Fake Silk

Paul David Blanc

Fake Silk

THE LETHAL HISTORY OF VISCOSE RAYON

Yale UNIVERSITY PRESS
New Haven &
London

Published with assistance from the foundation established in memory of Philip Hamilton McMillan of the Class of 1894, Yale College.

Copyright © 2016 by Paul David Blanc.
All rights reserved.
This book may not be reproduced, in whole or in part, including illustrations, in any form (beyond that copying permitted by Sections 107 and 108 of the U.S. Copyright Law and except by reviewers for the public press), without written permission from the publishers.

Yale University Press books may be purchased in quantity for educational, business, or promotional use. For information, please e-mail sales.press@yale.edu (U.S. office) or sales@yaleup.co.uk (U.K. office).

Set in Postscript Sabon type by Westchester Publishing Group.
Printed in the United States of America.

ISBN 978-0-300-20466-7
Library of Congress Control Number: 2016937699
A catalogue record for this book is available from the British Library.

This paper meets the requirements of ANSI/NISO Z39.48–1992 (Permanence of Paper).

10 9 8 7 6 5 4 3 2 1

Contents

Preface   vii

1 In the Beginning   1

2 The Crazy Years   30

3 Wrapped Up in Cellophane   57

4 Body Count   78

5 Rayon Goes to War   110

6 The Heart of the Matter   152

7 Rayon Will Be with Us   193

Notes   221

Acknowledgments   289

Index   293

Preface

When I say that I have been writing a book on rayon, the initial response frequently is, "Radon? That's radiation, isn't it?"

My attempt at a clarifying follow-up and the ensuing exchange typically runs: "Not radon, rayon, the synthetic textile."

"Oh, didn't they stop making that years ago—it was invented for parachutes or something in World War II, right?"

"Actually, rayon has been around since the turn of the last century—about 1900—and it's still being made. Maybe you know it better as viscose?"

"Viscose? I didn't know that was rayon. I thought viscose was a green product, not a synthetic."

"Viscose rayon is based on cellulose. That part may be 'green,' but the chemical used to make the viscose isn't. It's a toxic chemical called carbon disulfide."

"Does that mean viscose isn't safe to wear? I'd better go through my wardrobe!"

"No," I reassure at this point. "It's only the workers who make it that suffer, and maybe the surrounding environment." Consumer angst allayed, the conversation usually turns to some other topic.

I understand fully. Occupational disease is not the standard stuff of casual conversation. Admittedly, viscose is pretty far from central to almost anyone's thoughts. Moreover, carbon disulfide, the toxic agent perquisite to the making of viscose, is an unknown entity to anyone but a practicing chemist. Even most physicians have never heard of carbon disulfide unless they happen to remember it from an organic chemistry lab class they were forced to suffer through in premed. But the story of viscose manufacturing and viscose-caused disease, by rights, should not be obscure. It deserves to be every bit as familiar as the cautionary tales of asbestos insulation, leaded paint, or the mercury-tainted seafood in Minimata Bay.

Throughout most of the twentieth century, viscose rayon manufacturing was inextricably linked to widespread, severe, and often lethal illness among those employed in making it. Viscose is behind another product closely related to rayon—cellophane—and both rely on carbon disulfide as their key manufacturing constituent. Viscose, a technological innovation in its day, once was a very big business. In fact, it was one of the first truly multinational corporate enterprises, having achieved this status in the period just before World War I. A bit later in the twentieth century, during the Great Depression, the viscose business did not suffer appreciably. Rather, it flourished. Viscose went on to assume a highly profitable position as a strategic matériel on both sides in World War II.

Peace finally came; viscose went from strength to strength. For the Courtaulds company (phoenix of the postimperial British textile industry) and for the behemoth state-owned enterprises of the Eastern bloc, rayon was pivotal. During the same period in the United States, DuPont held on to its lucrative near monopoly on cellophane, fighting an antitrust ruling all the way to the Supreme Court. After the industry's midcentury apogee, viscose manufacturing found itself in the vanguard of those hazardous industrial processes exported to the developing world, starting in the 1960s and continuing through the decades that followed. Even now, viscose is still very much with us. Its successful rebranding as a renewable, eco-friendly product cleverly sidesteps the inconvenient reality that carbon disulfide, whether mixed with soft wood pulp or bamboo or straw, is anything but green.

The basic industrial manufacturing steps employed in making viscose never have been much of a trade secret. Cellulose wood pulp is treated with caustic soda at a high pH; carbon disulfide is added to that solution; the mix is churned, allowed to "ripen," and then mixed with more caustics to form a syrupy semiliquid that is the eponymous viscose of the process. The viscose syrup is forced through tiny spinning nozzles submerged in a bath of sulfuric acid, like sprinklers irrigating a Hadean garden. It is in this unkind environment that the extruded filaments of viscose rayon fibers coagulate and grow. Replace the tiny spinning holes with a long, thin slit, and one produces viscose-based film (that is, cellophane). Along the way there have been variations on this theme, but the basic story line has stayed the same.

Carbon disulfide has remained a constant in the mix. This chemical has the nearly unique ability to engage cellulose molecules, lining them up for guidance into a new form. Then, at just the right moment, the carbon disulfide lets go of the cellulose. Unless the process is engineered with care, the place where the "carbon disulfide lets go" is directly into the factory air breathed by viscose workers, with the rest wafted out into the surrounding environment.

Just as the basic process for making viscose is well established, the most dramatic effect of carbon disulfide on humans has been long appreciated. For more than 150 years, considerably in advance of rayon's invention, the chemical's potent and special toxicity has been clearly recognized. Carbon disulfide's industrial debut was as a vulcanizing agent in the rubber trade, back in the middle of the nineteenth century. Trouble was noted very soon after the chemical was first introduced. The effect was hard to miss: carbon disulfide exposure led to acute insanity in those it poisoned.

In the many years that followed, more and more medical evidence documented in exquisite detail the many ways in which carbon disulfide adversely affects the nervous system. Besides frank insanity, poisoning can be manifested in subtler personality changes. Carbon disulfide causes toxic degenerative brain disease and acts by damaging the sensory capacity of nerves (including those responsible for vision). After years of exposure, even more insidious carbon disulfide damage appears through increased risk of heart disease and stroke. Only in recent decades have these latter effects been established conclusively. Sophisticated epidemiological investigation was required to confirm the unusual pattern of individual cases that were occurring among viscose workers: disease that was happening both too frequently and among those at too young an age typically to suffer from these problems.

Despite its shocking legacy, viscose's history is almost completely unknown. Even those otherwise well versed in issues of public health are largely unaware of this story. Yet it is a history hiding in plain sight. In *Fake Silk*, I want to shine a bright light on this terrible story. In large part, I am motivated by a desire to memorialize the terrible suffering that has occurred. Almost everyone not only knows about radon—unlike rayon—and many can even name a specific victim: Madame Curie, who succumbed to leukemia, almost assuredly due to her occupational exposure to radiation. I want those who paid the full price for carbon disulfide's use to be as well remembered.

And I intend to name names. This may not be possible for the unrecognized thousands whom carbon disulfide made ill, although here and there personal traces of individuals can be detected. It is far easier, however, to identify the perpetrators. Indeed, some of them are still in business today, albeit after having been renamed, acquired, spun off, and then remerged through a string of new, also-known-as names and business acronyms. I also believe that this past history is highly relevant to other manufacturing innovations of today and tomorrow, processes that, like viscose production, may endanger worker health, threaten to degrade environmental quality, or compromise consumer product safety.

As I dug deeper into the story of viscose, one question that I was forced to consider over and over again was this: when a new technology leads to disease or even death, how high does the body count have to be before any protective steps are taken? One could imagine that slowness to act might simply be due to delayed recognition. After all, if no one suspects that there is a cause-and-effect relationship between an exposure and a new outbreak of disease, why would any prevention be undertaken?

It did not require extensive research for me to find out that when it came to carbon disulfide, lack of knowledge about its dangers has never been the problem—far from it. From nearly the earliest days of its discovery and first use, the toxic actions of carbon disulfide were observed, documented, and disseminated widely. This record was primarily recorded through medical reports and monographs appearing initially in French, then in German and English, and later in Italian, Japanese, and any number of other languages. Wherever and whenever carbon disulfide was used, reports of its adverse effects did not take long to appear. Occasionally, notice would spill out from medical reports into the popular media, including newspapers and newsmagazines.

While following this long chain of evidence, I encountered a curious pattern. The path was not linear, but rather ran almost in circles. There were places where the trail went cold, only later to emerge again as a new outbreak of illness caught the attention of a medical practitioner or a public health worker previously unfamiliar with carbon disulfide's toxicity. Each time, the neophyte seemed to rediscover the problem all over again, going back to square one. It was as if a kind of cyclical amnesia had come into play in which all that had been learned was soon forgotten, or nearly so, and the knowledge had to be reconstructed. These recurrent memory lapses were particularly prolonged just as carbon disulfide use fell off in the rubber industry and began to pick up again with the new technology of viscose.

By the early twentieth century, a considerable body of shared knowledge, acquired in fits and starts, delineated the dangers of carbon disulfide. Even so, this had virtually no effect on preventing exposure in the ever expanding viscose industry. Effective action on the part of the labor force most affected could not have been expected. Workers have always had limited leverage to protect themselves from health hazards on the job. Carbon disulfide exposure in the nineteenth-century rubber industry preceded the existence of organized labor, and over most of the twentieth century viscose manufacturing workers have been either nonunionized (especially in the United States and in many locales where the industry was exported overseas), under state control, or, at best, in a position far less advantageous than that of their employers. The globalized, multinational corporate nature of viscose manufacturing almost since its inception further exaggerated such power imbalances.

In all this, governmental "agency" was missing in action for most of the time. For example, when the plight of rayon workers became a topic of parliamentary debate in Great Britain, ministerial disingenuousness dampened any hopes for corrective action. Fifty years later, British governmental experts were still meeting, discussing compensation for injury to rayon workers, and reaching the predetermined outcome that it was premature to expand case eligibility to encompass newer medical knowledge on carbon disulfide–induced risk of heart attack. Meanwhile in the United States, the Occupational Safety and Health Administration made a feeble attempt at imposing somewhat tighter exposure controls that would have brought its standards nearer to the norms of other industrialized nations, and failed even at that. Needless to say, regulatory oversight is limited in many of the places where the viscose industry thrives today, including Indonesia, China, Thailand, and India.

My journey in attempting to unravel this complex story has led me in a number of directions, many with unanticipated findings. I began simply, working backward from current biomedical and other technical texts on the nature of viscose manufacturing and the known or suspected toxic effects of carbon disulfide. Such sources have footnotes. The footnotes identify papers, monographs, and book chapters, and all these have their own citations as well.

At a certain point, the citations repeated themselves to the point that there seemed to be nothing left to mine in a particular vein of inquiry. This seeming convergence, I learned the hard way, can be deceptive. A certain citation may be referred to repeatedly because it is indeed the definitive report. But more often than not, another rich source relevant to the very same matter may have gone ignored. This phenomenon, which I have encountered many times, was brought home to me again just as I thought I had closed out every major avenue of such bibliographic investigation for this book. A colleague called my attention to a reference in a 1960s Russian textbook on neurological illness, alluding to an outbreak of disease due to carbon disulfide in a prerevolutionary Russian viscose factory. I tracked down the original paper. It turned out to be an important early report of worker ill health and only the second one specific to the nascent rayon industry, just as the process first was coming online (the earliest outbreak was documented in a small American factory).

Valuable as biomedical and other scientific journal publications are, they can tell only part of the story. Archival and other documentary evidence provides another piece of the puzzle. And it is surprising how personal some of these old records are. Sitting in the British national archives in Kew and opening a large-format governmental ledger of industrial illness certifications for carbon disulfide poisoning, I was struck by the realization that the inked-in names and symptoms, and especially the ages, all belonged to real people. This sense of connection across time is still there even if one is holding a photocopy or a scanned image that has been sent from a remote source: a telegram, pages from a pocket diary, or photographic images, even if only a postcard showing a factory site long out of business.

At a certain point, it became clear that even these objects and images, despite their richness, were not sufficient for the full story I needed to tell. I wanted to have more direct contact in some way. In the end, this book includes information I gathered across a range of sources, including other clinicians and researchers as well as rayon workers (mostly retired) and their families. All these sources have been incredibly generous with their time, knowledge, and memories.

Especially among those who worked in viscose, a shared value I encountered time and again was *not to forget*. This was brought home to me most clearly through a blog hosted by the BBC North Wales on which former rayon workers could post their memories. Courtaulds' rayon mills in Flintshire, Wales, once the largest concentration of such production in Great Britain, have long been shuttered and all their workers laid off, or "made redundant," in British parlance. I began an e-mail correspondence with one of the bloggers, a woman named Vicky Perfect, who had worked at Courtaulds for more than ten years. At one of the works, she remembered, women could perm their hair simply by being on the job because of the high levels of sulfur fumes in the workroom atmosphere.

Viscose is so deeply imbedded in the history of the twentieth century that wherever one follows its many ramifications, they seem to lead out, cross over, and interconnect, unexpectedly linking together any number of otherwise disparate events and people. Even the word "rayon" itself carries meanings within meanings. It first denoted (in the late sixteenth century) a ray of light. As centuries passed, "rayon" also came to mean a military radius defined by fortifications and, in Eastern European areas under Russian influence, an administrative region. Rayon, as a 1920s trade name for the textile synthetic, is presumed to be a riff on light shimmering off the fabric, but the word also was chosen for its subtle but intentional allusion to "radon," so moderne at the time. No wonder people so often misunderstand the topic I have been working on.

The global interconnectedness of the viscose industry was borne home to me again when I finally had the opportunity to visit rayon plants still in operation. All were outside the United States. One of the smaller factories I visited is located in the Czech Republic, nestled in the Bohemian countryside north of Prague, in a town called Lovosice. It had been the site of a major eighteenth-century military engagement between Prussia and the Austrian Empire, the Battle of Lobositz. In the twentieth century, Lovosice was a quiet town with a modest industrial base, including rayon manufacturing. Along with many other state enterprises, the rayon plant in Lovosice was shut down after the Velvet Revolution of 1989, but then was resurrected with foreign investment.

Lovosice sits along the Elbe. By the time of my visit, I had already been to another viscose manufacturing site farther down river, at Wittenberge Elbe, in the former East Germany. The defunct Wittenberge facility, really a factory complex, had been the jewel in the crown of Deutsche Demokratische Republik state textile enterprises. Until its closure, the Wittenberge DDR operation had carried on the work of its immediate corporate predecessor, a Nazi-era component of the Phrix rayon conglomerate. I was particularly keen to visit Wittenberge because Phrix had used slave labor from the Neuengamme concentration camp. Delving into this dark corner of the viscose story, I began to come across more and more locations where similar arrangements had supplied forced labor to the rayon factories across the Reich. These victims suffered in double measure: first, from the terrible conditions general to their status, and on top of that, from the poisonous atmosphere of carbon disulfide–laden workrooms.

Lovosice lies in the Sudetenland, territory that was seized early by the Reich. I had not considered geography as I rode along a modern Czech highway for my day trip to the factory, accompanied by a fellow occupational medicine physician who had facilitated my entry to the rayon facility. We came to the exit for Lovosice. Another name was marked on the sign, too. This was also the turnoff for Terezín, where the Theresienstadt "model" concentration camp had operated: it was only eight kilometers from Lovosice. Follow-up correspondence with the archivist at the Terezín memorial confirmed that it, too, had supplied inmate laborers for rayon.

Fake Silk is most certainly about illness, about the disease and death that the viscose industry caused in its factories. And it is about technological innovation, an engine fueled by carbon disulfide that churned out novel products for an eager consumer public. *Fake Silk* is also about economics. After all, this was an industry that helped coin a new term, "duopoly." Referring to a market with only two sellers, duopoly characterized the comfortable arrangement between DuPont and Courtaulds' U.S. subsidiary in divvying up the lucrative American viscose business. More than that, economic profit has always been at the heart of viscose's power, whether that was parlayed by robber barons, war profiteers, state capitalists, or, in our own time, savvy players in a globalized market.

Fake Silk tracks cultural history as well. Sometimes this path runs through high culture, including avant-garde nods from Gertrude Stein and the Italian Futurist poet F. T. Marinetti. More frequently, however, rayon and cellophane are the stuff of mass culture, mirroring the ups and downs of shifting popular tastes. Giving something of a fairy-tale air to rayon's shifting role in fashion, for example, Genevieve Antoine Dariaux notes in *A Guide to Elegance* (1964), "Once upon a time, you only had to hint that a material contained rayon in order for the customers to turn up their noses. But by now all women are aware of the miraculous advantages of man-made fibers, and the former standards of quality have been adjusted to them."

Yet most of all, *Fake Silk* is about people. Some of them are truly despicable, but many were honorable women and men, and more than a few were quite heroic. In just over a century of viscose's history, five generations have toiled in it and, before that, another three worked with carbon disulfide–tainted rubber. This may not be the full ten generations that traditionally mark the passage of time from Adam to Noah, but it is still a significant slice of the human experience. I know that I cannot do full justice to that, but I hope never to lose sight of the real people who lived the story told in these pages.

Fake Silk

I

In the Beginning

Carbon disulfide is an elegant little molecule. Ethanol, for example, contains two carbon atoms. Carbon disulfide has just one. Like water, with its single oxygen and flanking hydrogen atoms, carbon disulfide is made of a one-plus-two combination, in this case a single carbon with two matching sulfur atoms, one on each side. Thus, the molecular formula for water is H_2O, and for carbon disulfide, CS_2 (for comparison, carbon dioxide, having a similar two-plus-one plan, can be rendered as CO_2).

To a modern practicing chemist, this formulaic shorthand of letters and numbers speaks volumes about the characteristics of the molecule in question. And even a basic science course would presume some fluency in such notations. This was not the case back when carbon disulfide was first synthesized, in 1796, by a German mining and metallurgical chemist named Wilhelm August Lampadius.[1] He announced his discovery by publishing a brief scientific report characterizing the novel substance that he had created by heating coal and mineral pyrite, a source of sulfur.[2] But beyond stating that the new material was a volatile liquid that seemed to be a sulfur-containing alcohol, which would have made it an unusual compound, Lampadius was not able to be more definitive about the structure of carbon disulfide.

Other chemists soon became interested, setting out to reproduce the experiment of Lampadius in a manner consistent with the basic scientific principle of first establishing consistent results before going further. Initially, though, finding a sulfur alcohol in the residue of carbon heated with sulfur proved to be elusive. Indeed, there were doubts whether the odd compound Lampadius claimed to have observed even existed. Then, when what seemed to be the same material was successfully re-created, arguments ensued about its structure: some French chemists got into a bit of a catfight over the precise elemental components of this new sulfur alcohol. Lampadius, along with both the groups in France, proved to be fundamentally wrong about the molecule's structure.[3]

The riddle of carbon disulfide was finally solved, not in Germany or France, but rather through a British-Swedish collaboration. In 1808, the British scientific leader Humphry Davy published the results of his initial inquiry into the nature of Lampadius's chemical.[4] Davy confirmed that the molecule contained sulfur and carbon, but he could not clearly determine its other constituents, in particular whether the oxygen and hydrogen atoms that might make it an alcohol were present. It was a scientific protégé of Davy, Alexander Marcet, who deciphered the structure of carbon disulfide. Marcet did a key part of this work in collaboration with the great Swedish chemist Jöns Jacob Berzelius in the summer of 1812 while Berzelius was on an extended working visit to London.[5] For example, in one experiment out of a series performed that summer, on 22 July (after first breakfasting together) Marcet and Berzelius were able to show that only when burned in the presence of external oxygen did carbon disulfide yield carbon dioxide as a major by-product.[6] On the same day in July, Wellington was fighting in the Battle of Salamanca, a victory that proved to be a turning point in the Peninsular campaign against the French.

When they announced their findings the following May in a paper subsequently published in the *Philosophical Transactions of the Royal Society,* Berzelius and Marcet proposed calling what had been thought of as a sulfur alcohol by the new name "sulfuret of carbon," consistent with having conclusively established that the compound was made up solely of a two-to-one combination of sulfur and carbon.[7] The absence of oxygen meant it could not be an alcohol. Other synonyms for the compound were coined as the nineteenth century progressed and as technical nomenclature and chemical spelling became standardized over the years that followed: bisulpheret of carbon, followed by carbon bisulphide, then finally carbon disulphide (and not until the twentieth century written as "carbon disulfide"). But the chemical composition of the molecule as determined by Marcet and Berzelius was never seriously questioned again.

It is not surprising that it took so many scientists, leading chemists of their day, to correctly discern the atomic makeup of carbon disulfide. In the first decades of the nineteenth century, working out the precise number and kind of atomic components in a molecule was a laborious and error-prone process. At the Royal Institution in London, where Davy and Marcet did their work, the laboratories could give a misimpression that this was a haphazard operation: there were crucibles and retorts in which test materials could be broken down by heat, acids, or caustics, along with apparatuses for condensing and distilling. In fact, Berzelius described the chaos of Davy's workspace admiringly, noting that "a tidy laboratory is the sign of a lazy chemist."[8]

To be successful, analytic work of the sort these chemists carried out required meticulous measurement and identification of the residuals left over from chemical and physical manipulation of the starting materials. In practice, such analyses also drew heavily on extrapolation from previous results of taking apart similar compounds with established formulae. The task is analogous to coming up with the recipe for a soufflé by scraping out and studying the remains from a charred ramekin. In the case of carbon disulfide, the experimenters were doing their work without ever having seen the metaphorical soufflé in question, let alone having tasted it. Nothing like carbon disulfide had ever been encountered before by the first scientists trying to figure out what it was.

Actually, carbon disulfide exists in nature, but it is characteristic of only a very few, quite particular niches. One favored environmental locale for carbon disulfide turns out to be volcanic formations, in particular fumaroles, which are geologic vents that release hot gaseous emissions. When such a formation is high in sulfur-containing compounds, carbon disulfide can be formed, and the fumarole is known as a solfatara. For example, the aptly named Solfatara volcano near Naples is famous for such fumaroles. It was at Solfatara (in an ancient sauna building) that scientists only recently isolated a type of bacteria that not only can metabolize carbon disulfide but also seems to thrive on the stuff.[9] Such organisms, happy in volcanic mudpots, living off what is toxic to most creatures, have been labeled "extremophilic" by the biologists who study them, as if such bacteria were the unicellular equivalents of a thrill-seeking sports enthusiast.

There are also a handful of tree species that naturally synthesize carbon disulfide. The chemical may serve them as a protective agent, given carbon disulfide's fungus-killing capabilities. One of the best-studied carbon disulfide producers is the noble valley oak (*Quercus lobata*), perhaps not coincidentally a very long-lived tree.[10] The valley oak, like a solfatara, is associated with species of bacteria that manifest the unusual ability to digest and be nurtured by the sustaining carbon disulfide that their host so generously provides.

The chemists working in the Royal Institution early in the nineteenth century knew nothing of bisulphuret of carbon being released by fumaroles or produced by oak trees. Nor did they speculate on any direct practical applications for the odd substance they had finally characterized chemically. Lacking any real use, carbon disulfide remained little more than a chemical curiosity in the initial decades that followed its discovery. Indeed, even for the chemically curious, carbon disulfide was an arcane substance.

An excellent measure of carbon disulfide's marginal status can be taken from the tutorials of young Miss Emily by Mrs. B., the protagonists in the hugely successful *Conversations on Chemistry* written by Jane Marcet. Although her husband, Berzelius's collaborator Alexander Marcet, was a well-respected and highly successful chemist-physician, Jane Marcet came to be far more widely known. Her fame derived from the groundbreaking educational texts that she authored, each structured as a dialogue between a kindly, very smart tutor and her eager female charges. The publication that led off the series was *Conversations on Chemistry*. It first appeared in 1806, well before her husband's work on carbon disulfide, and went through many editions. The author added new topics as she felt they were warranted (for instance, in the tenth edition Marcet introduced material on the steam engine), but carbon disulfide, it seems, did not merit inclusion among such updates.

Conversations on Chemistry, beyond the editions overseen by Marcet herself, also served as the basis for minimally altered "adaptations" published in America under the cover of other authorships. It is in one of these texts that carbon disulfide finally makes a brief appearance. Dr. Thomas P. Jones, in *New Conversations on Chemistry* (1831), lifted verbatim many sections of the original Marcet text and abridged others. One of Marcet's revisions had addressed Humphry Davy's invention of the safety lamp for coal miners (which had followed not long after her husband's work on carbon disulfide). Jones appended to this an added monologue put into the mouth of Mrs. B. as she patiently instructs Emily: "There is a curious compound of sulphur and carbon, which may receive passing notice.... This sulphuret, or rather *bisulphuret of carbon* is very inflammable, acrid to the taste, and has a very offensive odor. It may be formed by passing the vapour of sulphur over fragments of red-hot coal. The fact of the existence of such a combination is all that would at present interest you in regard to it."[11]

Although they did not directly consider practical uses for carbon disulfide, Marcet and Berzelius commented on the chemical's high volatility, noting that it was even greater than that of ether, which was well appreciated for its rapid vaporization. Marcet even went on to carry out additional experiments demonstrating the powerful cooling effect of carbon disulfide as it quickly evaporated.[12] Thirty years later, when ether anesthesia was first introduced in 1846, the next generation of experimenters quickly turned to other similarly volatile substances to test their therapeutic potential.[13] By 1847, James Simpson at the University of Edinburgh had introduced chloroform anesthesia, to great renown. He went on to study a number of other candidate anesthetics from among recognized volatile substances, publishing his results the following year. Among the test agents he investigated was carbon disulfide: "I have breathed the vapour of bisulphuret of carbon, and exhibited [that is, exposed] it to about twenty other individuals, and it is certainly a very rapid and powerful anaesthetic. One or two stated that they found it even more pleasant than chloroform; but in several it produced depressing and disagreeable visions and was followed for some hours by headache and giddiness, even when given only in small doses."[14]

When Simpson tried out carbon disulfide in clinical applications, it was even more unsatisfactory. In the end, he recommended against any clinical use for carbon disulfide, adding that yet another drawback was the very unpleasant odor of the substance, characterized as the stench of putrid cabbage.

Simpson's was the first human experimentation with carbon disulfide; it was followed soon thereafter by animal data. Two months after Simpson's report, Dr. John Snow, another key figure in the early history anesthesiology, published his experimental results from exposing mice to inhaled carbon disulfide. The mouse that was subjected to the lowest dose manifested violent tremors until taken out of its exposure jar, after which "it flinched on being pinched; attempted to walk but fell over on its side: it had no appreciation of danger at first, but it quickly recovered." Additional test mice, exposed to higher concentrations, fared less well. The final test mouse, "after running about for a minute, fell down, and stretched itself violently out, and died." Snow was impressed by the dangerously powerful effects of inhaled carbon disulfide, marked by profound toxicity combined with its volatility. And as he was to show six years later when famously investigating a London cholera outbreak attributable to contaminated drinking water, Snow did not shy away from the human health implications of what he had observed with carbon disulfide: "Indeed, I feel convinced that, if a person were to draw a deep inspiration of air, saturated with its vapour at a summer temperature, instant death would be the result."[15]

Unfortunately, it was not only the potent anesthetic effects of carbon disulfide that were beginning to be recognized. Carbon disulfide's powerful solvent capabilities, too, came to wider attention in the 1840s. In 1845, William Gregory's influential *Outlines of Chemistry, for the Use of Students* appeared. He provides instructions for how to synthesize bisulphuret of carbon, describes its physical properties, and ends by delineating its utility as an effective solvent:

> It dissolves sulphur and phosphorous readily; and these solutions, by spontaneous evaporation, yield fine crystals of those elements. It also dissolves camphor, essential, oils, and resins.
>
> Sulphuret of carbon is occasionally used as an external application in burns; and it promises to be useful as a solvent for resins, many of which it dissolves readily, and thus forms varnishes, which from its great volatility, dry very rapidly.[16]

Thereafter, the solvent properties of carbon disulfide drove its success. There had never been another solvent like it.

Before the discovery of carbon disulfide, scientists had only a few choices when it came to dissolving things. Water, the fluid of life, was always first. It fills the inland sea of every cell in a microscopic recapitulation of the primordial oceans in which biological systems first evolved. Each cell's viability depends on that fluid, rich in the salts and nutrients dissolved within it. Water is not the only solvent in nature. Alcohols of all sorts, and that includes consumable ethanol, do a pretty good job of getting substances to go into a solution. Whatever its other virtues, wine is as good a solvent as water, and usually even better. Turpentine is another ancient solvent, since antiquity obtained from the resinous discharges of various trees.[17] Turpentine is a natural mix of multiple related chemicals that are grouped together as terpenes. These molecules are more complex than either water or ethanol, with the great advantage of being better than alcohol at dissolving oils (water being of nearly no use at all for that purpose).

For the first millennia of human history, water, alcohol, and turpentine were it when it came to solvents. The early alchemists struggled against the limitations imposed by their relative weak dissolving powers. The very term "solvent," derived from a root meaning "to loosen or dissolve," reflects the roadblock this presented. The alchemists came up with a way around this problem in the form of a universal solvent, referred to as the "alkahest."[18] Chemistry, in the modern sense, had not yet come into being: the alkahest was purely a product of conjecture, not experiment; but the alchemists that promoted its virtues were not dissuaded by the absence of any confirmation of its existence. The term "alkahest," a pseudo-Arabic formulation meant to echo the names of chemical entities that did exist (including alcohol), was as much a fiction as the solvent it claimed to describe. It could not slice or dice, but the alkahest was said to be able to dissolve anything and everything and yet not destroy the basic essence of the substance it put into solution.

The absurdity of a universal solvent was clear. After all, what vessel would be capable of holding the alkahest without itself disintegrating? The new chemistry, as it began to emerge, sought to discover real rather than mythical new solvents. When distilled wine was mixed with sulfuric acid, the result was a novel substance initially called sweet oil of vitriol, first produced in the sixteenth century.[19] Two hundred years later, in the early 1700s, this solvent, finally having been characterized in laboratory experiments, was renamed "ether."[20]

Carbon disulfide was a far more powerful dissolving agent than ether, and in many ways a more potent solvent than newer agents that soon followed. It was the unusual ability of carbon disulfide to dissolve both phosphorous and sulfur that led to its initial commercial applications. The first such use for carbon disulfide was in a major emerging technology of the day, electroplating. This process, introduced at the beginning of the 1840s, applied exceedingly thin layers of precious gold or silver to objects made of less expensive base metals. Electroplating was able to displace a far more labor intensive and technically limited process called Sheffield plate, which mechanically layered silver on copper and then fused the hybrid with heat. Electroplating moved the industry from its former center at Sheffield to Birmingham, England, where Elkington, Mason and Company held all the key patents for this new and highly lucrative technology.

A brilliant young chemist and inventor named Alexander Parkes had gone to work for Elkington. The key breakthrough that made electroplating commercially viable was the discovery that metals could be put into solution using cyanide and then drawn by electrical current and deposited on a target metal object. Parkes took the basic Elkington process as his starting point. To it, he added a special flourish that got around the fundamental limitation of requiring a metal object of some sort to serve as the electrically charged recipient of the superimposed metal coating. Parkes figured out a way to electroplate an organic, nonmetallic object through a sequential process in which he first dipped it in a solution of phosphorous dissolved in carbon disulfide, followed by a silver nitrate coating. A metallic silver coating resulted, serving as the base metal layer that could then be plated with additional silver, gold, or copper.[21]

Elkington's stock-in-trade included tea services, tureens, and many other objects for both special occasions and everyday use. Harriet Martineau, a prime defender of the Victorian cultural and political high ground, glowingly described Elkington's operation in its heyday. Martineau was most impressed by some of the special-order items she saw, including a decorated commemorative object: "The group of palm-tree and oak, overshadowing the sick Hindoo, and the soldier-surgeon stopping over him, lancet in hand; the piece of testimonial plate presented to the surgeon of a regiment."[22]

Within the broader Elkington inventory, Parkes's unique metal-plated natural objects essentially served as technological vanity items meant to trumpet the seemingly unlimited manufacturing horizons stretching out around the Birmingham factory. Thus, on 29 November 1843, when Prince Albert paid a high-profile visit to a series of Birmingham industrial sites, the day ended with an extended tour of the Elkington facility.[23] A high point of the visit was the presentation to the Prince Consort of a delicate object of virtue plated using the Parkes method. This event even became something of an urban legend, purporting that what had been presented was nothing less than a silvered spider web. Indeed, such an object would have achieved the ultimate in refined delicacy, marrying art and science in a way that could only have very much pleased Prince Albert. Harriet Martineau, following up on this story with a reporter's nose for fact-checking, determined that the object had actually been a silvered rosebud, which happed to have attached to it a bit of cobweb unintentionally plated at the same time.[24] In any case, the prince was charmed.

Parkes's use of carbon disulfide to dissolve phosphorous essentially used one chemical curiosity to produce another. But his next invention based on the dissolving properties of the compound proved to be much more far-reaching in its commercial impact. On 25 March 1846, Alexander Parkes filed British patent no. 11,147 for "the change upon caoutchouc, gutta percha, and their compounds, by employing agents in a state of solution capable of producing such change."[25] The key step that Parkes included in his patent was that of dissolving sulfur in carbon disulfide and then immersing in that solution any objects intended to be treated, thus causing the desired "change."

By "change upon caoutchouc" (that is, India rubber), Parkes meant vulcanization, but he could not have known of that term in March 1846. The first known appearance of "vulcanization" was in another British patent, filed by Thomas Hancock of Charles Macintosh and Company just one week before that of Parkes. The Hancock patent was for a further refinement of a very different sulfur treatment process for rubber, one that Hancock had developed a few years earlier and that was also for changing the character of caoutchouc, but ultimately called vulcanization.

Whether or not in plating a rose Parkes was simply painting the lily (in fact, gilding it), such frivolities were never going to be a major part of the electroplating business. In contrast, any successful method for vulcanization had the potential to be a very big deal indeed. Without vulcanization of one sort or another, natural rubber is next to useless for most applications, too gooey or too brittle by turns. But once changed, rubber can be transformed into any number of highly marketable products. Because the financial stakes were so high, at the time of Parkes's invention Hancock and Charles Goodyear were already waging a bitter transatlantic fight over who held primacy to the original vulcanization patent.

Parkes's simple technique for dissolving sulfur in carbon disulfide and then dipping rubber-made goods into the solution was ingenious and novel. Unlike the Hancock-Goodyear method, Parkes's did not use heat and pressure to introduce sulfur into the final product. For that reason, his method came to be known as cold-process vulcanization in order to differentiate it from the Hancock-Goodyear hot-process technique. Given the obvious originality of Parkes's discovery, Hancock had no basis for fighting its patent, so he negotiated to take over the sole rights to its use.

Hot-process-vulcanization establishments were large-scale mega-factories of their day, with huge mixers churning together, under pressure and heat, rubber combined with sulfur and other additives, and then squeezing this mix through massive calenders working like a group of huge, synchronized rolling pins. Constructing and running such plants required large capital investments and significant technical engineering know-how. In contrast, Parkes's cold-process technology was easily adaptable to small-scale operations where any number of handy rubber items could be produced easily, quickly, and relatively cheaply, feeding a growing consumer appetite for such objects.

In 1847, shortly after adding Parkes's cold-process invention to his vulcanization portfolio, Thomas Hancock published *Personal Narrative of the Origin and Progress of the Caoutchouc or India-rubber Manufacture in England,* noting at the outset: "In writing a *personal* narration it is impossible to escape the very disagreeable necessity of frequently repeating the pronoun I,—my readers must excuse this unavoidable egotism" (emphasis in the original).[26]

One of the plates that illustrates Hancock's narrative, titled "Domestic Articles," shows a wide range of objects that seem to have changed little in the last one hundred and fifty years (although plastic may have since replaced natural rubber).[27] These include bathing caps (one for men and another for women), an infant pacifier, a playing ball, and even home exercise equipment in the form of stretchable "chest expanders." Other rubber paraphernalia shown are less timeless, such as an "invalid cushion" and a handy "sponge bag" (and a matching rubber soap bag). But the one consumer item that arguably was the biggest beneficiary of mass production using the Parkes process for dipping small rubber objects is not illustrated among Hancock's potpourri of domestic household objects: the India rubber condom.

The dipped-rubber consumer goods industry, restricted by patent limitations, did not become widespread in Great Britain. Across the Channel and beyond the reach of such patents, however, France proved to be fertile ground for this type of manufacturing, well suited to precisely the sort of small-scale workshops that had long characterized production in and around Paris. The first English-language appearance of the term "French letter" as slang for "condom," which dates from this period, may derive from the geographic concentration of this product of the cold-process-vulcanization industry.[28] What is certain is that such manufacturing grew dramatically in that time and that place.

The rapid expansion in carbon disulfide vulcanization in France can be gauged by the attention it began to receive in technical texts. Even as late as 1849, when a major text on applied industrial chemistry appeared—a two-volume, more-than-six-hundred-page work by the French chemist Anselme Payen—carbon disulfide was relegated to a brief reference in a single footnote.[29] Only two years later, the 1851 updated edition of the same text highlighted on its title page that the book had been augmented with chapters covering a number of new chemicals and their applications, specifically noting among them carbon disulfide in the rubber industry. In his revised text, Payen warns that carbon disulfide's use in cold vulcanization can be not only inconvenient, but even dangerous to workers on account of the chemical vapors present, cautioning that such work was best done in an open space with good ventilation.[30]

Payen was a major presence in applied chemistry in mid-nineteenth-century France. Among his many accomplishments, in 1839 he became the first to chemically name cellulose, identifying it as a principal constituent of wood.[31] This discovery would prove critical in later applications of carbon disulfide. And although Payen may have been an authoritative voice in many technical aspects of industrial chemistry, his cautions regarding the potential dangers of handling carbon disulfide went unheeded.

In 1853, two years after the new edition of Payen's text appeared, no less a medical figure than Guillaume Duchenne de Boulogne provided the first published notice of carbon disulfide toxicity to the human nervous system. Duchenne, after whom a major form of muscular dystrophy would later be named, made a presentation to the Medical-Surgical Society of Paris on muscle atrophy, a core interest of his, in relation to various neurological and psychiatric disease manifestations. Although Duchenne gave only brief mention to carbon disulfide exposure among vulcanization workers, his observation was ominous: the chemical appeared to cause a disease whose symptoms resembled general paresis of the insane, that is, the mental derangement seen in end-stage syphilis. Duchenne underscored the point by stating that he had been presented a recent case of this syndrome by a colleague at the Hôpital de la Charité in Paris.[32]

Duchenne's patient zero did not remain an isolated case. Early in 1856, another Parisian physician, Auguste Delpech, gave a brief report to an afternoon meeting of the French Academy of Medicine, describing his clinical experience with cases of carbon disulfide–caused disease.[33] The unfortunates he examined suffered from a frightening range of symptoms. They experienced agitated sleep with vivid and disturbing dreams when able to sleep at all, followed the next day with sleepiness and a sense of inertia. Overall, the patients complained of compromised memory, confusion, and, in extreme cases, extremely abnormal behavior that could be characterized as maniacal. Other complaints induced by carbon disulfide included headache, muscle weakness, and numbness. End-stage syphilis symptoms would have been bad enough—what Delpech was describing seemed to be even worse.

Delpech was thirty-seven years old when he made this first report to the academy, but he was already a professor of medicine associated with the University of Paris, and he had a long-held interest in neurological manifestations of disease.[34] Carbon disulfide became the central focus of his work for years to come. Continuing to sound the alarm on the dangers of carbon disulfide, Delpech published the details of one of his cases in a medical newspaper of the time, *L'Union Medicale*.[35] Twenty-seven-year-old Victor Delacroix, Delpech reports, after three months of using carbon disulfide as a solvent to patch and repair rubber objects, began to manifest nervous system disease. By the time Delpech examined him, the patient appeared aged beyond his years, a broken man whose "sexual desire and erections were abolished." Delacroix experienced marginal improvement after a convalescence completely removed from further exposure. But the impotence remained.

In another publication that appeared the same month as the case report on Delacroix, Delpech detailed animal experiments that he had performed with carbon disulfide.[36] Apparently unaware of earlier animal experimentation by the British anesthesiologist Snow, Delpech had rapidly dispatched two pigeons, but was able to preserve a test rabbit long enough for it to display paralysis. Later the same year, Delpech published a further expansion of his work as a seventy-nine-page freestanding monograph.[37]

Delpech did not leave off there. He continued to amass clinical experience with carbon disulfide poisoning over the next seven years, finally publishing in 1863 his definitive work on the subject.[38] In a scientific paper more than a hundred pages long, Delpech extensively details twenty-four case histories of cold-vulcanization workers specifically engaged in what he terms the "inflated rubber industry." This employment involved the mechanical distension of rubber objects that were then dipped into a carbon disulfide solution before being dried in the open air of the crowded workrooms where this manufacturing typically was done. The two principal products of this trade, Delpech reports, were colored balloons and condoms, the latter with a "special destiny for export."

To this day, Delpech's 1863 opus holds its place as the classic descriptive account of carbon disulfide poisoning. The two dozen cases reinforced his initial findings of years before: the intoxication manifested wide-ranging, devastating neurological symptoms, both physical and psychological. Delpech's wealth of clinical experience allowed him to further refine his assessment of the natural history of the disease at hand. He was able to demarcate between two distinct phases of carbon disulfide intoxication: first came a period of illness marked by mental disturbances that could be quite profound; a later phase was associated with distal nervous system manifestations, including muscle weakness and numbness of the arms and legs.

Sexual disturbance played a prominent role in the syndrome that Delpech described. In the first flush of illness, he documents multiple cases of inappropriate male sexual arousal. Case 10 was a prime example: a twenty-year-old troubled with constant erections. Nor was this genre of toxic effect limited to men. For Madame D., case 19, carbon disulfide induced genital agitation, aphrodisiacal excitation, and abnormal menstrual bleeding. In the second phase of carbon disulfide poisoning, the primary sexual manifestation that Delpech remarked upon was impotence, the same problem that troubled the very first patient he had described.

Two of Delpech's cases in the 1863 series manifested frank exposure-related insanity. There was also an even more disturbed, twenty-fifth case that fell outside the cohort because Delpech never had the opportunity to directly examine her. She had become progressively deranged, and in the end she intentionally self-asphyxiated with carbon disulfide vapor. It was a death by acute inhalation of exactly the sort that John Snow had predicted based on his laboratory studies when he rejected carbon disulfide as an anesthetic for human use.

Importantly, Auguste Delpech did not see himself as simply a passive chronicler of the industrial plague he was witnessing. In his earliest reports, he recommended that carbon disulfide vulcanization be completely forbidden in small rooms, and that even in larger work spaces, direct contact with the toxic substance should be reduced if it could not be eliminated entirely. Becoming more sophisticated in his proposed interventions, by 1863 he was promoting the adoption of a protective device originally devised by one of his patients. It was a kind of glove box apparatus that presaged the equipment used in the next century to handle radioactive substances. Delpech reports that the worker's invention, despite its simplicity and potential efficacy, was derided as nothing more than a "magic lantern."[39]

Delpech was not alone in his early concerns over the hazards of carbon disulfide or even in striving for better workplace protections. The French Academy of Sciences, which awarded an annual prize on the topic of unhealthy work activity, received a paper in 1858 that proposed placing open wooden boxes of quicklime on the floors of workshops using carbon disulfide, in order to absorb the toxic fumes.[40] Still, little was done to improve the lot of the workers. Delpech noted that there had been in his experience only a single case in which a worker injured by carbon disulfide took legal action against his boss, winning in court thanks to Delpech's expert opinion.

Medical reports of carbon disulfide toxicity continued to appear. A medical thesis for the Paris Faculty of Medicine in 1874, for example, included five new observations of cases of workers poisoned by carbon disulfide.[41] Three of them were only fifteen years of age. Another medical thesis, published two years later, added four more new cases, all seen in the hospital service of Dr. Delpech, including Louis Herbunot, who worked making balloons; Arthur Bahin, who made balloons and condoms; and a patient named A. Surtout, who did the same work. The fourth patient, Mlle. Louise Genet, was discreetly described as simply working in vulcanization.[42] The year was 1876, the twentieth anniversary of Delpech's first poisoning report.

Delpech, who died in 1880, devoted the better part of his career to the study of carbon disulfide's deleterious effects. This toxic substance caused distinct neurological deficits that could be easily assessed by direct physical examination, but it also induced far less easily quantifiable impairments in mental perceptions. Thus, in this work, Delpech was the forerunner of an evolving discipline, the novel science of the mind.[43]

In the decade that followed, the French neurologist Jean-Martin Charcot was widely acknowledged as that new science's rising sun. In particular, it was out of Charcot's promotion of hysteria as a medical construct that modern precepts of psychiatry began to crystallize. As laid down by Charcot, the clinical presentation of hysteria combined multiple physical complaints that seemed to be features of a neurological syndrome. Muscle weakness, which could be as extreme as paralysis, was common. Often, the manifested pattern of weakness did not conform to any known anatomic lesion or nervous system pathway. Delving deeper, the astute clinician could uncover many other problems that also commonly troubled the classic hysteric. Sexual dysfunction, in particular, was a notable complaint. Needless to say, hysteria was a disease that especially afflicted women, although male exceptions to the norm were noted.

A century later, when hysteria as a diagnostic entity had long since been abandoned, a discordance between symptoms and physical findings (or laboratory data) would come to be labeled in pejorative medical slang as the presence of a "high serum porcelain level" (that is, the patient is a "crock").[44] But in late-nineteenth-century France—and thanks to Charcot's wide influence—hysteria was taken to be an established medical fact throughout the allopathic medical world.

It stands to reason that carbon disulfide, an intoxicating agent capable of both physical and mental effects, including prominent psychosexual manifestations, would be of interest to Charcot. Late in 1888, in one of his famous "Tuesday lessons," Charcot took up the question. With the afflicted case present, the demonstration began: "You have without doubt, gentlemen, more or less heard talk of the carbon disulfide industry. This industry includes the preparation of carbon disulfide itself, as well as subordinate industries, among which one must cite the example of the fabrication of vulcanized rubber. Hygienists and clinicians are concerned with these industries because of certain accidents, principally neurological, to which its workers are subject. . . . The patient you have before your eyes offers a perfect example of this genre."[45]

In this lesson, Charcot acknowledged the previous contributions to the field of the late Dr. Delpech and the current work of Charcot's junior colleague Dr. Pierre Marie, who had called the case at hand to the professor's attention. Charcot described the patient's pre-morbid status: always sober (never drinking to excess) and tranquil of manner. Charcot also provided a detailed and revealing occupational history. The patient had worked in the rubber trade for seventeen years, but in the few months before his illness he had begun performing a new and particularly odious task: manually cleaning out vulcanization vats laden with carbon disulfide. Six weeks before the lesson, the patient had collapsed on the job, literally anesthetized by the high level of carbon disulfide vapors to which he was subjected.

The prodromal symptom before collapsing, Charcot was keen to comment, was a sensation of burning in the scrotum. It was half an hour before the rubber worker could be aroused from his comatose state, and then, stuporous, he had to be carried home. He remained bedridden for the following two days. Recurrent nightmares of fantastic and terrible animals characterized his convalescence, if it could be called that: he continued to be weak, suffered from twitching, and experienced vision loss.

This constellation of mixed neurological complaints could signify only one syndrome according to the formula of Charcot—hysteria. Male hysteria was uncommon, Charcot himself admitted in the lesson, but not unheard of. The patient's despondent mood sealed the deal: female hysterics (typically petit bourgeois) were detached in regard to their condition, whereas males, commonly working class, whose hysteria was often linked to a traumatic event, were not indifferent to their debility.[46]

In fact, this was not hysteria. The patient's acute response to high levels of carbon disulfide fit with what the British anesthesiologists, Simpson and Snow, had shown in human and animal experiments. The patient's residual illness was in every way consistent with the chemical toxicity of carbon disulfide, which Delpech had documented so well. Yet despite this, Charcot took this case to be a confirmation—rather than refutation—of the diagnostic entity of hysteria, which he championed. So strong was the authority of his opinion on this matter that even years after the general diagnostic label of hysteria fell out of fashion, "Charcot's carbon-disulfide hysteria" persisted in use as an industrial-medical diagnostic label.[47]

After Charcot, carbon disulfide poisoning continued to fascinate the next generation of French neurologists, many of them leaders in the field. The neurologist who had provided Charcot with his teaching case, Dr. Pierre Marie (one of the namesakes of Charcot-Marie-Tooth disease) wrote a full paper on carbon disulfide, further expanding upon the case presented in Charcot's lesson.[48] Dr. Georges Gilles de la Tourette (of Tourette's syndrome) wrote on the subject of carbon disulfide.[49] In the same period, Georges Guillain (of Guillain-Barré syndrome) did as well.[50] As with Charcot (and Delpech before him), these French neurologists consistently alluded to frequent genital involvement associated with carbon disulfide intoxication, either excitation or impotence or both, sequentially.

Driven by the relatively large number of exposed cold-process-vulcanization workers in France, the clinical investigation of carbon disulfide toxicity was largely, but not entirely, a Gallic enterprise up until the 1890s. As time progressed and the industry spread geographically, physicians elsewhere also began to pay attention to the problem. In Germany in particular, scientific papers and entries in medical texts on the subject of carbon disulfide began to appear with increasing frequency in the later nineteenth century. This new interest coincided with a shift in the center of gravity for psychiatric practice from France to Central Europe.

In 1899 there appeared an exhaustive publication marking the culmination of that transition. *Die Schwefelkohlenstoff-Vergiftung der Gummi-Arbeiter unter besonderer Berücksichtigung der psychischen und nervösen Störungen und der Gewerbe-Hygiene* [Carbon Disulfide Poisoning in Rubber Workers, with Special Consideration of Mental and Neurological Disorders and Industrial

Hygiene] is a dense monograph more than two hundred pages long.[51] Its author, Dr. Rudolf Laudenheimer, presents the clinical details of no fewer than forty-two cases of carbon disulfide intoxication. Sets of patients fall into various subgroups (cases 18 through 22, notably, suffered from mania).

Laudenheimer was not a generalist with an interest in the nervous system, as Delpech had been (Laudenheimer eclipsed Delpech's previous record number of twenty-four cases with this publication), nor was Laudenheimer a neurologist championing hysteria, on the model of Charcot. Laudenheimer was a full-fledged psychiatrist. After his foray into this subject, he apparently left poisoned factory workers far behind. Two years after publishing his magnum opus on carbon disulfide, he opened an exclusive spa-sanatorium where his patients included prominent artists and intellectuals of Munich's fin de siècle avant-garde.[52]

In contrast to France and Germany, and despite being the original homeland of Parkes and the cold process for vulcanization, Great Britain showed scant interest in carbon disulfide toxicity during these years. There was a brief flurry of concern when, in 1863, the ponderously named Commission of Inquiry into the Employment of Children and Young Persons in Trades and Manufactures not Already Regulated by Law visited the Charles Macintosh India rubber manufactory. The commission was reassured that even though carbon disulfide was used by Macintosh, no adverse effects from it had been noted by the employer.[53]

Despite the denials of Macintosh and the commission's glossing over the problem, working conditions in the British rubber industry in the late nineteenth century were not appreciably better than those in France or Germany. Indeed, the *Oxford English Dictionary,* under its entry for "gas" as an inflected verb, gives "to be gassed: to be poisoned by a gas," citing as the earliest usage in that sense an 1889 appearance in the *Liverpool Daily Post:* "'Gassed' was the term used in the india-rubber business, and it meant dazed."[54]

Not all British clinicians accepted at face value the claim that all was well in the India rubber curing rooms. Those treating the poisoning cases that resulted from carbon disulfide exposure were aware of the problem and took notice in particular of the compound's capacity to damage the nerves of sensation, especially the faculty of vision. Indeed, multiple British medical publications on the subject of optic nerve damage (a condition known as amblyopia) appeared in this period.[55]

Moreover, in addressing amblyopia, doctors catalogued other nervous system deficits caused by carbon disulfide as well. For example, when Dr. James Ross of the Manchester Infirmary in 1886 treated a twenty-four-year-old (identified as J.N.) poisoned by carbon disulfide in rubber vulcanization, he was careful to include a detailed mapping of the patient's color vision deficits. In addition, Ross included a complete neurological profile that, in a nod to the French literature on the subject, also addressed J.N.'s impairment in sexual function: "The patient lost all sexual desire a few weeks after he began work in the curing-room, and even at the present time he never has any erections. The loss of this function was not preceded by a stage of sexual excitement."[56]

Benjamin Ward Richardson was the most notable British physician with an interest in carbon disulfide during those years. He was a proponent of an eclectic mix of public health interventions, at one point even promoting tricycle riding for its exercise benefits. Richardson was also an early animal rights advocate; being familiar with the early anesthesia literature, he saw carbon disulfide as an ideal vehicle for the painless euthanasia of unwanted pets.[57] But Richardson also cared about humans. In 1879, he wrote a tract on the general subject of occupational health, intended not for his medical peers, but for workers themselves; it was written in a plain style and published by the Society for Promoting Christian Knowledge. In this pocket-size book that could easily fit in a work smock, Richardson includes specific warnings on the dangers of carbon disulfide, cautioning that those exposed to it (he singled out balloon makers) could be "rendered imbecile and insane."[58]

Poisoning from carbon disulfide continued to be endemic in the British rubber trade. As late as 1902, Dr. Thomas Oliver, later knighted in recognition of his prominence as a physician expert in occupational disease, described the deplorable current state of affairs in the British rubber trade. Oliver wrote of factory girls who, at a work shift's end, "simply staggered home," only to wake in the morning to a headache that could be relieved solely by "renewed inhalation of the vapour." Oliver described fits of insanity leading to suicide: "Sad as this state of things is, it is nothing to the extremely violent maniacal condition into which some of the workers, male and female, are known to have been thrown. Some of them have become the victims of acute insanity, and in their frenzy have precipitated themselves from the top rooms of the factory to the ground. In consequence of bisulphide of carbon being extremely explosive, vulcanization by means of it has generally to be carried on in rooms, one side of which is perfectly open. This open front is usually protected by iron bars."[59]

Reports from the United States were almost entirely absent from the medical literature on carbon disulfide poisoning in the rubber industry in the nineteenth century. One notable exception was a description of an outbreak of insanity among factory workers treated at the Hudson River State Hospital for the Insane.[60] The first case, a twenty-seven-year-old, was committed in April 1887, raving and incoherent. Twelve days later, a second case came in, a clinical match to the first. In fact, the two men worked at the same factory. In August of that year, a third co-worker appeared: "M.B., male, age thirty-one, married, a Hebrew, born in Austria," who had been "in a condition of great mental excitement, disturbing the neighborhood by loud noises and violent praying."

The treating physician for these cases, Dr. Frederick Peterson, was chief of clinic of the Nervous Department of the College of Physicians and Surgeons in New York. He could not accept that some kind of bizarre coincidence explained the identical employment histories of all three cases. When the workers regained their sense enough to describe their work, Peterson learned that their rubber factory jobs exposed them to carbon disulfide. Peterson believed that it was important to make it known that this poison, so well recognized in Europe, had come to America. Eventually, he reported these cases in the prestigious *Boston Medical and Surgical Journal* (later renamed the *New England Journal of Medicine*). He attributed the delay in the dissemination of his findings to a lack of needed cooperation. In effect, Peterson was stonewalled, and he did not hesitate to state the reason: "I have delayed in publishing these cases for some years, thinking that I might hear of other similar ones, or that I might acquire more information from the owners of the factory or from doctors in attendance upon their employees, but it is astonishing what a large amount of ignorance and secretiveness develops among the authorities connected with any factory, when questions arise as to the unhealthful conditions under which the operatives pursue their vocations."[61]

Amblyopia was the other manifestation of carbon disulfide poisoning in the rubber industry documented in the United States, perhaps because of the interest in such optic nerve damage in Britain. The first U.S.-reported case worked in a job at the intersection of two growth industries: the rubber trade and bicycle manufacturing. Dr. F. C. Heath described his patient:

> Miss T., visited my office early in the summer of 1900, complaining of weak eyes, saying there was a fog before them. Her sight was then about normal, far and near, but the eyes soon tired and the pupils were a little dilated. She had worked in the Indiana Rubber Company from Jan. 1, 1900, to the last of March, 1900, when she was unable to continue. She spliced the inner tubes of bicycle tires which she washed with bisulphide of carbon and chloride of sulfur solution, from open cans except the last two weeks, then from stoppered cans. She and her mother described her symptoms as great nervousness, excitability, irritability of temper.[62]

Dr. Heath is surprisingly forthcoming in his report. Over and above naming the employer, he recounts that he was later called to testify in a legal suit brought by the patient against Indiana Rubber for $5,000 in damages. Heath admits that he was not a strong witness; since the patient's eyesight had not been quantitatively impaired, he did not feel that he could state for the record that she had amblyopia. After being out for two days, the trial jury came back with a compromise award reduced to $500, one-tenth of the original claim. When Heath later saw the patient in medical follow-up, her pupils were more dilated, her eyesight was worse, she had lost peripheral vision, and there was pallor of optic disc (the optic nerve ending in the eye), all findings consistent with amblyopia, but this was after the trial.

Rubber treatment was the dominant but not the sole market for carbon disulfide in these years. For example, even as Delpech was reporting his first cases of poisoning in Paris in 1856, the chief military pharmacist for France in Algeria, Dr. Auguste-Nicolas-Eugène Millon, was promoting the use of carbon disulfide to extract the essence of perfumes from the native flowers of North Africa, a potentially lucrative enterprise.[63] Other uses of the chemical for its solvent properties also led to occasional poisonings, for example, in two young men in Australia who used it to extract oil from shale rock[64] and in a horse groom who tried to do away with himself by drinking two ounces of carbon disulfide, which was used to clean harnesses.[65] Carbon disulfide even found a place on the druggist's shelf as a compounding agent for selected cure-alls, including a combination with alcohol promoted as a treatment for rheumatism.[66] Nonetheless, all of these applications were inconsequential as sources of large-scale exposure.

But a major use for carbon disulfide outside of vulcanization emerged in the 1870s. It became the first widely promoted synthetic chemical pesticide. Beginning in the middle of the nineteenth century, a new scourge nearly obliterated viticulture in Europe. A small, aphid-like insect attacked the roots of grape vines, initiating a process that choked off nutrients and led to crop failures across the Continent. This disaster, of biblical proportions for the wine industry, came to be known as the "phylloxera plague." There seemed to be no cure and no way to halt the progression of an infestation that marched inexorably from vineyard to vineyard. Then, in 1873, a seemingly miraculous intervention began to be promoted.[67] Carbon disulfide, like a deus ex machina, arrived on the scene. Injected into the soil with pumps or even tilled in with specially designed plows, carbon disulfide in liberal amounts was reported to control the dread infestation. If the disease was far gone, the affected plots could be sterilized with even more carbon disulfide, enough to kill not only the insects but the grape vines as well, at least saving any nearby untainted vineyards. This destroy-the-village-in-order-to-save-it strategy was known at the time as carbon disulfide extinction treatment.

The grape vines of America were threatened as well, but in the end the U.S. grape crop never succumbed to widespread phylloxera loss. This was not because of chemical treatments, but thanks to naturally resistant rootstock. When this rootstock was eventually transplanted to Europe, it proved to be a cure for the disease. But before then, hundreds of thousands of pounds of carbon disulfide were used in a failed attempt to control phylloxera (the standard treatment was at least 150 pounds of the chemical per acre). Yet there seems to be no documentation of what happened to the agricultural workers who used it. After all, they were far from cities and unlikely to come under the care of a neurologist.

The same was true for those who may have become ill from other uses of carbon disulfide as a pesticide. Some of these applications have long histories. Even before the phylloxera bonanza for the carbon disulfide market, the toxic chemical had been proposed as a grain fumigant. Carbon disulfide also had established itself as a preferred extermination agent for gophers and ground squirrels because its vapors, heavier than air, sank down effectively into animal burrows.

We do have the record of one rural poisoning incident, from the casebook of Dr. C. L. Bard of Ventura, California. His report reads like one of the Grimms' tales:

> In the year 1882 there resided twenty-five miles distant from Ventura, two brothers, Alois and Ludwig Albrecht by name, who occupied a small cabin on a government claim. Honest, industrious and genial, they existed on the best of terms with their neighbors, who entertained for their good qualities the highest appreciation. Early one morning Alois sallied from the cabin and proceeded to the home of Robert Stocks, his nearest neighbor, with whom he never had had the slightest difficulty, or misunderstanding. Finding him in the barn, he entered into conversation by saying neither he nor his brother had felt well for some days, and finally, after accusing him of having poisoned them, drew out a pistol and shot him in the breast. Stocks fell to the ground and the assassin believing that he had killed him, started to town to report the matter to a friend.
>
> I was summoned to attend Stocks, and on the way met the German, and I shall never forget the spectacle that he presented. Mounted on a horse which he urged to the greatest speed, he reminded me of Don Quixote in his memorable crusade against the windmills. Arriving in town, he was persuaded to go to the sheriff, by whom he was locked up for safe keeping. He expressed no regret for the deed; said that before shooting Stocks, he saw poison on his hands; and that it would be useless to search for his body, as he saw the Devil carry it away.[68]

Dr. Bard goes on to tell how, when Ludwig arrived in town, he declared that his brother had not shot anyone and was being detained as a part of a Masonic conspiracy. A justice of the peace let Alois out on bail the next day. Soon after, he "removed his coat, made a pillow out of it, laid down in the hall, and placing a pistol to his head, blew out his brains." Ludwig returned, kicking the corpse and saying it was only a wax model. Ludwig was sent back to Germany, but eventually returned, according to Dr. Bard, once more "mentally sound."

Dr. Bard continued to care for Stocks, who recovered, and took advantage of a house call to inspect the nearby Albrecht cabin. The living arrangement had the brothers sleeping in the same low-to-the-ground bed, separated by a loose partition from a second room. What Bard found there explained the seeming folie à deux: "In the rear room, on a bench which stood close to the partition, was a fifty-pound can of bisulphide of carbon, which had never been opened, but which had been leaking for some time through a small hole in the bottom and but little of its contents remained. An odor, similar to that of decayed cabbage, pervaded the cabin, and was due to the escaping liquid. The can stood two feet above the pillows of the sleepers, and the vapor of the bisulphide of carbon, heavier than the atmospheric air, would necessarily descend to their faces."[69]

The gophers and ground squirrels nearly had their revenge on at least one California poison manufacturer. In the spring and summer of 1883, advertisements carried in the *Los Angeles Times* claimed: "Read and Foster's Bisulphide is safe, cheap and effective, and can be applied by anyone at any time without the least danger. The vapor of this material is heavier than air, and by its pressure reaches quickly every crack, crevice or department in the hole, destroying immediately all animals and insects contained therein."[70] For good measure, the advertisement also included a printer's fist [☞] highlight, "A boy can use it without danger." To get the message out even more clearly, the advertising copy began to be shortened with the simple header "Death to Squirrels and Gophers!"[71] One such advertisement even continued to run on 14 July 1883, the day after Charles Foster, the Foster of Read & Foster, was arrested for the attempted murder of A. H. Judson, a prominent Los Angeles business figure. The case quickly went to trial, and Foster was acquitted. His successful defense: temporary insanity due to the inhalation of carbon disulfide fumes at his own factory.[72]

The use of carbon disulfide as a pesticide continued into the twentieth century, but never again on as grand a scale as in the midst of the phylloxera crisis. At the same time, carbon disulfide was also falling out of favor in the rubber industry. Limited legal controls to protect workers finally came into play, first in Britain and later in Germany and France.[73] More important, carbon disulfide was being chemically engineered out of the vulcanization process. A new group of chemical compounds, called rubber accelerators, became the preferred agents. Although some of these chemicals were synthesized from carbon disulfide, thus still exposing chemical-manufacturing workers to its dangers, in the rubber trade the old hazard was largely reduced, albeit not eliminated.

Here and there, industrial poisonings still occurred in the rubber industry, especially when carbon disulfide was used as a solvent to transform rubber into a malleable spread or even to turn it into glue. In one such outbreak, a physician in France asked, "Is there a 'Leather Madness'?" when reporting on eight out of thirty employees who became deranged while working in one department of a single factory.[74] There seemed to be no explanation for the epidemic. Although senior colleagues told the doctor simply to mark it up to a "curious coincidence," he pursued the question, eventually discovering that carbon disulfide was being used in the workroom as a solvent to make rubber glue for leather.

In the twentieth century, carbon disulfide poisoning would no longer be an endemic human by-product of rubber vulcanization. Yet even as this source of carbon disulfide poisoning was finally on the wane, after producing fifty years of madness and blindness and paralysis and impotence, a novel and every bit as dangerous industrial practice was being introduced. This new manufacturing process had certain commonalities with the rubber business. Most saliently, it started with a natural raw material and treated it in a chemically intensive manner. The chemical manipulation of rubber started with a natural raw material that could be obtained from a limited variety of trees. In practical terms, in fact, natural latex rubber could be commercially derived from only a single New World species, *Hevea brasiliensis*.[75] Successful cultivation was achieved in plantations distant from the tree's native range, most prominently in Southeast Asia, but geographic limitations always played a major role in the politics and profits of the natural rubber business. Moreover, the commercial market for rubber products still had a limited scope, even with the sales boom provided by rubber tires first for bicycles and then for automobiles.

The new industry that came to depend on carbon disulfide had a far grander vision. Its potential feedstock stretched to the horizon, and its target consumer was not an adult who desired a condom or a child who fancied a balloon or even a baby who might need a new pair of (rubber-soled) shoes. Rather, the potential user of this new product was every man, woman, and child who had to be clothed. This was a technology that sought a revolutionary break with the past. It meant to leave behind earlier traditions of combing wool, scutching flax, ginning cotton, and, most importantly of all, reeling silk. A new synthetic textile industry, unprecedented in human history, sought to accomplish nothing less than taking cellulose, the abundant vegetable building block of every stalk and trunk, and transforming it into man-made silk. Noble man would be able to accomplish what heretofore had been the purview of the lowly caterpillar or the lurking spider.

Scientists and inventors had speculated about this sort of breakthrough years before it finally took on even a semblance of technical feasibility. Yet finally, in the last decades of the nineteenth century, artificial silk began to be made on an industrial scale. From the very beginning of synthetic fiber making, multiple manufacturing techniques came online nearly simultaneously. They differed from one another in key, separately patentable points. But all were predicated on the use of chemical processes to transform cellulose into filaments that could be twisted, wound, and spun. The thread and yarn produced could then be woven or knit in any way that far more expensive natural silk might be used—cheaply enough, in fact, to compete with other textiles, even cotton.

The first important artificial-silk-making process introduced was patented in the mid-1880s. In 1905, the *Journal of the Society of Dyers and Colourists,* a traditional organ of the British textile industry, published a three-part review of the topic, titled "Different Imitations of Natural Silk," that tallied no fewer than seven distinct processes to transform cellulose.[76] Of these, four survived a start-up phase and went on to become commercially viable.

The first method introduced, and initially the dominant process, took cellulose and modified it with nitric acid. The defining characteristic of this fiber was its nitrocellulose formulation; it was commonly known after the name of its French inventor as Chardonnet silk. The molecular makeup of the nitrocellulose, unfortunately, was similar to that of explosive guncotton. Moreover, its production required a number of relatively expensive materials, including a final treatment to reduce the material's inherent flammability. Or as the *Society of Dyers and Colourists* succinctly noted, the multistep processes required to produce nitrocellulose "show two distinct drawbacks; they are dangerous and very costly."

Two of the other types of artificial silk that emerged at this time were not at first as prominent as Chardonnet silk, but did go on to find their own market niches. One of these new inventions, first patented in France soon after nitrocellulose, originally was known as Parisian artificial silk, the better to differentiate it from Chardonnet's product, since he hailed from Besançon, far from the capital. When the Parisian patent lapsed, German inventors refiled it, going into production with what eventually became known Bemberg silk, after its major corporate producer. The manufacturing technique for this product, technically the cuprammonium process, put cellulose into a solution of copper and ammonia and then passed this through tubes shooting into a sulfuric acid bath. Cuprammonium was less dangerous to fabricate than nitrocellulose, but even more expensive.

The other new product, cellulose acetate, took a very different approach. Instead of manipulating cellulose without making any fundamental changes to the molecule, its production required the addition of a new organic chemical that formed a repeating side group bound to the long chain of the parent cellulose. The initial chemical characterization and manufacturing of this product was carried out in Germany. But unlike either nitrocellulose or cuprammonium silk, there was also an important American presence in the nascent cellulose acetate industry. This Yankee know-how was personified by an industrial chemist named Arthur Dehon Little.

Long before building his A. D. Little consulting empire, Little worked as a chemist in the paper industry, where his attention was drawn to the potential applications of modified cellulose-based fibers.[77] Little's partner was a fellow chemist named Roger B. Griffin. Little handled the clients of their chemical firm, and Griffin worked at the lab bench doing analyses. It was there that Griffin suffered from fatal burns in a laboratory fire in 1893.[78] Little filed his first cellulose acetate patent one year later and continued to be actively engaged in further technical innovations in acetate-silk-related business ventures in the decades that followed.[79]

But Little was also deeply involved as a key technical consultant to the American start-up of yet another artificial-silk process, one whose market would go on to dwarf that of the competing technologies of nitrocellulose, cuprammonium, and cellulose acetate combined. On 24 June 1899, at the instruction of his client, a lawyer named Daniel C. Spruance, Little sailed to England. Little's assignment was to meet with Dr. Charles Cross, co-inventor with Edward Bevan of the viscose method for producing artificial silk fiber. Further, Little was to directly assess the manufacturing process at facilities in England and on the Continent. Then, having fully explored the territory, he was to report back to Spruance on the best approach for exercising his options to purchase the exclusive rights for U.S. domestic commercialization of viscose.[80]

Impressed with what he learned during his sojourn, Little detailed his experiences in a typewritten account over four hundred pages long: "Report to Daniel C. Spruance, esq. on the Technical development of Viscose on the Continent of Europe and in Great Britain." It included a precise description of the multistep process involved in viscose making. Little describes the core features that define viscose and highlights the key chemical constituent that characterizes its manufacture to this day—carbon disulfide.

What Cross and Bevan had discovered and patented in 1892 was that carbon disulfide, small and elegant, was uniquely capable of liquefying cellulose without fundamentally changing its structure. The cellulose–carbon disulfide solution, a syrupy concoction whose consistency allowed it to be forced through fine apertures, yielded filaments amenable to thread making and then to weaving. There were tricks to the trade, needless to say. The cellulose starting material first had to be treated aggressively with caustic solutions, and the mixture of carbon disulfide and cellulose needed to mature a bit, ripening in industrial vats like a not-so-fine wine, in order to create the ideally viscous solution. That syrup was called xanthate. At the end of the process, to resolidify the xanthate into filaments, extrusion into a bath of sulfuric acid was required. Doing so niftily forced out the carbon disulfide, regenerating the native cellulose. The process was both efficient and less costly than competing artificial-silk-making methods. Caustics and acids were downright cheap; carbon disulfide was relatively inexpensive as well. In fact, its cost was low enough that there was no real economic impetus to methodically recapture the carbon disulfide coming from the extrusion baths. Arthur Little was unequivocal in his recommendations to Daniel Spruance: "I regard the rights possessed by the American Viscose Co. and of which you may obtain control under your main option as an extremely valuable piece of property which is probably worth many times the price named in that option and it is certainly well worth acquiring at the figure given. Assuming that you should see fit to act upon this advice I would recommend the immediate equipment of a small, well-located factory to be utilized mainly for the purpose of developing the process and extending its applications."[81]

Little, in his recommendations, offered no cautionary advice about any potential safety or health complications of working with carbon disulfide, a chemical for which there was a half century of accumulated misadventure in the rubber trades. English manufacturers certainly understood the risk of carbon disulfide explosion, but also were aware of at least some of the adverse health effects of viscose manufacturing, too. One of Cross and Bevan's earliest viscose laboratory workers, a chemist named Edwin Beer, kept a journal during his early years of employment. More than once Beer documented explosions, including one that occurred in 1901 when his co-worker tried to solder a piece of equipment meant to draw off carbon disulfide. Fortuitously, the door to the lab had been left open, and Beer was blown out into the street without serious injury. Beer also documented persistent eye troubles in multiple journal entries, noting as early as 1898 an April visit for the problem to a Harley Street (London) consultant. In 1900, the colleague who later caused the explosion wrote off his own loss of a billiards championship to disturbed vision, stating, "[I] would have won, but I saw *two* balls and hit the one that wasn't there." Beer noted later, in hindsight, that carbon disulfide could cause "intoxication, and even madness," and also believed that exposure to it could lead to hair loss, perhaps contributing to his own baldness.[82]

Although ignoring the hazards of carbon disulfide, Little otherwise was very attentive to details in his report; an appendix lists all the individuals and corporate entities involved with the American Viscose Company and Cross and Bevan's patent rights.[83] In this list, Little acknowledges the thirty-five shares held by Mary Griffin, the widow of the former partner who had died in the line of duty. Commenting on carbon disulfide as a poison may have been beyond the scope of the technical report that Little was engaged to compose. And after all, he was a chemist, not a medic. But having lost a partner to a laboratory conflagration, Little might have been more sensitive to the dangers of carbon disulfide explosion, at least.

That small factory that Little so strongly recommended Spruance to organize soon came to into being, well located in Lansdowne Borough, Pennsylvania, not far from Philadelphia. It did not take long for trouble to emerge. On 8 February 1904, a young man identified as H.R., who worked at the new factory, turned up at the University of Pennsylvania Hospital Dispensary complaining of weakness, especially in his legs, but his troubles went far beyond that:

> His work kept him in the bisulphide room all day. When he would first go into the fumes he would feel exhilarated, became loquacious, and passed for a jolly fellow. When he cleaned casks he would be almost overcome, and he would have to come out frequently for fresh air. After he had been at work for about a month he began to have headache. . . . He was depressed, taciturn, and irritable when he would go home; his memory was poor, and his appetite failed, for he tasted and smelt the carbon bisulphide constantly. He slept poorly. These symptoms continually became worse. Later the weakness extended to the legs until they became worse than the arms, and he was so debilitated that he had great difficulty in getting up the stairs. His sexual power was lost.[84]

A few weeks later the same physicians treated a second worker from the Lansdowne factory. This man came in complaining of nervousness and, rather than impotence, sexual excitement. At nearly the same time, a third worker from Lansdowne presented for care at the University of Pennsylvania, suffering from "weakness and mental depression after leaving the works, although while there he was often exhilarated"; he was cared for by a doctor who did not know of the other two cases.[85] These treating physicians had no trouble clearly identifying carbon disulfide as the cause of the illness, and both publications documenting the cases were careful to cite relevant works that thoroughly described the chemical's poisonous effects. Indeed, in the activist tradition of Delpech, the physicians involved in the first two cases did not shy away from recommending preventive measures: "Masks and nose-pieces seem to be of no use. A vacuum hood placed over the open vats would draw off a good deal of the gas; moreover, if the workmen were rotated between this work and some other, there would be less of the poisoning gas."[86]

A Philadelphia suburb may have marked the far western border of what was becoming a rapidly expanding international presence for viscose manufacturing, and its easternmost outpost was established in 1909 in the town of Mytishchi, near Moscow.[87] An early Russian interest in viscose had been spearheaded by no less a personage than Dmitri Mendeleev himself, of periodic table fame. In 1900, Mendeleev prophesied:

> The work on viscose just started in 1899 and now is in its first, or embryonic, stage of development and therefore it is best to talk about it with caution. But ... the victory of viscose will be a new triumph of science: just as the discovery of beet sugar freed the world of its dependence on the tropics, the discovery of viscose may free the world in relation to cotton.... Russia with its heartland of forests and grass could, with the production of viscose, provide the entire world with a colossal amount of fiber.[88]

The Mytishchi Viskosa factory, however, was no realization of Mendeleev's arcadia. Not long after the factory opened, the regional sanitary inspector (a man named Lebedev), the plant's director, a factory mechanic, and another worker were all killed in a carbon disulfide explosion. But conditions went from bad to worse when, in 1911, the enclosure that had covered the manufacturing line was removed to accommodate new equipment. By 1913, Dr. Vasily K. Khoroshko of Moscow's Imperial Medial University reported four cases of carbon disulfide poisoning in the factory's small workforce of six production laborers.[89] Khoroshko, who had a prominent career as a neurologist in prerevolutionary Russia and later in the Soviet Union, detailed the impairments in workers' nervous system function, including a thirty-three-year-old named Sareicheff whose vision was so adversely affected that he could not differentiate "a cow from a horse." In his paper, Khoroshko also thoroughly reviewed the medical literature on carbon disulfide, from Delpech through Laudenheimer, including recent German publications on the rubber trade. But unaware of the U.S. cases in Lansdowne, Khoroshko incorrectly assumed that his were the first cases of carbon disulfide poisoning to be reported in the new viscose industry.

Meanwhile business in Lansdowne had not been going well: originally operating as the General Artificial Silk Company, the facility was not as effective in generating profits as in producing disease. From its founder, Spruance, the business was taken over by his major shareholder, Silas Petit, and its name was contracted to Genasco. By 1908, Silas was dead; his son, John Read Petit, replaced him. "Genasco" may have been pithier than the original moniker, but the business still sputtered. Not so for Cross and Bevan's original operation in England. After being taken over by Samuel Courtauld and Company, it became integral to a powerful manufacturing concern. Courtauld and Company made John Read Petit an offer that he did not refuse—Courtauld bought back the patent rights that Petit held and set about establishing its own American subsidiary operation.[90]

Lansdowne was abandoned, and a new facility was constructed in another Pennsylvania town, Marcus Hook. Production began there in 1911 as the American Viscose Company, the same name as the entity for which Arthur D. Little had done his original consultation a decade before. Despite the old name, this was meant to be a new beginning for the industry. Indeed, the 1911 founding at Marcus Hook is the date most often claimed as the establishment of the viscose industry in America.

From the start, under Courtaulds sheltering wings, the American Viscose Company had a clear vision of how it meant to present itself. This included nothing less than a model industrial village for its workers, designed by Emile G. Perrot, an American architect, who first went to England to study similar projects there. In *Discussion on Garden Cities,* a small, hand-sewn Arts and Crafts–style pamphlet promoting the project, Perrot sings the praises of his patrons: "It is to the credit of the owners that they have deviated from the stereotyped American practice of building houses in rows, on straight streets, in which no attempt is made to relieve the monotony of design, nor conduce to the general uplift of the surroundings, instead they have taken into account the physical and intellectual needs of the workers and created an Industrial Village on Garden City Lines."[91]

An auditorium, a gymnasium for boys and one for girls, and a library were part of the plan. There were to be no outhouses allowed. It was a new day for viscose and America, and the industrial village at Marcus Hook was meant to be its equivalent of a shining city on a hill. Lansdowne was forgotten, utterly.

Most importantly, no further mention of ill health within the U.S. rayon industry would be made, at least publicly, for another twenty years. There were no follow-up investigations from the University of Pennsylvania, and Khoroshko's Russian-language report languished in obscurity, never to be cited in the West. Treating cellulose with large quantities of carbon disulfide, which emerged as downstream fumes when the liquid viscose was sprayed out into baths of acid, was a process every bit as dangerous as dipping rubber condoms in vats of the same poisonous solvent. To think otherwise in the face of what was already well known about carbon disulfide was little more than an exercise in make-believe. And yet physicians and hygienists in the United States as well as in Britain, France, Germany, and all the other European viscose-producing countries ignored the obvious hazards of this process, and the owners and managers of the viscose factories, who were most familiar with the facts on the ground, were more than satisfied to maintain this convenient fiction. All were silent, but carbon disulfide, the old rubber industry hazard, was back with a vengeance.

2

The Crazy Years

On 15 March 1928, W. T. Kelly stood up in Parliament to pose a question. Addressing the home secretary, the MP asked pointedly whether "the Home Office has received any Reports as to the conditions of health of workpeople employed in the artificial silk factories; and, if so, will he make known to the House his decision on these Reports?"

Sir William Joynson-Hicks responded, "I have received reports by the medical and other Inspectors of Factories who have been visiting these works. It appears from those reports that the conditions generally are satisfactory but cases of conjunctivitis have occurred at one or two works, and there have also been some cases of dermatitis. Suitable precautions are being taken in each case and the hon. Member may be assured that the conditions will continue to receive the special attention of the medical staff."[1]

This exchange was hardly the stuff of high drama of the sort that we have come to expect from parliamentary debates. Indeed, within the year George Arliss set the benchmark for this sort of thing, cinematographically at least, with his Oscar-winning portrayal in 1929 of Disraeli. Nonetheless, the repartee between Kelly and Joynson-Hicks was duly noted in the "Parliamentary Intelligence" section of the medical journal the *Lancet,* although considerably more column space was allotted to the question of "invaliding from the Navy" (dealt with by the parliamentary secretary to the Admiralty).[2] Even so, for artificial silk manufacturing, any mention at all of possible hazards in a medical journal—and in the *Lancet,* no less—marked something of a turning point. The industry was being given a clean bill of health, but the fact that it had to be done publicly and at such a high level was significant.

In the early twentieth century, the production of synthetic silk fiber was little more than a start-up industry based on small manufacturing units, often carried out on a prototype scale. By the time of this 1928 parliamentary exchange, however, the manufacturing of viscose-process artificial silk was a very big business indeed. In Great Britain, that industry was nearly synonymous with a single entity: Samuel Courtauld and Company.[3] In addition to flourishing at home, Courtaulds was reaping tremendous returns through its wholly owned foreign subsidiary, the American Viscose Company.

Thus, when Sir Joynson-Hicks replied to MP Kelly's question on the health hazards of artificial silk production, he had every reason to consider his response carefully, because of the size and importance of this industry. Moreover, Joynson-Hicks had a personal connection with the textile trade. In 1895, he married the daughter of Richard Hampson Joynson, a silk manufacturer of Bowden, subsequently assuming the family name of his prosperous in-laws.[4] Joynson-Hicks, also known by the moniker "Jix," was first elected to Parliament as a Conservative from Manchester in 1908, defeating Winston Churchill (who stood for the Liberal Party) in large measure by Jew-baiting him by association.[5] After losing that seat in a follow-up election, Jix later reentered Parliament from Middlesex. Part of the "Die Hard" faction (Tea Partiers of their day), Joynson-Hicks became the Conservative Party's home secretary in 1924, making a name for himself through anti-immigrant policies.[6] Although widely seen as a xenophobe and an anti-Semite, he was hardly a friend to organized labor either. In 1926, Joynson-Hicks played a prominent role in the government's repressive response to the general strike of that year.

What the *Lancet*'s "Parliamentary Intelligence" column did not include was MP Kelly's follow-up. Pressing the home secretary for more specifics, Kelly asked, "Have the reports been received as affecting Messrs. Courtaulds, the Spondon Factory of British Celanese, and the Belfast Factory?" Joynson-Hicks, in good political form, temporized: "Before I answer on definite cases, the hon. Member must give me notice."[7]

At this point, his sidestepping crossed over the line to falsehood. In fact, the home secretary already *had* received prior notice from another MP. In the previous House of Commons session, Labour MP James Maxton, one of the Glasgow "Red Clydesiders," transmitted a query whether "the attention of the factory inspectors has been drawn to the need for special vigilance in the examination of artificial silk factories to safeguard the health of workers in that industry?"[8] Joynson-Hicks's written reply, dated 23 February 1928, and thus preceding by three weeks the Kelly exchange, was an unequivocal denial of any problem: "Yes, Sir, the attention of the inspectors has been directed to these works, and will continue to be. There is no evidence of any serious risk to health, but certain precautions are required and these are being enforced."[9]

Less than a week later (29 February) the secretary of labor (Sir Arthur Steel-Maitland) had to respond to a series of parliamentary questions also posed by MPs Maxton and Kelly on working conditions (as well as wages) in the artificial silk industry.[10] Steel-Maitland, passing the buck, suggested that some of the issues raised would be better addressed by the home secretary.

Disagreeable as fielding these labor health questions may have been for Jix, this duty came with his job. As his colleague the secretary of labor understood well enough, the home secretary was the one directly responsible for enforcement of the Factory Act, which covered worker exposure to harmful agents such as might be encountered in artificial silk manufacture. Thus, Jix would have been the minister to bring forward any legislative amendments to the Factory Act. Since it had not been substantively modified since 1901 (when there was not an artificial silk industry), it was by the time of these parliamentary exchanges a quarter century out of date. In fact, in the 1927–28 legislative session, Joynson-Hicks and the Conservatives were under particular pressure to introduce reform legislation to address deficiencies in the Factory Act. The home secretary promised action, although in the end none came about, at least on his watch.[11]

Even in its outdated state, the 1901 version of the act allowed for periodic updates to its list of hazardous intoxications that were recognized as occupational illnesses for which reporting to the authorities was legally mandated. Indeed, under Joynson-Hicks's aegis, a new group of such conditions was put on the books. Carbon disulfide poisoning was among them. The inclusion of carbon disulfide was something of a holdover from its past use in rubber vulcanizing, but it nonetheless applied to any industry in which it was found, including artificial silk manufacturing.

Joynson-Hicks had responded to a formal request on the topic well before he ever had had to deal with Kelly and Maxton. In December 1925, MP William Cornforth Robinson (Labour, Elland in Yorkshire) asked Joynson-Hicks to tell the Commons how many cases of poisoning due to carbon disulfide and the other newly listed toxins had been reported in that year (the first in which this was mandated). Joynson-Hicks acknowledged that such new reports included, among others, "two cases of carbon bisulphide poisoning."[12] He misreported the number of cases from that year. In fact, there were three, not two. And there was a related omission: Joynson-Hicks did not say a word about artificial silk manufacturing, the industry from which that third, unmentioned case had been reported.

The sanitized accounting by Joynson-Hicks made it into the pages of the *British Journal of Medicine* (*BMJ*), which, like the *Lancet*, had a regular column highlighting parliamentary health-related items ("Medical Notes in Parliament").[13] But almost no one who might have seen that small notice in the *BMJ* would have been in a position to know or even suspect that something in it was amiss. No one, that is, except the man who reported directly to the home secretary on health conditions in the factories, Dr. Thomas Legge.

By the end of 1925, Sir Thomas Legge (knighted earlier that year) had served for some years as the senior medical inspector of factories, enforcing industrial health provisions of the Factory Acts. Legge's background was traditional, but with a twist. He was born in Hong Kong, the son of a reverend who later became the first professor of Chinese at Oxford. His impeccable training took place at his father's institution, Oxford, then at St. Bartholomew's Hospital, and later during a de rigueur medical stint in Vienna. His *BMJ* obituary notice would summarize him as "a man of charming personality, with a cultivated mind and a deep sense of social responsibility."[14]

As senior medical inspector, Legge oversaw production of *The Annual Report of the Chief Inspector of Factories and Workshops*, published six months after the year being covered. The first (and lengthiest) chapter usually took the form a "general report" covering questions of sanitation and employment. Another chapter presented a first-person narrative summary, in the voice of the senior inspector, of problems in industrial health, organized by sections that addressed each reportable occupational disease.

The *Report* for 1925 was the first to capture carbon disulfide poisoning as a reportable disease. This new condition would have been of particular interest to Legge, and so he focused on it in some detail:

> *Carbon Bisulphide.*—Two cases occurred in the cold cure process in the rubber industry and the other—a very acute case, the symptoms being headache, vomiting, delirium, loss of muscular power, and almost complete anaesthesia (loss of sensation) from which he has made only slight improvement—in the manufacture of artificial silk. Dr. Henry and I shortly afterwards (on different occasions) examined some 12 men similarly employed in artificial silk and noted in several of them symptoms of absorption, such as headache, gastric disturbance, distaste for food, lassitude and depression. In the factory in question considerable attention had been given to the question of ventilation, but at the machines where carbon bisulphide was used this had not been sufficiently locally applied. This has since been remedied.

> Other effects on health in this very important new industry have been noted—e.g. irritation of the eyes from minute proportions of hydrogen sulphide gas present in the air in spinning rooms, and the effects on skin from the acid used in spinning.[15]

Based on Legge's synopsis, there were fifteen cases of carbon disulfide toxicity in that year. The "very acute case" manifested the hallmarks of severe carbon disulfide nervous system toxicity that had been so well established in the rubber industry, most ominously delirium and neuropathy (that is, a dying off of the nerves supplying muscles and peripheral sensation). The other twelve cases suffered from less specific complaints, but their symptoms were entirely consistent with potentially dangerous carbon disulfide effects. In particular, "lassitude" and "depression" were red flags for the ominous psychiatric effects of the toxic chemical, with the attendant risks of manic psychosis and even suicide. Legge pointed out that eye and skin injuries were two other problems troubling an unspecified number of workers. Although toxic damage to the optic nerve, leading to vision loss, due to carbon disulfide had been well documented in the rubber trade, eye and skin irritation in the viscose industry, Legge emphasized, were likely attributable to other chemicals. Indeed, the question of eye damage in the viscose industry was to grow into a major and persistent problem, in Britain and elsewhere.

Despite Legge's tally, the official tabulation in the *Report* for 1925 lists only three carbon disulfide cases in total: two from the rubber industry and the single "notified" case from an artificial silk factory.[16] One has to go back and realize that "symptoms of absorption," in governmental phraseology, can be taken to mean a condition just short of classifiable as "poisoning." Poisoning would have required an official report and would have made it in to the tabulation. Less severe symptoms seem to have gotten a pass as merely reflecting absorption of the poison. Of course, three cases still add up to one key case more than the two that Joynson-Hicks was prepared to acknowledge.

As was typical in the *Reports*, the industrial manufacturer visited by Drs. Legge and Henry is not named. In his memoir, *Industrial Maladies* (published posthumously in 1934), however, Legge lets the cat out of the bag. He characterizes the manufacturer as being "two large works at Flint" (Wales). He goes on to state, "In 1925 I visited, I think with Dr. Bridge, these works because of severe poisoning from carbon bisulphide."[17] There were only two viscose plants in Flint at that time, the Aber and Castle works. They were both major Courtaulds facilities.

The 1925 *Report*, the first one to include carbon disulfide poisoning as a notifiable disease, was the last one overseen by Legge. In 1926, he abruptly and quite publicly resigned his position. Legge, truly a man with a "deep sense of social responsibility," as his obituary later noted, did so in protest against an egregious policy decision by the British government: it reneged on an international treaty that it had committed to and that Legge had helped negotiate. The treaty was intended to ban the use of lead in interior paints.

The change in factory inspector leadership was clear by the time that the next *Report,* for the year 1926, appeared. Dr. John C. Bridge (whom Legge said had been with him in the visit to Flint), in his first report as senior inspector role, noted, "By far the most extensive use of carbon bisulphide at the present time is in the manufacture of artificial silk by the 'Viscose' process. Many new and large works have been erected during the past two or three years, but no case of carbon bisulphide poisoning from this manufacture has been reported."[18] Bridge went on to describe the viscose process in some detail, emphasizing the importance of good ventilation, a fact "universally recognized" by the manufacturers.

Universal recognition, however, was not the story told by the factory inspector's field records. Over a five-year period through 1927, eight inspection reports on the viscose churn room at the Courtaulds facility in Flint contain no comment whatsoever on ventilation. Many other "interventions" were recommended explicitly (guaranteeing an adequate number of coat hooks was one notable concern). In parts of the factory beyond the churn room, ventilation was mentioned only a handful of times, such as in the "viscose cave," where xanthated cellulose ripened before being spun into filaments, and in the spinning room itself.[19]

All in all, things seemed to be going fairly well for the British viscose industry, which is to say, things were going gangbusters for Courtaulds. No cases of carbon disulfide intoxication were officially notified through the chief inspector of factories for 1927.[20] Legge was out of the picture. Joynson-Hicks remained a stalwart at the Home Office. True, there were a few troublemaker Labour MPs, but Jix could handle them.

Most conveniently of all, medical science (in Britain and elsewhere) seemed to be in some kind of stupor when it came to noticing any potential for carbon disulfide toxicity in viscose manufacturing. This chemical, after garnering extensive medical attention in the last half of the nineteenth century, and being clearly recognized as a terrible hazard in the rubber trade, was virtually ignored in the first decades of the twentieth. And after all, what need was there to be concerned? Such, at least, was certainly the view of the dominant figure in British occupational medicine at the time, Sir Thomas Oliver. He had early sounded the alarm on the hazards of carbon disulfide in the rubber trade, was the author of several textbooks, and spoke with the voice of authority. Underscoring his preeminence, in the fall of 1925 he managed to publish nearly identical commentaries in the *Lancet* and the *British Medical Journal*. This was the medical publication equivalent of achieving a "full Ginsburg" on a single Sunday-morning round-robin of TV talk show appearances.[21] In "Some of the Achievements of Industrial Legislation and Hygiene," extracted from an address he had delivered in Amsterdam to an international medical congress, Oliver trumpeted the near elimination of carbon disulfide poisoning through successful controls in the rubber trade. But he wrote not a word about artificial silk manufacturing.[22] The major British occupational medicine textbook of the period, published in 1923, at least acknowledged that carbon disulfide was being used "in the preparation of cellulose for artificial silk," even though it does not refer to any cases of disease in this industry.[23]

Ironically, the one independent hygienic investigation on the subject that seems to have been carried out in Britain in this period was never properly published and, to the extent that it was ever cited at all, was summarized only in Italian. Late in 1925, under the heading "Cause of poisoning in the artificial silk industry," *La Medicina del Lavoro* published a very brief notice of a British report on carbon disulfide exposure.[24] This eighty-five-word synopsis stated that Dr. Arnold Renshaw, in a session of the Institute of Chemistry in Manchester, presented a study of industrial poisoning, in particular the dangers represented by the use of carbon disulfide among artificial silk workers. Despite the abstract's brevity, it includes a specific, quantifiable warning: "More than one milligram of carbon disulfide per liter of air produces chronic poisoning and . . . this limit is easily achieved even in ventilated rooms."[25] The first part of this statement—that chronic exposure to a concentration above one milligram per liter of carbon disulfide is toxic—allows for a terribly high air level of the chemical. Indeed, this cutoff point was far too permissive, even based on what was known in the 1920s. Renshaw may have derived this dangerous benchmark from earlier work by one of the most prominent toxicologists of the era, Dr. Karl B. Lehmann.

Lehmann's classic approach was to expose test animals to lethal and sublethal concentrations of a test chemical and then to try out far lower levels on a few humans, typically his own research doctoral trainees. In 1894, Lehmann published data on carbon disulfide inhalation based on work in his laboratory in Würzburg, Germany.[26] He carried out a series of experiments subjecting to exposure cats, rabbits, and guinea pigs as well as Drs. Sigmund Rosenblatt and Michael Hertel. Dr. Rosenblatt did pretty well after a few hours of exposure up to a concentration of 0.76 milligrams per liter of carbon disulfide. Headache and other neurological symptoms started to kick in at higher levels, especially at 2 milligrams (or higher) per liter. Dr. Hertel, in contradistinction, had trouble getting past 1 milligram per liter without symptoms. One milligram per liter of air equates to more than 300 parts (molecules) carbon disulfide per 1 million parts air (parts per million, or, more commonly, ppm).

Lehmann subjected Rosenblatt and Hertel to exposures that lasted only three to four hours at most. Prolonged exposure to carbon disulfide, day in and day out, was an entirely different toxic scenario. For example, the 1927 edition of the handbook *Noxious Gases,* for many years the bible of industrial-hygiene exposure control, recapitulated Lehmann's cutoff point, but with a clear admonition: "The concentrations given here are for acute effects; exposure of several hours a day to concentrations lower than those mentioned leads in a short time to chronic poisoning."[27] It was precisely because Renshaw allowed for such a high threshold of carbon disulfide that the other half of his equation was even more telling: the statement that levels as high as this were "easily achieved" in the British industry—in other words, easily achieved and thus commonly exceeded.

Arnold Renshaw was a prominent Manchester forensic pathologist with a particular interest in industrial toxicology early in his career.[28] The *Journal and Proceedings of the Institute of Chemistry of Great Britain and Ireland* confirms that on 2 March 1925 he gave a lecture in the rooms of the Manchester Literary and Philosophical Society, titled "Chemical Poisoning Occurring Amongst Industrial Workers."[29] Although it is noted that the lecture covered both organic and inorganic poisons and that an "interesting discussion" followed, in which "Mssrs. Brightman, Ellsdon, Hannay, Herbert, Rogers and others took part," no indication is given that carbon disulfide was specifically addressed. The minutes of the meeting in question confirm the same information, but do not provide any further details of the presentation.[30] Apparently, no scientific paper based on this presentation, including any measurements of carbon disulfide levels that Renshaw may have made, was ever published, nor is a text of the presentation preserved in either the archives of the Royal Institute of Chemistry, which subsumed the Institute of Chemistry, or the Manchester Literary and Philosophical Society. In later life, Renshaw became known as a collector of Constable and other English landscape painters.[31] In the end, his important early work on carbon disulfide appears to have been fleeting and ethereal, as if summoning up the spirit of one of Constable's cloud studies.

The pathway by which the Renshaw report came to the attention of the editors of *La Medicina del Lavoro* remains obscure, but it is not surprising that they found it noteworthy once they knew about it. The viscose industry was beginning to be of particular interest to the discipline of occupational medicine in Italy in the mid-1920s. In the first decades of the twentieth century, Italian occupational medicine was one of the most advanced centers of that discipline, holding world-class status. Its flagship institution was the Clinica del Lavoro, established in 1910 as a groundbreaking medical facility specifically dedicated to the study and treatment of occupational diseases. Formally renamed years later for its founding director, Dr. Luigi Devoto, the Clinica still functions today on its original site, surrounded by a complex of later buildings that form a modern academic medical center in central Milan.

The Clinica, in addition to its clinical activities, also oversaw the publication of *La Medicina del Lavoro,* the premier Italian specialty medical journal in occupational medicine. *La Medicina del Lavoro* has been published for more than a century. A complete set can be found in the ground-floor library of the Clinica, along with an encyclopedic collection of Italian occupational health reports, proceedings, and regional publications.

It is on those dusty shelves, and this is no mere metaphor, that one can find the seminal publications first sounding the alarm on carbon disulfide in the Italian artificial silk industry. The first such report appeared on 31 March 1925 in *Minerva Medica* (subtitled *Gazetta per il Medico Pratico*).[32] The journal was published in Turin, an industrial city then at the epicenter of the new and expanding Italian artificial silk industry. The article was a clinical review of causes of polyneuritis, a clinical dysfunction involving multiple peripheral nerves. It was written by Professor Angelo Ceconi, director of the University of Turin's Institute of Special Pathology and Demonstrative Medicine. The latter is an arcane term that once was used to refer to clinical medical instruction of the sort epitomized by bedside patient rounds. It has been preserved in Russia and the former Soviet Republics as meaning something akin to "evidence-based medicine"—hence, the Kyrgyzstan state medical school's Department of Pharmacology and Demonstrative Medicine.

Consistent with this didactic context, Ceconi's article is more than a general and theoretical overview of the subject—it includes illustrative cases too. Two of these involve neuropathy caused by carbon disulfide. One was a patient employed as a rubber worker—the classic scenario of carbon disulfide exposure of the past. Notably, the other case was employed in an artificial silk factory. The symptoms in this latter case included weakness in the limbs and impaired gait. The patient also recounted that the symptoms first started with headache, loss of appetite, vertigo, insomnia, and a tingling sensation. All these symptoms are consistent with fairly heavy exposure to carbon disulfide. Ominously, the patient reported that such symptoms were commonplace among his co-workers, noting that when someone was removed from exposure early in the course of such illness, the condition would abate. Ceconi did not mince words in his assessment of what needed to be done: "It is hoped that the authority responsible for the welfare and hygiene of work, as it addresses this issue, establishes labor standards aimed at reducing to a minimum, if not to suppressing altogether, any danger to these workers."[33]

It did not take long for a representative of the "authority" to weigh in. The chief medical inspector for the Italian Ministry of Labor, Giovanni Loriga, published a report on the subject two months later, in the May 1925 issue of the governmental *Bollettino del Lavoro*.[34] This was about when Legge, more or less in a parallel governmental position to that of Loriga, was traveling up from London to Flint, Wales.

Loriga's focus was reflected in the title of his report: "Hygienic conditions in the artificial silk industry." He covered, in excruciating detail, the technical aspects of viscose production, going through each phase of the manufacturing process, in sections lettered *A* through *L*, albeit skipping the nonexistent Italian *J*. Loriga methodically highlighted the principal sources of carbon disulfide exposure and also addressed the adverse health effects of the chemical in the Italian viscose industry: "In one of these [factories], an outbreak of an epidemic of nervous disease was noted affecting almost all of the workers in the [fiber] twisting process, which was diagnosed by some physicians as collective hysteria, while by others it was judged to be due to carbon disulfide."[35]

In addition, Loriga summarized the findings from a recent clinical-pathological report of the death of a young artificial silk worker. The case had been presented earlier in 1925 to the Società Medico-Chirurgica di Pavia by a clinician named Piero Redaelli. In retrospect, Redaelli's case is not particularly germane to the toxicity of carbon disulfide for the nervous system, since it concerns gastric pathology. Nonetheless, it underscores the contemporary general Italian medical interest in the potential for carbon disulfide–related toxicity in the rapidly growing workforce of the viscose industry. Indeed, Loriga cited the report before it appeared in print, indicating that it had been called to his attention through other means. He apparently was unaware, however, of Ceconi's far more relevant case report.

The pace of Italian medical reports of carbon disulfide illness picked up rapidly. Loriga's review was guaranteed a wider exposure after an extended summary was published later in 1925 in *La Medicina del Lavoro* (in an issue preceding the one summarizing the Renshaw report).[36] Redaelli's case appeared in full in the *Bollettino della Società Medico-Chirurgica, Pavia*, along with a companion publication by M. Arezzi reporting the findings from animal studies of carbon disulfide exposure.[37]

Later in the year, another, far more comprehensive report on the health of artificial silk workers appeared in the same Pavia journal. The focus of that article, which acknowledged the work of both Redaelli and Arezzi, was the effects on the blood of carbon disulfide exposure among eleven workers.[38] Brief case histories were presented for each, and five of the eleven (those with the most direct carbon disulfide exposure) manifested neurological symptoms of toxicity. The first subject, twenty-two years old, well-nourished, a modest smoker, and nondrinker, had worked for only a month in what the author referred to as the "carbon disulfide department" of the artificial silk factory. Although not further specified in the article, this most likely meant viscose churning. The job certainly involved heavy exposure, since the subject displayed symptoms that the author recognized as corresponding to the initial "excitation period" of carbon disulfide intoxication, first described by Delpech more than three-quarters of a century earlier among the Parisian condom makers who were dipping rubber into liquid carbon disulfide. In addition to dizziness, unsteady gait, tingling, and lower limb weakness, the young Italian man also experienced "exaggerated emotions."

In 1923, two years before the first flurry of publications specific to the Italian artificial silk industry appeared, a case report described more than just "exaggerated emotions" in a carbon disulfide worker—it documented acute psychosis in an employee of a primary carbon disulfide chemical production plant in Sicily.[39] We are told that "Giuseppe Bonacc . . . ," from Giarre (in southeastern Sicily), first experienced hallucinations and paranoid delusions that later worsened until his supervening aggressive outbursts directed at his neighbors ultimately led to a psychiatric hospitalization.

It is not surprising that Italian carbon disulfide manufacturing was based in Sicily. The crucial starting material, mineral sulfur, was a prime natural resource of that region. Indeed, the sulfur mines of Sicily were famous or, better put, infamous. This was clearly the view of Sir Thomas Oliver, otherwise such a booster of the carbon sulfide success story. In 1910, Oliver made a visit to see firsthand the sulfur mines of Sicily. Among the various horrific conditions that he described, the most heartrending was the fate of the child laborers hauling up from underground the mined sulfur ore, carried on their backs: "The ore got by the men is carried on the shoulders of barefooted, scantily-clad boys up the steep and worn steps to the surface. As these *carussi* are not always given lights, the journeys up and down are made in the dark. Many are the sad accidents which have taken place owing to the *carussi* slipping. The boys and their burden roll down the steps, entangling in their decent other *carussi* who may be ascending."[40]

Oliver provides extensive details on the sulfur mining operations in Sicily, but he does not mention the carbon disulfide that then was manufactured from the raw material garnered at such a high human cost. The factory making the chemical was online at the time of Oliver's tour. In fact, the author of the case of Giuseppe B. from Giarre makes it a point to state that as of his writing in 1923, the plant in question had been in operation for thirty years. (It had in fact been thirty-one years.) Of course, Giuseppe B. was identified by name, even if only partially, while the factory was kept completely anonymous: long before U.S. health care laws protected patients' privacy, there was already an international medical commitment to the privacy of brand names and the protection of corporate identities.

Fortunately, the Internet can be used to determine the identity of the only factory in Sicily making carbon disulfide at that time. Called Insulare, it was founded in 1892 by a British engineer named Robert Trewhella. He had made his fortune by building railroads for the Sicilian sulfur mines, including the operations in the area that Oliver visited. Trewhella then expanded into sulfur refining. The Insulare was originally located in Catania proper and then relocated in 1901 to the nearby seafront area of Ognina.[41] At the time, there had been a good deal of popular protest against this expansion because of the noxious fumes the factory brought with it. The new and expanded carbon disulfide factory was built anyway.

Italian medical reporting played a critical role in virtually rediscovering that carbon disulfide was still a hazard in the mid-1920s, undermining the prevailing "mission accomplished" British view championed by Oliver in his 1925 *Lancet/BMJ* pronouncements. One clear contributor to the Italian alacrity in calling attention to this problem was the overall strength of the country's occupational medicine discipline at the time; equally important was the rapid expansion of the Italian artificial silk industry.

Economic growth parallel to that in Italy was taking place in the German artificial silk industry. Although at a slower pace than in other parts of Europe, carbon disulfide–related disease also was coming to light there. The first two German publications on viscose-associated poisoning from this period appear to report the same outbreak affecting five workers, one of whom committed suicide in a state of carbon disulfide–induced mania. The first is merely a brief notice of a presentation made at a meeting of an occupational medicine professional society in 1924;[42] the second is a more detailed report published in the following year, although also originating from a 1924 medical conference.[43] It took several more years, however, for additional information to disseminate in the German medical literature, and this improved reporting was largely driven by secondary reports of the Italian experience.[44]

Beyond simply expanding production, the artificial silk industry in the 1920s was characterized by important technological innovations. Above all, a key viscose-manufacturing advance, initially led by German and Italian producers, not only changed the way much of the fiber was made but also increased the health hazards of those who worked by producing it. This modified manufacturing process created a new viscose product called "staple." This novel form of short-length artificial silk filaments could be admixed with natural fibers such as cotton or wool and then woven into a variety of materials for which previously standard "continuous" (unbroken) filament was poorly adaptable. Although there had been attempts to accomplish this by cutting continuous filaments post hoc or even by using broken fiber waste, the commercial results of such earlier methods had always been unsatisfactory.

The new staple-making process entailed additional steps to viscose processing: the extruded fibers were gathered into a "tow," stretched, sliced into short lengths, and then baled. In the continuous-fiber process that had characterized artificial silk production up until then, the key manufacturing production points at which the worst carbon disulfide overexposure was endemic were the churning of the liquid viscose and then the filament-fiber extrusion in the spinning baths. Later steps in the process involved potential carbon disulfide exposure, too, but to a substantially lower degree. Staple manufacturing was likewise vexed by overexposure to carbon disulfide in churning and extrusion, but also introduced heavier exposure to carbon disulfide off-gassing further downstream in the towing, cutting, and even baling operations.

The commercial birth of staple fiber took place in Germany, where, in 1920, the Premnitz factory of Koln-Rottweil AG began commercial production of a new product, "Vistra" fiber.[45] Although there were other German producers of viscose continuous fiber at the time (the largest being Vereinigte Glanzstoff), staple became something of a niche for the manufacturer of Vistra. By 1922 the new product was already commanding attention across the Atlantic. *Textile World* announced on its front page a lead story headlined "Status of German Fibre Substitutes": "The new Koln-Rottweil Aktiengesellschaft has so far produced exceptionally good artificial silk, but it has also brought out a different artificial fibre bearing the name 'Vistra.' This is particularly suitable for articles of clothing, carpets, rugs, decorating material, underwear, knit goods, etc."[46]

As others also began to make staple not branded as Vistra, the Germans adopted a generic term for the product, *zellwolle* (essentially, "cellulose wool"). This differentiated it not only from standard, continuous-fiber artificial silk, referred to as *kunstseide* ("artsilk"), but also from an earlier, unsuccessful product that was called staple fiber (*stapelfaser*).[47]

In Italy, the corporation SNIA was the most prominent artificial silk producer of the day. SNIA's name originally came from the Società di Navigazione Italo-Americana (later "Italo-American" was replaced by "Industria e Commercio," but the acronym went unchanged). SNIA had come on the viscose scene in the early 1920s, acquiring controlling interests in existing producers in Pavia and Turin, the locales of the first Italian medical reports of carbon disulfide toxicity.[48] By 1925, SNIA had announced that it had developed an important new viscose product to be marketed as "Sniafil."[49]

Even though French medicine had been the first to identify and characterize carbon disulfide poisoning in the rubber trade in the mid–nineteenth century, publications in the mid-1920s make it clear that the French came late to an appreciation that viscose manufacturing was emerging as a key new source of exposure to this previously recognized toxic substance.[50] The first original report from France on carbon disulfide toxicity in the artificial silk industry appeared in 1927, but it was a medical doctoral thesis coming out of Lyon, a viscose manufacturing center. Its opening sentence is classic: "In the course of the last six months on Dr. Nordman's service at the Hospital Saint-Etienne, we had occasion to admit an ill young worker for observation that was identical in all major respects to a workmate of his who was on our service two months previously."[51]

This erudite medical thesis cites all the medical case reports of carbon disulfide illnesses in viscose manufacturing workers that had recently appeared in Italian and German publications. But it was not able to point to any original sources from the French medical literature specific to this question, because there simply weren't any. Nor did the findings of the thesis ever see a wider dissemination through publication in a French biomedical journal.

In Japan, the viscose industry was first established in 1916, and national production increased rapidly over the next decade. An ever more industrialized Japan meant that industrial safety was a growing problem there, too. Japanese injury prevention posters of the period, similar to those in the West, emphasize individual workers' responsibility to avoid physical injury. One striking poster from 1928, however, does concern toxic fumes relevant to rayon and other chemical-intensive manufacturing: in the upper left, a samurai in traditional armor is shown, and below a worker wears a cartridge face-and-nose mask as well as goggles, with the caption, "For this job, protective armor!"[52] In the same year as the poster, an outbreak of work-related disease in the artificial silk industry was documented in a domestic Japanese publication, and information on the Japanese experience with carbon disulfide toxicity was being shared internationally by 1931.[53]

While the Italians were amassing important new information on the health risks of viscose, and as the Germans were stirring, the French sputtering, the British debating, and even the Japanese beginning to suspect that something was going on, what on earth was happening in America? The answer: branding. In 1924, on the cusp of the emerging new knowledge from Europe, a representative committee of U.S. textile manufacturers held a meeting at the Vanderbilt Hotel in Manhattan. The committee was headed by S. A. Salvage of the Courtaulds subsidiary the Viscose Company. They did not discuss any health concerns that might affect their employees. Rather, the committee was formally voting to discard "glos" and adopt "rayon" as the commercial term of choice for artificial silk. Their considered recommendation was later adopted by the U.S. National Retail Dry Goods Association, overcoming initial opposition by the Silk Association of America, which had previously been stuck on "glos."[54]

Of course, more was going on in the U.S. rayon business than simple marketing maneuvers. In 1920, the DuPont Company acquired the American rights for viscose from the French Comptoir des Textiles Artificiels, breaking Courtaulds' lock on the U.S. market. DuPont quickly established and rapidly expanded the DuPont Fibersilk Company in Buffalo, New York, and soon after (1923) converted a munitions works in Old Hickory, Tennessee, into yet another viscose facility. These plants later were grouped together as the DuPont Rayon Company, in keeping with the new product name, along with a parallel DuPont Cellophane Company, which manufactured the viscose film product.[55]

Even up against DuPont, Courtaulds was so big that the Viscose Company was still dominant in America. Besides its flagship manufacturing site at Marcus Hook, by 1921 the Viscose Company had an additional plant in operation in Lewistown, Pennsylvania, and the Viscose Corporation of Virginia had started up a major facility in Roanoke. It was similarly held by Courtaulds, but set up separately to avoid taxes that would have accrued under the Viscose Company aegis. In 1922, faced with additional potential liabilities under new federal tax statutes, Courtaulds scrambled to create an overarching entity to act as the holding company for all its U.S. operations. The American Viscose Corporation (AVC) thus was born, its operations still generically referred to as the Viscose Company. Thanks to these maneuvers and despite new market challenges from DuPont and other smaller industrial concerns that soon entered the market, Courtaulds, through the American Viscose Corporation, was making more money than ever.[56] Collectively, counting AVC,

DuPont, and smaller operations, the United States was the unrivaled geographic leader in worldwide rayon production throughout this period.[57]

In the 1920s, the discipline of occupational medicine was far less developed in the United Sates than in Italy, Germany, Great Britain, or France. But though rudimentary, it was growing. Teaching and research programs were ongoing at several U.S. medical schools. A major U.S. specialty journal in the field (the *Journal of Industrial Hygiene*) was established in 1919; articles on occupational medicine regularly appeared in general medical journals (such as the *Journal of the American Medical Association*); and there were U.S. medical textbooks that focused on this topic.

One new textbook that appeared in 1924 carried an inconsistent message on carbon disulfide: one chapter alluded to problems in the new industry of artificial silk without providing specifics, while another focused on carbon disulfide in rubber vulcanization, but made no mention of viscose.[58] In the following year, however, a new U.S. occupational medicine textbook was loud and clear on the subject. In her seminal text *Industrial Poisons in the United States*, Dr. Alice Hamilton provided extensive details of two cases of carbon disulfide poisoning in the artificial silk industry, both of which she had seen personally.[59] The cases had been presented to her by a fellow physician, Dr. Richard Cameron, in 1923. Together they had gone to see one of the cases in the hospital—Hamilton's description is moving:

> He had no motor paralysis, but he was in a very emotional and nervous state, although trying hard to control himself. He seemed as if about to burst into tears at any moment and suffered keenly from mortification, over his betrayal of weakness, saying, "I was all right, doctor, till you came now and stirred me up again." He did not wish to talk about his symptoms but to be left alone. Only unwillingly would he admit that he had been much depressed, and that he slept heavily, but his wife said that he jumped and jerked all night, that he was always drowsy and that latterly he had had queer things happen,—for instance that when he was looking at a thermometer it seemed to leave the wall and come towards him. Suddenly, while talking to us, his self-control gave way and he buried his head in the pillow and burst into tears.[60]

Hamilton did not name the city in which these poisonings arose, nor did she ever separately publish this report beyond her textbook entry (this is not surprising, since she was not directly involved in the care of the two patients). In fact, it was rather unusual to include the details of such unpublished case reports in a modern medical textbook. That atypicality underscores the importance that Hamilton placed on this novel information. She also discreetly avoided mention of a third physician who also accompanied her and Cameron on that hospital visit. The facility in question, also unnamed in Hamilton's textbook, was DuPont's original Fibersilk plant. In her pocket diary for 1923, Hamilton noted in the entry for Thursday, 12 April: "Buffalo with Wade Wright & Rich. Cameron to Du Pont artificial silk then to hosp. to see CS_2 cases."[61]

Hamilton's two Buffalo cases of carbon disulfide poisoning were corroborated in print shortly thereafter by Cameron, who provided some additional details in a brief report that appeared in a New York State Department of Labor publication in the summer of 1925. Cameron refers to the two affected workers as "Case X" and "Case Y," and although he is clear that they worked making artificial silk, he does not identify the employer or even indicate the city in which they worked.[62]

Wade Wright, the physician whom Hamilton did not bring into the story as it appeared in her textbook, was a young occupational medicine colleague of hers from Harvard. In 1924, Wright was recruited by the Metropolitan Life Insurance Company to become its assistant medical director.[63] Just a few years later, the Policyholders Service Bureau of Metropolitan Life took it upon itself to publish something of a paean to artificial silk, *Rayon: A New Influence in the Textile Industry*.[64] This thirty-one-page booklet, graced with a bold geometric cover in an Art Deco style, includes a brief history of the industry, an assessment of the market for rayon in the United Sates, and an analysis of the effect of rayon on the consumption of other fibers ("On the basis of available information it cannot be definitely stated that rayon has had either the effect of increasing or decreasing the consumption of cotton, silk or wool fibers").[65] In a section titled "Conditions Influencing Plant Location," Metropolitan Life seems to be aware of certain environmental considerations, although not in a positive way, the sixth of seven stated prerequisites for siting being: "6. Disposal of wastes. Site to adjoin a river with not less than 550 to 600 square miles drainage above the site."[66]

There is not one word in the entire text on any health risks to the workers manufacturing rayon. Among the entities acknowledged for contributing to the preparation of *Rayon: A New Influence in the Textile Industry* are both the DuPont Rayon Company and the Viscose Company.

Standing up in Parliament in 1925, Joynson-Hicks initially pleaded ignorance about whether there were any health problems in the artificial silk industry. That uncertainty was tenable when he was first asked about the issue of occupational disease in light of the new listing of carbon disulfide as a reportable cause of work-related poisoning. Aside from his convenient "omission" of the single case of poisoning in the artificial silk industry, documented in that same reporting scheme, Joynson-Hicks at that time might very well have been ill informed about carbon disulfide's dangers. After all, reports of illness due to artificial silk work, both Hamilton's and a spate of new ones, were only just appearing in 1925. But increasingly from 1926 to 1927, and certainly by 1928, the situation in the industry was far clearer to those under Joynson-Hicks's command, even as he continued to minimize the carbon disulfide problem to Kelly and his fellow opposition MPs. Jix's deniability was no longer plausible.

Meanwhile, MP William Kelly was like a dog with a bone, and with good motivation. He was not only a Labour MP, but also a Workers' Union official, and moreover, there was an artificial silk viscose plant located in his Littleborough parliamentary constituency—a Dutch-based competitor of Courtaulds called Breda Visada Ltd.[67] Kelly began to extend his search for answers beyond the factory door, pursuing the issue of chemical exposure in the local environs.

Kelly's environmental interest was timely. In July 1928, the British High Court of Justice had granted an injunction (later stayed) abating the release of nuisance fumes from a relatively small artificial silk producer and minor competitor of Courtaulds, the Rayon Manufacturing Company of Surrey.[68] Going beyond and around Joynson-Hicks and the Home Office, in 1928 Kelly repeatedly addressed questions to the Ministry of Health (at that time headed by Arthur Neville Chamberlain). This was the governmental authority with jurisdiction over neighborhood exposures by virtue of the Alkali Acts, a Victorian legislative holdover regulating industrial air pollution.[69]

Just as Kelly failed with Jix, he didn't get very far with Chamberlain either, who apparently was unwilling to practice appeasement, at least with someone that he probably considered little more than a Labour Party gadfly. Having been told repeatedly that any health issues inside the factories were trivial and that neighborhood issues would be dealt with in a report forthcoming at some future but unspecified date, Kelly in 1929 once more took up the matter of occupational disease with the home secretary. He received two written answers reiterating the absence of any ill effects beyond self-limited conjunctivitis and dermatitis, an exchange duly reported in the *British Medical Journal*.[70] Nor did Kelly limit his activity to parliamentary debating and ministerial inquiries. He worked with the Trades Union Congress (TUC) to try to expand, through legislation, workers' compensation coverage for artificial silk workers, and in early March 1929, he was part of a deputation to the home secretary to address that issue, emphasizing the eye problems that were becoming rampant in the industry.[71]

Still, like clockwork, the factory inspector's annual *Reports* continued to come out. The *Report* for 1929 included quite a bit on artificial silk. Its

"General Report" (chapter 1; section, "Dangerous Trades"; listing, "Artificial Silk") notes: "This growing industry has been associated with the definite risk of temporary eye trouble, notably in the spinning rooms, and with some risk of poisoning through the use of carbon bisulphide, which can only be prevented by an efficient, localized ventilation."[72]

The *Report* goes on to draw on Mr. H. R. Rogers's experience from "a large works in Coventry" (namely, Courtaulds), with praise for its successful prevention of eye injuries. Indeed, we are reassured by Rogers that not only were there no reported cases in that year, but there was also not a single visit to the factory first aid station for such a problem. A single case of carbon disulfide poisoning was officially reported that year, but the precise circumstances of this case were not stated in the *Report*.[73]

The situation was much different, and much worse, in the next annual *Report*. A walloping 230 cases of conjunctivitis had been recorded over three months, detected after a special monthly reporting program had been established in the artificial silk industry. But there were even more serious problems at hand. Dr. Bridge, in his chief medical officer's report, notes:

> In the early part of the year [1929] I was asked by Mr. Poore, the District Inspector, to visit an Artificial Silk Works with him, as, on the previous day when inspecting this factory he was concerned not only with the conditions existing in the churn room but also by the appearance of the men working there. I examined the three men engaged on the shift and found two of them to be suffering from well-marked symptoms of chronic carbon bisulphide poisoning. . . . Another man, who had been employed in the churn room until three weeks previously, was also examined and found to be more seriously affected. I also visited a man employed on the night shift who also exhibited marked symptoms.[74]

Bridge ordered a general inspection of the industry in response to this outbreak. Although no additional "definite" cases were identified, several other workers with "some indication of absorption" were found, but these were not assigned to a "definite" status. A short time later a case of full-blown, acute mania occurred.[75]

The official registry of reportable diseases for 1929 lists only those five cases that Bridge alluded to, every one of them a viscose churn-room operator from the same facility—none other than the Rayon Manufacturing Company of Surrey, the target of the court action against neighborhood pollution a year earlier.[76] The first two cases were reported by a certifying surgeon on 29 March, only a few weeks after Kelly's deputation to the home secretary; the symptoms of one of these included "failure of sight," a classic finding of toxic amblyopia (toxic optic nerve damage) caused by carbon disulfide. Amblyopia due to carbon disulfide exposure in rubber vulcanization had been well recognized in decades past; British ophthalmologists were medical leaders in documenting this problem in the nineteenth century.[77]

Damage to the surface of the eye, as opposed to the deeper injury of amblyopia, had not been a problem in the rubber industry, but rather emerged as a new malady peculiar to the viscose rayon industry—peculiar and very common. Outbreaks of eye troubles were linked especially to work near the viscose spinning baths. The condition came to be attributed to hydrogen sulfide or to a related irritant by-product released when carbon disulfide reacted with sulfuric acid in that operation, although the precise nature of the toxin was obscure. What was clear, very early on, was that viscose rayon manufacturing caused this sort of eye problem. Indeed, medical reports on this hazard began to appear several years before there was any consistent notice of carbon disulfide neurological poisoning.

As early as 1923, a brief report from Switzerland documented an outbreak of eye injuries among the spinning workers in the artificial silk industry due to exposure to "viscose fume[s]."[78] Dr. Strebel, the author of this paper, was a keen observer. By using a slit lamp examination rather than a simple clinical assessment alone, he correctly described the syndrome he diagnosed as a form of keratitis. Keratitis is a more serious injury than simple conjunctivitis: although it can still be relatively superficial, it goes deeper than the very uppermost layer of the eye's conjunctiva. The symptoms of keratitis are more severe and take longer to resolve than those associated with conjunctivitis. Strebel took note of the fact that the outbreak, which was located in the Emmenbrücke viscose factory, began in the winter months and was linked to a move into a new, more enclosed workspace, which might have facilitated exposure. He also presented an interesting discussion of possible eye protection by providing the viscose workers with specially designed safety goggles (*maskenbrille*) equipped with chemical filters. Unfortunately, though, Strebel did not seem to have a good grasp on the chemistry involved in viscose spinning. He not only fails to mention carbon disulfide but also misidentifies hydrogen sulfide as sulfur dioxide, a gas not emitted in viscose rayon spinning. Emmenbrücke, founded in 1906, was the main production facility of the Société de la Viscose Suisse.[79]

Virtually simultaneously with Strebel's 1923 report, a Dutch physician named Bakker described a similar outbreak of eye disease among artificial silk spinners in the Netherlands.[80] Detailed clinical assessments came to the same conclusion: this was not simply conjunctivitis, but keratitis, albeit superficial. The symptoms included swollen eyelids, painful light sensitivity, and constantly watering eyes, to the point that the tears simply dripped onto the ground. Once again, there was no mention of carbon disulfide. Considerable attention, however, was given to the potential role of hydrogen sulfide gas, although Bakker expressed reservations that it alone was the culprit. To back up his suspicion that more was going on, Bakker cited toxicological work on hydrogen sulfide that Karl Lehmann had performed decades before, in the same period when he was testing out carbon disulfide.

Bakker's conclusions did not sit well with Dr. W. R. H. Kranenburg, who was the consulting physician to the Dutch National Work Inspectorate. He wrote a letter to the editor critiquing Bakker's article, armed with a missive he had elicited from the great Lehmann himself. The master toxicologist stated, unequivocally, that hydrogen sulfide was obviously the source of the eye problems in the industry.[81] Two years later at the end of 1925, following close on the heels of Renshaw's report that 320 ppm of carbon disulfide was likely a commonplace air concentration in the viscose industry, Kranenburg teamed up with an industrial engineer to measure the atmosphere in the Dutch factories under his purview. His initial focus was on hydrogen sulfide, and in the first plant he studied, the 30–50 ppm concentration of hydrogen sulfide he measured outstripped by a factor of ten the 5 ppm of carbon disulfide found. Kranenburg was reassured by this latter finding, pointing out that, according to Lehmann, far higher levels of carbon disulfide were needed for toxicity.[82] Once again, as Renshaw had done in his report, Kranenburg seemed to confuse acute toxicity with chronic carbon disulfide poisoning.

At a second plant that Kranenburg inspected, however, carbon disulfide concentrations up to nearly 100 ppm were documented, and a number of workers were suffering adverse effects. They had complaints of headache, sleep disturbance, and tremor. One exposed employee had become confused and combative, leading to a psychiatric hospitalization. Finally, at a third factory Kranenburg documented carbon disulfide levels to a staggering level of nearly 1,000 ppm. It was a level of exposure that Renshaw had predicted based on his measurements in England, and its seriousness was such that even Lehmann would not have quibbled with it.

And yet, despite nearly five years of mounting medical evidence across Europe and even farther afield that serious disease was endemic in the international viscose industry, the British labor inspectorate seemed confident that the situation there was well controlled. Even the nagging issue of eye irritation was minimized, downgraded to conjunctivitis rather than officially recognized as the more serious keratitis; the problem was trivialized even further by Bridge as merely "temporary eye trouble." This willful official blindness to the problem flew in the face of experience in Switzerland, the Netherlands, and France. The French, despite their slowness to recognize deeper neurological disease in the industry, in 1928 documented over two hundred cases of keratitis due to viscose manufacturing.[83]

Not one of the 230 British viscose industry conjunctivitis cases that Bridge catalogued in the *Annual Report* for 1929 ever made it into the official tally of the Home Office's disease registry. This may have been a bureaucratic technicality: the illness in question was not due to carbon disulfide directly, but rather to a chemical by-product, and therefore was not considered a listed, reportable condition. Even if due to carbon disulfide, though, the temporary trouble simply did not meet the threshold of an illness that would lead to physician certification for the Home Office's purposes. A medical journal report documenting the British keratitis outbreak did eventually appear. It took until 1936, more than a decade after the fact. It was only through that long-delayed publication that the full extent of the problem became clear: in total, 1,598 cases.[84]

British officialdom minimized not only the risks from carbon disulfide to workers' health inside the factory, but also the potential ill effects of factory emissions on the surrounding community. Public complaints about water pollution and (especially) air pollution emanating from rayon manufacturing dated back to the earliest days of Courtaulds' Coventry operations, before World War I, but it was not until the spring of 1929 that the Alkali Acts chief inspector published a seven-page report on the question. The report wholly exonerated the industry: "There is no evidence that the health of the community has suffered from the existence of these works, although the amenities of every-day life may have been in certain places interfered with."[85] Just as keratitis, a serious eye condition, was officially downgraded to nothing more than a "temporary trouble," toxic environmental pollution was transformed into a mere interference with certain "amenities."

Following a Labour Party electoral sweep on 30 May 1929, that June's issue of *British Medical Journal* carried a brief opinion piece titled "Amblyopia and the Artificial Silk Industry," a scathing critique of continuing carbon disulfide hazards in artificial silk manufacturing.[86] Besides reviewing the history of carbon disulfide poisoning in the industry, it supplied a key, previously unpublished detail about one of the cases officially reported by the factory inspector. Although the viscose churns were sealed and under negative pressure, the worker became poisoned after one was opened because he was "compelled to put his head inside to remove the orange-coloured sticky mass." The piece concludes with the following admonition: "The artificial silk industry starts with a tremendous advantage in the experience already gained as to the effects of the toxic agents used or evolved. The day in, day out exposure to them by workpeople should never be forgotten."[87]

Although unsigned, the editorial was written by none other than Sir Thomas Legge. In fact, it was extracted verbatim from a longer presentation he had given a few months before.[88] Legge was up to date on certain aspects of the emerging scientific literature on carbon disulfide, having prepared lengthy briefing materials for the Trades Union Congress to be used in MP Kelley's deputation to the home secretary that year: a seven-page handwritten summary of governmental sources on eye and skin problems in the Dutch and German artificial silk industry, along with a typed, abstracted summary of a 1928 German medical article on the subject.[89]

Legge, however, had not reviewed for the TUC the medical reports that had been so rapidly accumulating since 1925 on neurological disease in the artificial silk industry, nor did the brief editorial format of the *British Medical Journal* provide an opportunity to bring such new information to the fore. Among the multiplying medical reports was a pivotal paper spotlighting the neurological hazards of carbon disulfide in the viscose industry. The paper, presented by Dr. Gustavo Quarelli of Turin at the Fifth International Congress for Work Related Diseases in Budapest in September 1928, included four clinical cases of neurotoxicity, along with an extensive review of the past (rubber industry) and current (artificial silk industry) medical literature on the subject.[90]

In April 1929, Quarelli presented two cases with manifestations of carbon disulfide toxicity different from the well-established toxicities of cognitive impairment and psychiatric derangement.[91] The first case was a fifty-five-year-old artificial silk spinner with "typical symptoms of a lesion of the corpus striatum," predominantly manifested by muscle rigidity. The corpus striatum is linked to Parkinson's disease, the implication of Quarelli's terminology being that the insult he was describing was similar. In describing the second case, age fifty-seven and also a spinning-room worker, Quarelli was more explicit. The disease present in this patient manifested as a "Parkinsonian syndrome," including typical symptoms of increased muscle tone and a fine resting tremor. In early October of that year, Quarelli revisited the topic, presenting another chronic neurological case history to that year's annual Italian National Conference of Occupational Medicine.[92] The patient was a twenty-eight-year-old artificial silk worker named Ferdinando, whom Quarelli had treated earlier that summer. Ferdinando was troubled by repeated spasmodic jerking motions of his arms and legs, which Quarelli hypothesized might reflect Parkinsonism, albeit in an earlier stage.

During the campaign preceding Labour's electoral sweep in 1929, the slogan of the losing Conservatives was, ironically, "Safety First," meant to convey surety through staying the course under Stanley Baldwin (the slogan also echoed a recent governmental road safety campaign).[93] Despite the change of government, it would be hard to argue that much had improved for the health of British artificial silk workers or their communities. The new home secretary, John Robert Clynes, despite his Labour bona fides, was not spared MP Kelly's questioning. Clynes, like Jix before him, focused on conjunctivitis as the main source of trouble in the industry and did not mention any more serious risks.[94] Joynson-Hicks had not run for reelection in 1929 (although his constituency remained a safe Conservative seat), yet he was still in Parliament. Having been of no small service to the state, Jix was created 1st Viscount Brentford that July and was thereby elevated to the House of Lords.

The May election was known as the "Flapper Election," the first in Britain in which women had the vote. Black Thursday was five months away. The stock market crash and ensuing economic failure mark the common end point of the Roaring Twenties, although given the degree to which neurological disease became endemic in the viscose industry during those years, the French term for the period, "*années folles*," has more resonance.

For eighty years, psychological illnesses had been known sequelae of central nervous system poisoning by carbon disulfide, even if international recognition of this hazard in artificial silk manufacturing had been slow in coming. By 1930, the evidence of this toxic nexus was irrefutable. The closing argument can be taken from a case series reported in that year by Dr. Karl Bonhoeffer. Bonhoeffer was world renowned as a founding father of modern psychiatry; he had originated a dichotomization of psychosis into diseases caused by endogenous factors and exogenous sources. For Bonhoeffer, psychiatric disease caused by carbon disulfide fit his concept of exogenous psychosis to a T. By 1930, he had had twenty-four years' experience with fourteen such cases, eight rubber workers, and six artificial silk spinners. Bonhoeffer did get one fundamental fact wrong: in the opening paragraph of his paper, he states that the substitution of other solvents for carbon disulfide in the spinning industry had brought the problem under control.[95]

Bonhoeffer did not consider Parkinsonism in his review, which is not surprising, since Quarelli had raised the specter of Parkinsonism in case presentations only in 1929. In early 1930, Quarelli published a fully developed paper more extensively documenting his findings on this syndrome in connection with carbon disulfide.[96] But it took another Italian to definitively establish the link. In November 1930, the leading French neurological journal of the time published Dr. Fedele Negro's "Les Syndromes parkinsoniens par intoxication sulfo-carbonée," the first major international paper on the subject.[97] Negro was a formidable neurologist and the doyen of Parkinson's researchers in Italy. The paper, although under his name, was largely based on the work of Quarelli and his colleagues, rather than containing original clinical observations by Negro himself. In the hierarchy of the academy, however, it was far more advantageous to have a figure such as Negro, rather than Quarelli, put forward the proposition that carbon disulfide could cause a Parkinson's-like syndrome.

Damage that mimicked Parkinson's disease, thus labeled "Parkinsonian" to differentiate it from the classic idiopathic condition, was both a new finding and an alarming one. By 1931, Quarelli had published a three-hundred-page general textbook of clinical occupational disease in which he not only addressed Parkinsonism in its brief section on carbon disulfide, but also took care to include a full-page photographic illustration of one of his cases suffering from torsion spasm. The worker, smartly dressed in a three-piece suit with a double-breasted vest and wearing a striped tie, is holding his contorted hand up, facing the camera; a somewhat macabre additional close-up view shows only the arm with its hand in spasm, as if disembodied.[98]

In the badly split British general elections of October 1931, MP Kelly lost his seat to a Conservative. One month later, Viscount Brentford rose to speak in the House of Lords, bemoaning an increase in foreign imports and supporting the need for protective tariffs, citing specifically the threat to "artificial silk tissues"; he made sure to note, "I have no interest whatsoever in manufacture or the commercial world myself."[99] Viscount Brentford's second son, Lancelot Joynson-Hicks, would himself later take a seat in Parliament, succeeding in a safe constituency previously held by John Courtauld, Samuel's brother.[100]

One of the last pieces Sir Thomas Legge wrote before his death, "An Industrial Danger," was on the subject of the artificial silk industry. An appeal to a general audience, it was published in the *Statesman and Nation* in August 1931.[101] He included a comment on the case of the Rayon Manufacturing Company of Surrey, tempered by the new economic hard times: "No one would wish to see a large factory closed down these days, but the issue recently of a writ of sequestration against a factory manufacturing silk by the viscose process for failure to remove the nuisance from fumes in a residential district must come as a shock to the industry. . . . The cause of the trouble lies in the use of carbon bisulphide in the 'churn' room, where, unless elaborate local exhaust ventilation is applied, serious consequences to the workmen will arise."[102]

Legge's critique, which seems so understated in retrospect, was taken by some to be both reckless and inflammatory. A pointed letter to the *Statesman and Nation* was penned by two of the leading British forensic toxicologists of the day: Sir William Willcox and his protégé, Gerald Roche Lynch, OBE.[103] Both served as consultants to the Home Office, albeit typically for homicide cases, not occupational disease.[104] Their letter makes clear that they had served as consultants and expert witnesses on behalf of the Rayon Manufacturing Company of Surrey. On this basis, they claimed special insights into the rayon industry, although they made it clear that they could not possibly comment on a case still before the court. But they did not hesitate to opine more generally:

> Our object in writing you is to correct the implications of Sir Thomas's article that the production of artificial silk occasions potential injury to the health of workers employed in it. . . .
>
> Any possible danger from carbon disulphide from fumes arising in the churn rooms of artificial silk factories has effectually been removed and is dealt with by departmental regulations. Beyond occasional transient and temporary irritation of the eyes (conjunctivitis) experienced by some workers in spinning rooms, who represent but a small proportion of the employees engaged in the production, manufacture and making up of the products of this important and growing industry, there is nothing to justify Sir Thomas's suggestion.[105]

Willcox and Roche Lynch went on to play what they clearly saw as their economic trump card:

> No one more fully than ourselves recognizes what Sir Thomas has done in securing measures to protect the health of industrial workers engaged in dangerous occupations; but, at this critical juncture in the industrial history of our country, we cannot but deplore that Sir Thomas should lend his authority, even remotely, to the crippling of a rising industry which, in addition to being a triumph of research and inventive genius, serves as a substantial source of direct revenue to the State by the duty imposed on the production of yarn and indirectly, affords employment to a vast number of people who might otherwise be unemployed.[106]

Two months after Legge's "An Industrial Danger" appeared and less than four weeks after the counterattack by Willcox and Roche Lynch was published, the case against Rayon Manufacturing was finally settled. In fact, the plant was allowed to continue its operations.[107] Within the year, Sir Thomas Legge was dead. Sir William Willcox continued being driven to his consultations by chauffeured Rolls-Royce, as was his practice.

Meanwhile, the Rayon Manufacturing Company shut down, despite its favorable court judgment. Its demise was due to price cutting rather than the hypothetical economic burden of the air pollution controls that it was never obliged to install. W. P. Dreaper, the company's technical director, a former Courtaulds worker who had gone out on his own, had pleaded personally with Samuel Courtauld to back off. "Dear Mr. Sam, Cannot we stop this nonsense of undercutting," he wrote to his old boss. The letter went unanswered.[108]

3

Wrapped Up in Cellophane

By the summer of 1931, the worldwide economic depression was well entrenched and getting worse. Improvement, when it finally came, was sputtering at best for much of the decade that followed. But things were far from grim for the viscose industry.[1] In the United States there was only a modest, brief downturn in rayon filament production from 1931 to 1932 and then considerable improvement during the next year. This was followed by steady growth through 1936, a booming 80 percent five-year increase overall. Production in Germany and Italy moved up every year after 1933. Japan simply shot up like a rocket from the start, surpassing the United States as the leading filament producer by 1935. For short-cut rayon staple, as opposed to the longer filament product, the trend from 1931 onward was even more dramatic. There were hundredfold, logarithmic increases in production worldwide. Germany and Italy led the pack, Japan caught up fast, and the United States, the United Kingdom, and France, albeit behind, marched in lockstep in a parallel, no-less-remarkable upward path of expansion.

The big players that hauled in impressive earnings even in the depth of the Depression were more or less the same ones that had sat at the table the decade before. It was Courtaulds in the United Kingdom and Canada. In the United States, it was Courtaulds yet again through its American subsidiary the Viscose Company (not yet the American Viscose Corporation), although DuPont also had a big piece of the action. In Germany, Glanzstoff was a dominant force; in Italy, SNIA; and the Comptoir des Textiles Artificiels (CTA) in France. In the Far East, the Teikoku Rayon Company, Ltd., around since 1918, was another old hand in the viscose rayon business.[2] In contrast, the Toyo Rayon Company (founded in Japan in 1926) was relatively new on the scene. Under the modest slogan of "No Victory Without Unity," Toray, the present-day corporate successor to Toyo, explains, "The design of the corporate logo used at the time—three interlocking rings placed inside a double circle, symbolized the opportunities given by heaven, the advantages of the land, and the harmony between human beings. It also represented the formation of the four letters of TOYO. Faced with the Great Depression and downturns in the fiber and textile market soon after its founding . . . Toyo Rayon overcame many hardships."[3]

An even more holistic view of the interlocking corporate spheres of rayon manufacturing interests informs Grace Hutchins's book *Labor and Silk* from 1929.[4] The picture that Hutchins paints is not a pastoral vision of heaven, land, and human harmony; she concretely illustrates the state of things with a full-page diagram.[5] Hutchins had done her homework. All of the large manufacturing countries are accounted for in the figure, along with many of the smaller ones too, each represented by a circle proportionally sized to the amount of national rayon production. Circles within circles illustrate separate corporate entities, also scaled by production volume. In addition, the diagram includes the minority of remaining producers that continued to manufacture rayon by other processes (nitrocellulose and cuprammonium rayon, as well as acetate silk, a related cellulose derivative). Solid or dotted lines connect corporate entities based on outright ownership of subsidiaries (solid) or the holding of an investment stake or some other transnational arrangement (dotted). Glanzstoff, for example, is shown with seven radiating ties, although several of these are dotted lines; Courtaulds manifests six interconnections, but all of its connecting lines are quite solid.

This bewilderingly complex web could easily come across as the product of some anticorporate conspiracy theorist spinning out of control. But if anything, Hutchins underestimated the true extent of the phenomenon. By late 1928, and unknown to her, a hidden layer of cross-ownership was operating through a transnational holding company, the Associated Rayon Corporation, based in Maryland and headed by a Dr. Fritz Blüthgen (conveniently of Glanzstoff), which tied together New York, Amsterdam, and Berlin through the banking concerns of Lehman Brothers, Teixeira de Mattos Brothers, and Lazard Speyer-Ellissen.[6]

Labor and Silk was one volume in the International Publishers' Labor and Industry series. Closely aligned politically with the Communist Party of the

United States, International Publishers had a special connection to worker safety and health through its chief, Alexander Trachtenberg, and its co-owner and angel investor, Abraham A. Heller.[7] Trachtenberg's graduate thesis work at Yale, eventually appearing through International Publishers, focused on the occupational hazards of coal miners.[8] The firm published fictional works, too, touching on the theme of industrial disease, one of the most notable of these being *The Way Things Are*.[9] This collection of short stories by Albert Maltz included "Man on the Road," which, when it first appeared in the magazine *New Masses*, brought to early public attention a massive outbreak of silicosis caused by a hydroelectric excavation project at Gauley Bridge in West Virginia. The corporate perpetrator at Gauley Bridge was Union Carbide. Ironically, Carbide might be said to have had a dotted-line connection with International Publishers, having purchased Abraham Heller's lucrative International Oxygen Company in 1930, the same year that Carbide's subsidiary began its deadly drilling in West Virginia.[10]

The Labor and Industry series was meant to educate the masses about the basics of several industrial sectors by taking a scientific approach, a kind of adult-proletarian precursor to the "All About" books (*All About Dinosaurs*, for instance), which began to appear two decades later. *Labor and Silk* covers the economics and working conditions of the entire U.S. silk textile industry, giving prominence to the Paterson and Passaic, New Jersey, silk strikes, but going far beyond local or regional questions. Rayon, as "artificial silk," was a part of that story. Hutchins acknowledges that the potential displacement of workers employed in natural-fiber weaving, silk included, might be counterbalanced by the new production of mixed blends (for example, rayon-silk or rayon-cotton). The more salient point was that the rayon industry was characterized by low wages, long hours, and underunionization relative to the overall textile-manufacturing labor force. Most of all, however, Hutchins emphasizes that the rayon industry operated as the preeminent international industrial cartel.

Unhealthy working conditions arising from the manufacture of artificial silk are given rather short shrift in *Labor and Silk*, even though elsewhere in the book Hutchins shows that she was quite attuned to industrial disease in the natural silk industry (including the special chemical hazards of textile dyeing).[11] She mentions eye and lung problems attributed to acid exposure among British rayon workers, also noting, "The atmosphere of the rayon spinning rooms is described by workers as 'etherized.'"[12] Yet the specific anesthetic vapor in question, carbon disulfide, goes unnamed.

Hutchins did her own research, but she relied on a talented artist collaborator named Esther Shemitz to craft the complex diagrammatic representation of the industry that is the centerpiece of the rayon chapter. Shemitz also provided more figurative line drawings that illustrate other parts of *Labor and Silk;* her contribution to the work is acknowledged prominently on the title page. Hutchins maintained her commitment to social justice and was highly productive (she subsequently published *Youth in Industry,* 1932; *Children Under Capitalism,* 1933; *Women Who Work,* 1934). She worked alongside her lifetime partner, Anna Rochester, who also wrote for International Publishers' Labor and Industry series (*Labor and Coal,* 1931). Esther Shemitz traveled another path, going on to marry Whittaker Chambers in 1930 (Hutchins and Rochester were the only two witnesses at the wedding).[13] Shemitz stood by her man and was herself a figure in the Alger Hiss Cold War affair, in which Chambers later was to level accusations of spying against the former U.S. State Department official.[14]

Labor and Silk threw a wide net meant to capture the international rayon behemoth, but not surprisingly, the nascent Soviet viscose industry slipped through. Hutchins did comment, rather glowingly, on working conditions in Russian silk factories and in the textile industry there generally, based on a personal visit that she had made to the Soviet Union a few years earlier.[15] She may have been unaware of rayon production in the USSR, which went back to the pre–October Revolution establishment of the factory sited in Mytishchi (near Moscow).[16] War and revolution had interrupted production, but the Mytishchi Viskosa factory was back up and running again by 1924, reborn under the aegis of the Moscow Council of the National Economy.

The production methods at Mytishchi Viskosa remained primitive, allowing ample opportunity for carbon disulfide overexposure. The process solution of cellulose liquefied with carbon disulfide was "aged in small boxes moved along on handcarts. . . . The spinning bath contained not only sulfuric acid, sodium sulfate, and zinc sulfate but also glucose. When the plant was short of glucose it used edible sugar instead. The fresh-spun cakes [were] carried on wooden trays."[17] Such was the boom market for viscose that even in such a technologically limited milieu, beyond the pale of Courtaulds, three other Soviet plants had been online by 1934 (including one for rayon staple), and national production increased by 200 percent over a ten-year interval.[18]

Independent of the otherwise dominant international cartel, the atypical Soviet artificial fiber enterprise of the early 1930s could be invoked as the exception proving the rule. But rayon's sister industry, cellophane, served as self-evident confirmation that concentrated corporate power was axiomatic of the worldwide viscose trade. Making transparent film, as opposed to longer or shorter filaments, was an obvious offshoot of the basic manufacturing process; its possibilities occurred to the earliest viscose tinkerers on both sides of the Channel. The British at first, however, did not exploit this opportunity commercially; uncharacteristically, Courtaulds was not the alpha male of the pack this time. So often forced to back down in rayon confrontations with its British competitors in the first decades of the twentieth century, it was the French Comptoir des Textiles Artificiels that took the lead, no longer the underdog. Before the outbreak of World War I, the CTA had established a new, separate manufacturing concern with its own corporate identity. Based in France and Switzerland, the sole mission of this new enterprise was to make and market transparent viscose film.[19]

It took a quarter century from the inception of viscose fiber for its marketers to give birth to the name "rayon." Further, the designation "rayon" had to be shared with other cellulose-based fibers made through other competing, albeit commercially marginalized, processes (that is, nitrocellulose, cuprammonium, and acetate silk). In short, rayon lacked both trademark protection and a brand identity. So what was the CTA to name its viscose film? Here is where the true genius of the company was manifested. It took "cell" from "cellulose" and combined this with "phane," a suffix derived from "diaphane" (that is, diaphanous) and voila!—La Cellophane SA was born.[20] With that company's name came one of the most enduring commercial eponyms of the twentieth century, and, unlike "rayon," it was trademarked. Thus, even though producing transparent viscose film did not necessarily mean having to purchase the patents for the specific industrial process owned by La Cellophane (there were other production options available), successfully marketing such a product virtually required purchasing the right to use the name "cellophane."

DuPont grasped this quickly, signing an agreement with La Cellophane on 9 June 1923 for the exclusive U.S. rights to make and sell cellophane. DuPont had brought manufacturing online by April 1924, side by side with its Buffalo, New York, flagship viscose filament plant ("Fibersilk" in those heady pre-rayon days) that had been in operation since 1921.[21] On the Continent, Hoechst had acquired from the CTA the rights to cellophane; with the creation of I. G. Farben in 1925, Hoechst was absorbed, along with its subsidiaries, one of which, Kalle and Company, became the leading German producer.[22]

Courtaulds, on the other hand, was not so fast on the uptake. And the pressure was on. La Cellophane had the audacity to start a British subsidiary in 1927 to market its imported product. To add insult to injury, in 1928 a home-grown, Courtaulds-independent operation was launched, Transparent Paper (TP) Limited. The underwriting was huge, thanks to a fairly speculative stock offering that allowed TP to acquire patents, purchase equipment, and bag its biggest trophy, a freehold lease to the Bridge Hall Mills of Bury, Lancashire.[23]

Bridge Hall Mills was a paper factory with a long pedigree, going back to the eighteenth century; it expanded into a major industrial operation in the nineteenth century before sputtering to a pre-Depression economic collapse and a 1925 shutdown. The TP stock prospectus, in a boast as empty as the abandoned and stripped mill buildings, put forth that the plant was "in a good state of repair."[24] Viscose film was now on the British map, but the address was not Courtaulds.

Of the other extant production processes for transparent viscose film independent of La Cellophane's patents, one was in the hands of a German manufacturer that, with some success, marketed it as Transparit. Even more prominent was a Belgian concern called the Société Industrielle de la Cellulose. In 1930, that company established a foothold in Virginia. The new entity, which was clearly in direct competition with DuPont, was named Sylvania (it was forced to market its cellophane product as Sylphrap).[25]

Courtaulds believed that it, too, could escape cellophane's brand-name chokehold. A new product called Viscacelle was Courtaulds' big transparent wrap hope. Indeed, in 1933, Courtaulds' C. P. Atkinson was very upbeat on Viscacelle,

> Containers of metal, glass, cardboard, and paper . . . will be superseded to a large extent, however, by the most modern of all wrapping materials, viz. transparent cellulose paper, which has additional advantages. Courtaulds Ltd. have developed and are producing large quantities of this class of paper under the name of Viscacelle. This paper has excellent pliability which enables it to conform to articles of all shapes, whilst the appearance of an article for sale can often be improved by wrapping it in coloured Viscacelle. . . . A glance in any shop window shows the ubiquity of its applications. Viscacelle does not catch dust. It is impervious to grease, oil, and alcohol.[26]

Viscacelle, as it turned out, was not around long enough to catch much dust. Even if impervious to grease and alcohol, apparently it was not resistant to the pressure of anemic profits. The product ultimately was abandoned in favor of an alternative strategy, far more in keeping with the way business was done in the rayon trade. Courtaulds simply made a compact with the devil, in 1935 establishing a wholly new entity, British Cellophane. It was a win-win of sorts: Courtaulds put up all the money for the new company, and La Cellophane provided the manufacturing patents and sales rights.[27]

Even with this large new investment, cellophane was still a side business for Courtaulds. Synthetic textiles remained its core enterprise, but an important lesson from the Viscacelle episode could be applied to rayon. The name could be branded more effectively, even if the appellation "rayon" was not trademark protected. Commencing on 18 May, National Rayon Week was to be the "great merchandising event of 1936," or as Courtaulds put forth in its advertising prospectus: "Spotlight on rayon. Rayon . . . that has brought colour and beauty into people's lives. Rayon . . . that has given to women to-day wardrobes that only a few years ago were reserved for the exclusive few."[28]

Not national, but rather imperial National Rayon Week was more like it: as far afield as Singapore, the *Straits Times* ran a piece headlined " 'Say Rayon' Week; Many Uses and Endless Variety."[29] Not that the Courtaulds campaign was eagerly embraced everywhere. From the beginning, rayon filament was a potential threat to the primary producers of natural silk, especially in the Far East. By the 1930s, however, a new phase of competition had come into play. Innovations in short-cut rayon staple allowed the mixing of far greater quantities of the viscose synthetic with wool or cotton, the resulting fiber to be respun as blends in weaving applications; rayon staple also could be used in nonwoven, felt-like applications that previously had been the exclusive bailiwick of natural fibers. So whether branded or not, viscose had grown into an industry so expansive that it had come to be seen as a threat not simply to natural silk feedstock, but to other fibers as well. The *Adelaide Advertiser*, voicing support for a countercampaign in behalf of the nation's native wool, noted that "Britain is on the eve of 'national rayon week' ": "There is too much reason to regard artificial silk, or rayon, as Australia's 'Public Enemy No. 1' in the markets of the world."[30]

Courtaulds' sloganeering of the time yielded "The greatest name in rayon" and "As good as it is beautiful," a bit staid perhaps, but nonetheless better than the cryptic advertising copy of Glanzstoff's U.S. subsidiary, "Tag always gets 'em"; cellophane was light-years ahead, though, with an apotheosis of pithiness in "Reveals what it seals" and "Shows what it protects."[31] Arguably, however, the most successful slogan ever associated with viscose was popularized not by paid advertisement, but rather through a partisan one-liner. In the election year 1936, the American Liberty League, with large underwriting from the du Pont family (personified by Irénée du Pont, former president of E. I. du Pont de Nemours and Company and, by that point, vice chairman of its board), was a major conduit for channeling anti–New Deal counterreaction in the period leading up to FDR's campaign for a second term. James Aloysius Farley, Roosevelt's postmaster general, Democratic National Committee Chair, and quintessential political operative, was acutely aware that to effectively counter the American Liberty League, offense was the best defense.

The perfect opportunity presented itself on 22 February 1936, when Farley took the podium to address the Washington Day Banquet of the Kansas

Democratic Club in Topeka. The first part of his speech emphasized gains made during the first Roosevelt term, with a local, farm-belt emphasis on FDR's actions to address the plight of agriculture. But then he came to the heart of the matter:

> The constant and unsportsmanlike campaign of distortion directed at the Roosevelt administration is so intense, bitter and biased that I frankly confess many honest and upright American citizens are confused and bewildered.... Let us reflect not so much on what is said *but who says it*.
>
> First let us take the miscalled American Liberty League, an organization of multi-millionaires which is run as a subsidiary of the Republican National Committee. They think alike, act alike, and their leaders are in constant heavy conference in Washington figuring out ways to destroy President Roosevelt's influence with the people. A brilliant editorial writer said it ought to be called the American Cellophane League and he gave two good reasons. He said first, it's a duPont product and, second, you can see right through it.[32]

The speech was widely reported at the time.[33] The Cellophane League aphorism is often misattributed to Farley as its originator, rather than to the unnamed editorial writer that he credited as his source. The original, thirty-word item on the *New York Post*'s editorial page titled "Wrapped Up in Cellophane" begins, "The *Post* suggests that the American Liberty League be renamed."[34] Farley had a long political career.[35] Even more enduring, the cellophane metaphor has gone on to enter the American political lexicon. The ill-fated Mitt Romney campaign of 2011–12, for example, was deemed particularly well suited for revisiting the allusion, either obliquely, as when Gail Collins wrote in December 2011 that Romney was "generally kept so far from one-on-one interviews that he might as well be wrapped in cellophane,"[36] or more directly, as when Kevin Kruse in January 2012 drew parallels between Romney's recent comments on "one nation under God" and the Liberty League's employment of a religious minister to speechify on its behalf (parenthetically, Kruse also invoked the Farley line on the American Cellophane League).[37]

Irénée du Pont may have been sorely vexed by the New Deal and frustrated that his investment in the American Liberty League was all for naught, given FDR's landslide victory in the 1936 presidential race (the classic Farleyism on that election being, "As goes Maine so goes Vermont").[38] But at least du Pont could take comfort in the continued and even growing strength of the viscose market in those years. There had been challenges early in the Depression: imports threatened the profits of U.S.-based production, and the inventory of unsold, locally produced fibers grew because of declining domestic purchases. These challenges were met through a marriage of convenience with DuPont's major U.S. rayon competitor, Courtaulds' Viscose Company. Their two representatives, Samuel Agar Salvage for the Viscose Company and Leonard A. Yerkes for DuPont, served as president and vice president of the Rayon and Synthetic Yarn Association. Founded in 1929, this organization had the superficial trappings of a standard trade association: it included most U.S. rayon manufacturers of the day under one big tent, predominantly viscose producers, but also what remained of those still using other methods, such as cuprammonium.

Salvage and Yerkes called the shots. In the battle over foreign imports, Salvage spearheaded a successful lobbying effort to maintain trade protections (already well established). This was accomplished by the inclusion of fairly hefty duties on imported rayon in the Smoot-Hawley tariff legislation winding its way through Congress in the spring of 1930. Domestically, DuPont and the Viscose Company had already agreed to curtail production in order to stabilize prices through reduced supply, but this effort was being undermined by suppliers that cut prices to increase their share of the reduced market. In July 1931, when the other companies would not fall in line, Salvage and Yerkes agreed to cut their own prices. The Viscose Company and DuPont abruptly left the Rayon and Synthetic Yarn Association, effectively disbanding the group that they had been instrumental in creating.[39]

Price cutting was coupled with a technical change in how rayon yarn was supplied to the knitting factories that were secondary users of the product. Rather than bundling in skeins, the yarn was wound on cones and oiled to make it ready for immediate knitting. The problem was that when the cones sat unused for more than three months, the oil tended to go rancid, compromising the quality of the yarn. As stocks of unsold product built up in a depressed market, the Viscose Company and DuPont sold off their defective product at a huge discount, although acknowledging that it came without any of the usual product guarantees. But this fix was temporary: the competitors started to sell off better-quality yarn at the same heavy discount or to sell poor-quality yarn but claiming it to be of high quality. Push came to shove. In October 1931, representatives of the rayon companies convened at the Union Club in New York City. A preset price cut for standard-quality yarn would come into force, along with an across-the-board cut in production; to ensure compliance, independent auditors would check the books of each firm.[40] The Union Club continued to meet over the following months, and although the entente collapsed in mid-1932, by then the worst of the market crisis had passed. And most saliently, DuPont and the Viscose Company were still on top.

Moreover, Irénée du Pont could take personal satisfaction in another DuPont–Viscose Company alliance that was contemporaneous with the

Salvage-Yerkes machinations. On 3 January 1931, Mr. and Mrs. Irénée du Pont "made known" (it was not a formal announcement) the engagement of their daughter Eleanor to Philip G. Rust. Eleanor was a graduate of the Baldwin School, active in the Junior League, and, needless to say, a horsewoman; Philip was a graduate of MIT and "connected with the Viscose Company."[41] The *New York Times* carried news of the engagement in January, dutifully reported Eleanor's choice of her bridal attendants in April,[42] and then covered the May wedding in detail: she held orchids and bridal roses, her veil was of old family lace, and her gown was of egg-shell tone satin—no art-silk for Eleanor, to be sure.[43] We are never told more about Philip's Viscose Company "connection," which, since it was of a technical nature, may have been considered a bit too hands-on. The patent for his invention of a new means of treating rayon yarn as a way of "freeing the thread from the excess of chemicals used in forming the thread" was granted in late October 1931, a few months after the wedding.[44] It is not specified where, exactly, through Rust's invention, the chemicals (prominently including carbon disulfide at this process step) were being freed up to go. No matter: Philip was moving up and out in any event, first into finance and later as a major cattle rancher on a "plantation" in Georgia.[45]

The prominent wedding of the du Pont daughter would surely have caught the attention of any regular reader of the society page. It is even likely that included among the wedding guests were members of the Union Club, and thus, it is theoretically possible that one of them happened to be at the club on a day when the Rayon and Synthetic Yarn Association was having one of its meetings there. Yet even allowing for this coincidence, he (no women were allowed as members) would have found little special in it. This does not mean, however, that neither the Viscose Company–DuPont sweetheart deal on tariffs nor their production volume and price shenanigans went unnoticed.

In August 1931, Frank William Taussig and Harry Dexter White published, "Rayon and the Tariff: The Nature of an Industrial Prodigy."[46] This thirty-three-page essay was no mere polemic on the financial advantages the industry had garnered through tariff protections. In it, Taussig and White also systematically review the history of the industry in Europe and the United States, technological aspects of production, and the general economic policy implications of rayon specifically. They also prominently feature the term "duopoly," meant to capture the essence of the DuPont–Viscose Company relationship within a small group of corporate entities dominating a large industry: "Most interesting to the economist, we have a case where, in some respects at least, there seems to have been successful application of protection to young industries; yet also conditions of production on a huge scale, monopoly or 'duopoly,' international combinations and agreements. Here we meet the problems most characteristic of the trends in modern industry, the most baffling and most difficult of accurate analysis, the least easy to interpret in those terms of competitive industry on which modern economic reasoning has chiefly relied."[47]

Although "Rayon and the Tariff" first appeared in the *Quarterly Journal of Economics* (Taussig was the journal's editor), this article was, in substance, a new chapter for a revised edition of Taussig's classic text, *Some Aspects of the Tariff Question*.[48] Frank Taussig was one of the leading U.S. economists in the first half of the twentieth century. Long associated with Harvard University, he had particular expertise in tariff and trade policies, and was no stranger to governmental service.[49] Harry Dexter White, who did his doctoral work at Harvard under Taussig, was the listed co-author of the revised *Tariff Question*. White soon left academia to serve in the Department of the Treasury (going on to lead the Bretton Woods Conference, which laid the groundwork for the International Monetary Fund).[50]

Taussig and White were surprisingly sympathetic to the possibility that earlier in the development of the U.S. domestic rayon industry, tariff support may have served a useful purpose. They were clear, however, that by 1931 such a time, if it had ever existed, had long passed. "Rayon and the Tariff" demolishes, one by one, the economic constructs that might be argued to support continued trade protection, including lack of local expertise, shortage of manufacturing equipment, high labor costs, and, above all, a large capital investment: "Allowance should, of course, be made for an adequate return on this capital, with regard not only to depreciation proper but also to risk and obsolescence. These in a chemical industry are important items, especially when the industry is new. Yet with all allowance, the evidence of the profits of the American Viscose indicates that earnings were brilliant—much more than can be thought necessary, by the wildest stretch of business imagination, to attract and keep capital in the industry."[51]

There was another contemporary, and no less keen, observer of the Viscose Company's legerdemain in manipulating the invisible hand of the marketplace—one with experiential insights. Isidor Reinhard owned and ran the Arcadia Knitting Mills, based in Allentown, Pennsylvania, with his two brothers, Samuel and David. Arcadia was, in fact, the Viscose Company's largest yarn buyer in the early 1930s. As a major client, Arcadia received a 5 percent discount on its purchases. This was standard in the industry, but emblematic of the sort of price inconsistency that the Rayon and Synthetic Yarn Association was working to do away with. Salvage told Arcadia that its discount would be falling to

3 percent. Arcadia refused the offer, countering with a proposal that the Viscose Corporation simply buy them out rather than squeeze them to death. On behalf of Viscose, Samuel Salvage declined.[52]

On 30 March 1932, Salvage cabled Henry Johnson, his direct boss at Courtaulds in England:

> HAD TALK WITH YERKES YESTERDAY WHO FEELS RAYON PRODUCERS WILL ULTIMATELY GO IN KNITTING BUSINESS AS KNITTED CLOTH MANUFACTURERS INCLUDING ARCADIA ARE JEOPARDIZING RAYON STOP . . . YERKES WILL INFORM ME BEFORE HE TAKES ANY STEP STOP ARCADIA WILL TAKE VERY LITTLE PROFIT THIS YEAR BUT ASIDE FROM THAT ITHINK [sic] IT IS TOO DANGEROUS TO BE FINANCIALLY ASSOCIATED AS TROUBLE WOULD IMMEDIATELY ARISE AND THEIR SMART JEW LAWYERS WOULD JUMP ON US STOP FURTHERMORE IT WOULD LEAD TO HAVING TO BREAK OUR ARRANGEMENTS WITH RAYON PRODUCERS STOP PLEASE SHOW THIS TO CHAIRMAN IN CASE THEY TRAP HIM AS I FEEL CONFIDENT I AM SPEAKING IN THE BEST INTERESTS OF THE COMPANY. SALVAGE[53]

As the situation was worsening, Reinhard was en route to England to make an end-run around Salvage. Reinhard radioed Henry Johnson in London from the SS *Europa* first on 11 August, then again on the 17th, and finally on the 19th, enumerating Arcadia's desperate need for viscose yarn (in total, 225,000 pounds a week). By the last message, his request had become a plaintive outpouring, "Please mister Johnson have pity on us and give us the amount of yarn we need otherwise I am afraid our health will break down mentally and physically from the pressure of our customers for their orders."[54]

But the story did not end there. In February 1934, the Federal Trade Commission (FTC) brought a complaint against ten rayon producers, with the Viscose Company and DuPont Rayon heading up the list, and also naming as respondents twenty-two members of Price, Waterhouse and Company.[55] This was the accounting firm that had been checking the manufacturers' books to make sure that standardized production levels and charges were adhered to by all.[56] The FTC believed that a better descriptor for that activity was illegal price-fixing. Hearings on the case began in May 1934 and did not conclude until three years, 6,500 pages of testimony, and 1,700 exhibits later, finally culminating with a 1937 federal cease and desist order against further price-fixing (although the practice, at that point, had already been discontinued among the viscose producers).[57] The proceedings against Price, Waterhouse were dismissed, thus preserving that firm's good name (for the meantime). Of special note, the future secretary of state Dean Acheson, of Covington, Burling, Rublee, Acheson & Shorb, served as co-counsel on the DuPont team.[58]

Just days before the judgment in the FTC case was handed down, the July 1937 issue of *Fortune* hit the stands. Its lead article was on Courtaulds and its American operations: "Mystery: The American Viscose Corp.; A U.S. Investment of $930,000 by Some Shrewd and Close-Mouthed Britishers Yields a Stupendous Net in Twenty-Six Years, Estimated by *Fortune* at $300,000,000-odd."[59] As the expansive title of the piece makes clear, its focus is on the huge profits American Viscose had gathered to date and was continuing to reap. Illustrated profusely in the *Fortune* style of the day, with striking photographs of plant operations, the article is also richly colored with corporate details. The competitors of American Viscose are reviewed, with a fitting emphasis on DuPont. Perhaps more surprisingly, however, the Arcadia Knitting Mills incident is covered extensively as well. And although the spin is tilted in favor of American Viscose, this part of the story concludes with a zinger. In *Fortune*'s retelling, Arcadia had at first undercut its competitors by taking advantage of the discounts it received:

> Then came the depression, and Arcadia Mills stopped buying yarn. By the time they wanted it again, Viscose had found other outlets.... The mills brought suit for $20,000,000, alleging that Viscose had violated an agreement to supply them with all the yarn they needed. The suit was settled out of court—for a nominal sum, so the rumor went in the trade. Meanwhile Viscose had found another struggling young business *to take up the yarn it didn't have for Arcadia*. The history of that business, Burlington Mills, [is] as amazing in its own way as Viscose. [emphasis added][60]

Fortune does not omit biographical details for Samuel Salvage, who is described as "well rewarded for his pains" as president of the American Viscose Company, a position from which he had recently retired, moving up to become chairman of the board:

> Now sixty-one, he keeps an apartment at the Sherry-Netherland Hotel in Manhattan and a country estate at Glen Head, Long Island. The house at Glen Head was built in 1928, just in time to entertain for Sir Esme Howard and 1,500 other Englishmen on Empire Day.... He has tried his hand at one hobby after another. Once he dabbled in show business. He has raised tulips that won the Holland Bulb Growers Association trophy four years in succession. And he keeps a 150-foot yacht, Colleen II, with a crew of 14 men, for short weekend cruises.[61]

Salvage came through the FTC crisis intact, though not unscathed. His affair with the Reinhard brothers had left a lingering taint. In the words of *Fortune*, "It was no secret that Viscose in 1931 had fallen to a new low level in the estimation of the rayon trade because of the Arcadia Mills episode."[62]

Moreover, the evidence at the heart of the FTC case comprised Salvage's cables to Courtaulds on the pricing and production tactics.[63] Perhaps not surprisingly, Salvage blamed Arcadia for calling out the FTC dogs upon him.[64] But as far as is known, he did not invoke the metaphor of Isidor taking a pound of flesh (or if he did, he at least did not memorialize the comment in a cablegram). Arcadia Knitting Mills, whose company motto was "Independence, Liberty, Virtue," stayed in operation through the 1930s despite being cut off by American Viscose—a photograph of the company's employee football team from 1932, with the management proudly standing in the back row, hangs today in the California office of Isidor Reinhard's son, Eli Reinhard, president of the enterprise that he named after his father's and uncles' business: Arcadia Development Company.[65]

Meanwhile, in the same years as the FTC episode, DuPont was fighting on a second legal front, going after Waxed Products Company, Inc., for the manner in which it sold viscose-based transparent film made by its chief competitor, the Sylvania Industrial Corporation. Although the cellophane business was dominated by DuPont, Sylvania used manufacturing processes covered by different patents and was thus independent, holding on to roughly a quarter of the market share (and it had the potential to expand in that niche, including by courting tobacco business—the Sylvania product performed well as a wrapper for Lorillard's chewing plugs).[66] This was a duopoly in the finest sense of the word. Moreover, Sylvania had been careful (in light of DuPont's aggressive stance) to avoid making any claims that its product was Cellophane. But that, apparently, was not good enough. It turns out that Waxed Products, when approached by customers wishing to purchase cellophane, did not clarify that what it was distributing did not have a capital C, as it were. DuPont sued for copyright infringement. In May 1934, a lengthy federal district court opinion found in favor of DuPont,[67] but two years later this was reversed by the U.S. Court of Appeals for the Second Circuit.[68] In a much more succinct decision, the Second Circuit determined that the term "cellophane" had come so much into the public use that trademark protection no longer held. DuPont twice asked the U.S. Supreme Court to intervene, and twice the petition was rejected; the second time, Dean Acheson's senior partner, J. Harry Covington, represented the company.[69]

The waning of cellophane's trademark protection, determined by mass-market-driven facts on the ground, was only a small part of a much bigger phenomenon. Over the 1930s, the ways in which perceptions of viscose rayon and cellophane played out were telling: both products, despite being chemically identical and produced by the same technology, traveled widely divergent paths in the public imagination. Cellophane became emblematic of something entertainingly novel, the camp chic of its day. Rayon, on the other hand, assumed a tawdry, even sinister aspect. In 1931, looking forward to what the future might hold for rayon, Taussig and White crossed over the border from the land of straight economics into the realm of cultural critique, writing in a Veblenesque tour de force:

> Fashion may change. It is more than possible that it will become less favorable to the rayon makers. Their product has been for a long time insistently in demand not only because it was like silk in softness and lustre, but also because, like silk, it had something of the glamor of display and distinction. So long as rayon was still a rare material emulating silk in softness and sheen, fabrics made from it catered to prestige and emulation. But as soon as it comes to be made cheaply in larger and larger quantities, and sold to a wider circle, they must lose some of their charm. "Silk" stockings are no longer insignia of riches and station; they are worn by the maid as well as the mistress.[70]

The maid in ersatz silk stockings that Taussig and White foresaw made her novelistic debut shortly thereafter, personified in Irmgard Keun's 1932 novel *Das kunstseidene Mädchen*, which appeared in Britain as *The Artificial Silk Girl* in the following year.[71] As the *Times Literary Supplement* noted in its December 1933 review, Doris (the first-person-voiced protagonist of the novel) is "a young lady of easy virtue in the distracted Germany of a year or two ago."[72] She goes to Berlin to make it in the big city; not surprisingly, Keun has been invoked as the Candace Bushnell of her day. If so, Doris is more Samantha Jones than Carrie Bradshaw. Although she prefers sheer silk, she will make do with rayon when necessary. But even in this Doris is discerning: she is particularly attracted to the cuprammonium form of nonviscose rayon known commercially as Bemberg silk (*Bembergseide*). In the 1930s, it had only a tiny share of the market, but even at a higher cost, it held on because it was desirably "super-silky"; and particularly in Germany, *Bembergseide* undergarments were considered especially hygienic.[73] In four places in the book Doris takes note of *Bembergseide* underwear (indeed, at one point pilfering a pair).[74] But aside from the novel's title, the more generic *kunstseide* (that is, viscose rayon) makes only a single appearance, and that as a caution: "Let me tell you, Herr Brenner, a woman should never wear artificial silk when she's with a man. It wrinkles too quickly, and what are you going to look like after seven real kisses? Only pure silk, I say."[75]

Keun was not alone in highlighting rayon accessories as key features in the wardrobe of the moderne social climber. An essay by the leftist German journalist Felix Stössinger, published in the same year that *Das kunstseidene Mädchen* appeared, critiqued the consumerist pretentions of the revitalized urban shopping landscape of the Berlin of his day. Stössinger noted, inter alia, that Tauenstzien Straße had been transformed into a cheap street carnival where the ribbon vendors had become "a temple to the God of rayon."[76] It is no coincidence that near the end of Keun's book, her material girl, Doris, walks dejectedly on the same Tauenstzien, down on her luck, no place to live, schlepping all her possessions. As she describes it: "I am carrying a suitcase—genuine vulcanized fiber—a suitcase with my Bemberg shirts and stuff, with my hard-earned Berlin things."[77]

In specifying that the vulcanized fiber of the carrying case was genuine (*echt* in the original German), the author sounds an ironic note not intended by Doris herself. Vulcanized fiber, like viscose, is a semisynthetic cellulose material made by a process (first patented in the 1870s) in which paper is hardened through a metallic-salt treatment.[78] Vulcanized cellulose fiber, vaguely leather-like, was particularly popular for use in making synthetic traveling cases that were less expensive than the real thing.

The "artificial silk girl," as a type, was far from uniquely German. Indeed, it is easy to see why Keun's novel might have resonated with British readers. Lux soap advertising copy in the 1920s pointed out how the female wage worker could aspire to pass as a member of the leisured class (an office assistant at an open filing cabinet, paneled with illustrations of each piece of garb and text about its laundry care): "HAPPILY THE BUSINESS GIRL OF to-day disproves that the only wear of efficiency is sombre hues of blacks and greys. Every morning she successfully vies with her more leisured sisters in charm and brightness . . . [Upper left panel] THE STOCKINGS—Artificial silk in the popular moonlight shade. Spare them from hot water harsh treatment—wash them in lukewarm Lux."[79]

But the blurring of social status was only part of the equation. Taussig and White posited that artificial silk might come to be viewed as something not only cheap but also cheapening. In the next issue of the *Quarterly Journal of Economics,* the publication in which their article had appeared, a physician with expertise in consumer product safety named Carl Alberg took this one step further. In "Economic Aspects of Adulteration and Imitation," he notes, "The word, *imitation,* also has an ignoble connotation which, however, does not imply harmfulness. The antagonism to imitations arises rather from the fact that their use confers the opposite of social prestige upon the user."[80]

Despite the downside to imitation, Alberg argues that at least it is more honest than the misnomer "synthetic":

> If the manufacturer is scrupulous yet anxious to avoid the word, *imitation,* he may use the word, *artificial,* or *synthetic,* instead. These words, however,

have meanings not identical with that of the word, *imitation*. An artificial article in the strict sense of the term is one in which chemical and physical properties is [sic] practically identical with some natural product, but produced by artificial means. Imitation leather is wrongly termed artificial leather, since it is not, chemically speaking, leather at all. This is also true of artificial silk, so-called, which is not silk at all but a cellulose derivative.[81]

The imitation in leather-like "vulcanized fiber" was twofold: a cellulose leather knockoff and a name that invoked vulcanization, but was actually based on a chemical manufacturing process nothing like the vulcanizing of rubber. Vulcanization introduced sulfur into the latex feedstock, which was why sulfur dissolved in carbon disulfide caused so much disease in rubber manufacturing. The lexicon of ersatz expanded further through another chemically modified substitute material, vulcanized vegetable oil. In this case, the vulcanization was real: sulfur was introduced into the vegetable oil (rapeseed was one of the preferred starting materials) to create a substitute rubber-like substance that could be mixed with the real stuff to get more product at a cheaper cost. One method of production used carbon disulfide to dissolve the sulfur.[82] This practice led to workplace poisoning, just as it had in cold vulcanization.[83] In the case of vulcanized vegetable oil, a new name for the material came close to admitting its imitation status: factice, initially introduced by the Germans and derived from Latin (as in "factitious").[84] Factice is still manufactured and marketed today for use in the rubber industry. Another appearance of the same word with a related etymology, but this time directly from the French *factice,* for "dummy" (used as in "sham" or "empty object"), is a popular eBay collectable referring to very large perfume bottles meant for display but without any real perfume therein. This is probably a good thing, since very old perfume might well contain traces of carbon disulfide, since the solvent was promoted very early on as a way to effectively extract the essence of roses and other flowers.[85]

The American Dry Goods Association, in 1925, had not been blind to the potential downside of "artificial"—or even worse, "imitation"—silk, and proposed the alternative brand name "rayon," which caught on right away in the United States. Even though the Brits took longer to warm to "rayon" (perhaps because it was perceived as an Americanism), they came around eventually, and "artsilk" eventually fell by the wayside. Even Samuel Augustine Courtauld could see the downside to "artificial." Chairing a 1926 meeting of the Royal Society of the Arts at which the chief designer of Courtaulds was reading a paper on synthetic fibers, Courtauld made introductory remarks (paraphrased in a third-person account) that were in sync with Alberg's argument against the term "artificial": "The name 'artificial silk' was unfortunate, because artificial silk was entirely different from silk both chemically and physically, and its properties were, in some ways, almost the reverse of silk. He supposed that it had originally been called artificial silk on account of its peculiar lustre; so far as the present invention went, it was the most lustrous fibre known. It had added, in reality, a new lustre to life."[86]

Non-English-language usage varied. The Italians embraced the word "rayon," using it widely in advertising throughout the 1930s (also applying the word *viscosa* frequently); the Japanese seemed somewhat open to "rayon" as well. The Germans, in contrast, stuck with "artificial silk" (*kunstseide*) when referring to filament, as did the French (*soie artificielle*). Whatever term was applied, the modifier "imitation" was universally avoided, or nearly so. In 1930, a traditional *cintu* ballad appeared in Tamil: "Imitesan cilkku celai cetamilc cintu."[87]

Taussig and White, and Alberg, easily might argue that such word preferences were not mere coincidences. But extrapolating beyond a negative market impact to a potentially sinister cultural-spiritual symbolism for imitation was beyond the ken of economic theory. The intersection of ersatz and modernity, however, was a theme well suited for treatment in dystopian fiction. The timing was ripe in early 1932 when, a few months before Kuen's novel appeared, Aldous Huxley published *Brave New World*.

If one comes at it from the angle of costume notes, *Brave New World* might as well be subtitled "The Artificial Silk Boy." Huxley very nearly perseverates on the topic of synthetic textiles. As with Keun's Bemberg silk, a nonviscose rayon gets prominent billing, in this case acetate silk (technically, a modified rather than a regenerated cellulose rayon; Huxley would have known cellulose acetate by its British trade name, Celanese). In the first chapter of *Brave New World*, the decanted embryos are trained to take to a hot climate, "predestined to emigrate to the tropics, to be miners and acetate silk spinners and steel workers."[88] Acetate textiles (including acetate-satin and plain acetate) appear many other times throughout the novel, including an impassioned "I thought I would never see a piece of real acetate silk again."[89] But viscose (Huxley's preferred term for rayon) makes as many appearances in the book as acetate (and maybe more), including as viscose-linen, viscose-wool, viscose fur, and viscose velveteen (the latter twice, in the form of sexy velveteen shorts worn by the novel's paramour, Lenina).[90] Juxtaposing viscose with the novel's title phrase underscores the key role played by the synthetic for Huxley: "He was thinking of Lenina, of an angel in bottle-green viscose, lustrous with youth and skin food, plump, benevolently smiling. His voice faltered. 'O brave new world,' he began."[91]

Rayon had traveled a long way, culturally, over a very few years, becoming transformed from the modern everyman's luxury of 1925 to a tawdry deceit by 1932, if only the short distance from "O, wonder!" to "O brave new world," separated by only a few lines of text in Shakespeare's *Tempest*.

Viscose even became the stuff of medical quackery, with a special flourish of imitation hucksterism. In the late 1920s, the Dr. Clason Viscose Company began to market by mail order a nearly miraculous cure for varicose veins called the "Viscose Treatment." The term "viscose" likely was co-opted in this scheme for its connotation of modern technology and for simultaneously operating as a near soundalike for "varicose." The viscose varicose treatment involved spreading a thick paste over the legs and wrapping it over with a bandage.

Within only a few years, the Viscose Treatment had mushroomed into a series of clinics called "Viscose Ambulatoriums" (never referred to in the company's advertising literature by the presumed Latin plural, *ambulatoria*).[92] The flagship operation was based on Alavarado Street in Los Angeles, whence it expanded to Chicago and other sites. Before long, doctors and patients around the country were writing in to the American Medical Association inquiring about the product's extravagant claims as a cure. This caught the attention of the AMA's Investigative Bureau, and it had the ear of the Federal Trade Commission. By 1933, the resulting FTC cease and desist order catalogued eight pages worth of false claims made by the Dr. Clason Viscose Company, heading the litany with: "Viscose Method stops pain from Varicose Veins, Milk Leg, Phlebitis, Poor Circulation; stops swelling, positively heals leg sores while you work."[93]

The AMA Investigative Bureau had carried out an independent laboratory analysis that showed the product had nothing to do with viscose and, thankfully, at least was not contaminated with carbon disulfide. It was no more than overpriced zinc oxide paste: "The material seems to be a little better than white carpenters' glue with glycerine. The name itself is misleading as Viscose, as ordinarily known, is a well recognized chemical substance known for years and used in making caps for covering bottles, for making artificial hog casings and in the manufacture of artificial silk."[94]

Remarkably, the contemporaneous image of cellophane was the polar opposite of that of artificial silk. Cellophane was newfangled but good-natured, contemporary and cute, hygienic yet not sterile, a lark, the cat's meow—in short, as Cole Porter immortalized it in 1934, *the top*, right up there with the National Gallery and Greta Garbo's salary.[95]

Musically, Cole Porter was not alone in composing odes to cellophane, although "You're the Top" has been perhaps more memorable than Edward Heyman and Richard Myers's "If Love Came Wrapped in Cellophane" (1935)[96] or Billie Madsen's "My Cellophane Baby" (1937).[97] Nor was the muse of song alone in pushing cellophane. The same year that *Anything Goes* opened, the defining avant-garde theatrical event of the decade took place, the 1934 stage production of *Four Saints in Three Acts*. The words were by Gertrude Stein, the music was composed by Virgil Thomson, choreography was by Frederick Ashton, and the production was staged by John Houseman. The costumes and scenery, the work of Florine Stettheimer, prominently featured cellophane (which Stettheimer had earlier incorporated into the curtains in her Manhattan apartment); the exuberance of the stage design led to its being dubbed "Botticellophane."[98] The souvenir program for *Four Saints in Three Acts* (introductory remarks by Carl Van Vechten, portrait photos by Lee Miller) includes special acknowledgment of the DuPont Cellophane Company.[99]

We have no documentation of any Algonquin Round Table conversations that included cellophane witticisms. But we do have Ring Lardner, in a piece in the *New Yorker* (19 December 1931), imagining the hotel's manager, Frank Case, interrupting a dancing male couple with the query, "Pardon me, Officer, but can either of you play the cellophane?"[100] The next month, cellophane even made it to the "Talk of the Town" (alongside and above the classic Thurber cartoon of a barking seal in the bedroom looking over the bedstead):

> It [cellophane] has all sorts of uses besides the familiar ones. They are making hats out of it, artificial grass and water scenes for the stage, raincoats for emergency use, costumes for burlesque-show choruses, and coverings to protect table tops from glass marks. Kosher cellophane is made to wrap up kosher products. You can make your own flypaper by smearing it with glue and laying it over something flies like to eat. It's very effective. The flies don't see it until they are stuck on it. In Oregon a fishing guide discovered that a little piece torn off a cigarette package and stuck on a hook fools trout and salmon. It looks like nothing to flies, but it looks like flies to salmon.[101]

The *New Yorker* returned to cellophane again in May 1933, in an essay by E. B. White, whose rapier wit is matched by an incisive critique of how things stood at the worst of the Depression and not long after FDR's inauguration. In "Alice Through the Cellophane," White remarks:

> The social revolution began with cellophane, which allowed people to see what they were buying instead of simply going ahead and buying it anyway. Consumers are now keenly interested in examining life through the cellophane film of paralyzed industry; they are intrigued with its new transparencies and convinced that some sort of order might be introduced somewhere. Some of the more common paradoxes of this transition period are wrapped in cello-

phane by economists, and we find people of shall I say average intelligence like myself seeing clearly that there can be a glut of food in a time of famine.[102]

Cellophane may have caught the notice of the elites, but this was because it had become so emblematic of the masses. As an instructional pamphlet produced by the Brooks Paper Company of St. Louis tells its readers, "Nineteen hundred thirty-three has set many new styles. *Cellophane* has surely taken its place as a leader in these. You will want to be in the swim. The newest thing is *Cellophane* Clubs, where the fashionable matron and miss join in making gifts, novelties and plainly useful things of Cellophane.... You should be the first in your neighborhood to start a *Cellophane* Club."[103] The pamphlet ends by linking cellophane to celebrity: "Hollywood's famous movie colony has established *Cellophane* as the anniversary symbol of the first six months of marriage. Already up-to-the-minute folks have grasped the idea and *Cellophane*-wrapped gifts and remembrances made of *Cellophane* honor the occasion."[104]

But it is the concluding paragraph of Dashiell Hammett's *The Thin Man* in 1934 that sums it up best: "Macaulay started to move. I did not wait to see what he meant to do, but slammed his chin with my left fist. The punch was all right, it landed solidly and dropped him, but I felt a burning sensation on my left side and knew I had torn the bullet-wound open. 'What do you want me to do?' I growled at Guild. 'Put him in cellophane for you?'"[105]

4

Body Count

On the morning of 11 March 1933, Western Union delivered a telegram to Dr. Alice Hamilton at the Harvard Medical School: "RAYON FACTORY HAVING EPIDEMIC OF MENTAL CASES LAID TO CARBON DISULPHIDE POISONING PHYSICIANS ANXIOUS TO GET GENERAL SYMPTOM OF DISEASE STOP BEST PROBABLE TREATMENT STOP CAUSES OF DISEASE PLEASE ANSWER QUICKLY."[1]

There could have been no more knowledgeable or better-placed recipient of such an urgent appeal. Alice Hamilton's work for the federal Bureau of Labor Statistics, back in 1915, made her a leading U.S. expert on the toxicity of carbon disulfide. In her erudite report on the subject, *Industrial Poisons Used in the Rubber Industry*, she provides a technical primer on the rubber manufacturing process, including the carbon disulfide–based cold curing method for vulcanization.[2] Even though falling increasingly out of favor in the industry, it was still in nine factories that she had inspected as part of the survey.

After detailing these technical aspects, Hamilton's report goes through, substance by substance, the toxic materials used in the rubber industry, including six pages devoted solely to carbon disulfide.[3] Hamilton summarized the scientific literature on the toxic effects of carbon disulfide, including many key medical reports in German. Besides being proficient in the language, she was personally acquainted with some of the leaders in German toxicology, having received training there early in her career. As she notes, the state of medical knowledge on the hazards of carbon disulfide at the time was fairly rudimentary in the United States compared to the expertise in Germany and other European centers. In Hamilton's rubber industry report, she puts forward the premonitory assessment that any disease in the United States caused by carbon disulfide might very well be missed because of a lack of medical recognition of the root cause: "It is certainly possible that the insane asylums have received cases of unrecognized carbon disulfide psychosis, since insane rubber workers are committed from these towns without any inquiry being made as to the exact occupation of the patient and the possible industrial source of his disease."[4]

In addition to reviewing the data available to her from published sources in 1915, Hamilton presented the results of her independent search for carbon disulfide poisoning cases. She did this through leads she had received from rubber factory foremen rather than from physicians, because, as she notes, the latter generally were ignorant of carbon disulfide's effects. Of sixteen separate cases Hamilton tallied (including an ill foreman), one worker had been briefly committed to an asylum and several others had experienced nervous system complaints, albeit they did not lead to hospitalization. One unfortunate had worked for only a month before he began to show signs of derangement: "He was Hungarian and spoke no English, and the foreman did not recognize his condition until he became very much excited and unmanageable. He was sent home, and his wife reported that he acted so strangely and was so uncontrollable that she took him to a doctor. When the latter asked him about his work he told a rambling tale of lumbering down a river, and could not be convinced that he had ever worked in a rubber factory."[5]

Although the cold curing of rubber vulcanization was of particular concern to Hamilton because it entailed the use of carbon disulfide, she also addressed other toxic substances used in hot rubber vulcanization, by far the dominant process in the U.S. industry. In the latter method, chemical accelerators typically are added to rubber. One of the additives widely used at that time, aniline, is a potent toxicant in its own right; it was of special concern because it poisons the blood, giving it a bluish tinge and causing a syndrome of acute oxygen deprivation. Hamilton reported that the problem was so endemic in one rubber plant in Akron, Ohio, that the workers there were referred to simply as the "blue boys."[6] But she also noted a relatively new rubber-manufacturing chemical additive introduced into the trade called thiocarbanilide.[7] This substance was a rubber vulcanization accelerant produced through the reaction of aniline with carbon disulfide, a combination of hazards that is something of a toxic perfect storm. When a *Journal of the American*

Medical Association editorial featuring the findings of *Industrial Poisons Used in the Rubber Industry* appeared early in 1916, carbon disulfide, aniline, and thiocarbanilide were each specifically acknowledged.[8]

In 1919, Hamilton paid a visit to the Marcus Hook, Pennsylvania, manufacturing site of a major U.S. producer of aniline and thiocarbanilide, the National Aniline and Chemical Company. She noted symptoms that suggested both aniline and carbon disulfide poisoning: "In making thiocarbinilid [sic] they have had a man suffer from intense headache and nervousness so that he could not sleep."[9] The production of thiocarbanilide likely led to a case of optic nerve damage (amblyopia) associated with carbon disulfide that was reported around this time. The patient's occupation was described as being a "dye worker" in contact with both aniline and carbon disulfide, which would have been consistent only with thiocarbanilide production. The treating physician for this case, a leading expert in amblyopia, was situated in Philadelphia, geographically positioned for a referral from Marcus Hook.[10] Ironically, although the Viscose Company's flagship artificial silk plant in Marcus Hook already was in full swing just down the road from National Aniline, viscose had not yet caught Hamilton's attention.

In the same year as her visit to Marcus Hook, Hamilton published a review titled "Inorganic Poisons, Other than Lead, in American Industries," but her comments on carbon disulfide were limited to a brief recapitulation of her 1915 Bureau of Labor Statistics findings in the rubber trade, still without mention of viscose manufacturing.[11] Six years later, however, when Hamilton returned to the subject of carbon disulfide in 1925 in her encyclopedic *Industrial Poisons in the United States*, she had important new information to transmit.[12] More than simply revisiting the hazards of the rubber industry, she considered emerging occupations that entailed carbon disulfide exposure. Chief among these was viscose, whose dangers she underscored with her firsthand description of the poisoning she had seen in two rayon workers in Buffalo.[13] In her textbook, Hamilton also came back to thiocarbanilide, highlighting its manufacture as another increasing source of carbon disulfide poisoning.[14]

Hamilton was an activist for worker protection on many fronts. Not limiting herself to governmental bulletins and medical texts, she made a point of writing about dangerous occupations for the general public. In a piece for *Harper's Magazine* in 1929, "Nineteen Years in the Poisonous Trades," she weaves her personal experiences in surveying hazardous factories together with a sharp policy critique of the kind of deficient worker protection that was the norm across the United States.[15] Hamilton singles out for special criticism the practice of trying out new chemicals without knowing ahead of time the danger they may pose to humans, or as she puts it, a scenario in which "the workers will serve as experimental animals." This practice, Hamilton argues, should be replaced with testing before use, and she cites as an example rubber accelerators, clearly with thiocarbanilide in mind: "If a new rubber accelerator is introduced the Bureau of Standards will test it and inform the manufacturer whether or not it will do the work; but nobody in the Governmental service is in a position to discover for him what effect the compound will have on his employees."[16]

Despite this allusion to thiocarbanilide, Hamilton never mentions carbon disulfide in the *Harper's* article. But building up to the piece's conclusion, in which she implicitly compares putting a premium on production over safety as being the equivalent to giving carte blanche to Fascism and justifying it because the trains would be more efficient, Hamilton singles out a location that later would prove critical in her work on carbon disulfide. Decrying a failed legal patchwork of worker protection, she notes that thirty-eight U.S. states (as of her writing in 1929) still had no legal provision whatsoever to compensate employees for work-related illness, although they might cover traumatic injuries, adding: "It is strange to think that a great industrial State like Pennsylvania should permit men to suffer and die of poisons against which they were helpless to protect themselves or even give them or their survivors what we call, rather curiously, 'compensation.' One wonders what would 'compensate' most of us for a ruined digestion, or for consumption, or paralysis, or pernicious anemia, or insanity."[17]

Thus, primed by a wealth of knowledge, personal experience, a newfound interest in the rayon industry, and, most of all, dedication to the goal of putting an end to a disease she believed easily could be prevented, Hamilton responded immediately to the telegram she received that March morning in 1933. She wrote back on the same day, providing a succinct review of carbon disulfide poisoning, then making clear her concerns:

> Such a condition as you describe is inexcusable under modern conditions. There is only one place where there is normally any danger of carbon disulphide fumes and that is in the churn room. Here the whole apparatus should be fume-tight, and never opened until all fumes have been drawn off by a closed exhaust system. As a usual thing these precautions are insisted upon by the fire insurance men because of the highly inflammable nature of carbon disulphide. I have not heard of any cases of poisoning in rayon manufacture in this country for many years.
>
> Will you not, in return for this, write me all the details possible about this occurrence? I am deeply interested in it.[18]

— At the bottom of the typed carbon copy Hamilton retained, she noted the following by hand: "No answer ever came."[19]

Alice Hamilton looked back on this episode in her memoir, *Exploring the Dangerous Trades*. Her recounting is intriguing for what it includes, what it alters, and what it omits altogether:

> I received a telegram, back in the early thirties, from an industrial nurse in a state which has practically no factory-inspection service. She wired "Epidemic of insanity has broken out in rayon plant. Doctors do not understand. Can you help?" I answered at once, with a full description of carbon disulphide and an offer of my services. I also begged for more information. No answer came to that or my other inquiries, for I wrote to physicians and to the nearest insane asylum. The veil of secrecy dropped and was never lifted, nor have I been allowed to visit that plant even after all these years. Once I was on the verge of being admitted. It so happened that I sat next to the president of that company at a dinner in Washington and got from him a cordial invitation to visit the plant as soon as some alterations were completed, but when the time came I received a formal, impersonal letter from the company saying that it had adopted a policy of not admitting outsiders.[20]

The telegram's text as reported by Hamilton is a condensed, but accurate, paraphrasing (albeit not literal, her use of quotation marks notwithstanding); the letter Hamilton promptly wrote back in reply is also correctly characterized. Oddly though, the telegram originated not from an industrial nurse, as Hamilton misstates, but rather from a Mrs. Petricha E. Manchester, who identifies her affiliation in the telegram's address line as being the Consumers' League of Delaware.[21]

The National Consumers' League, which is still active, was a nongovernmental organization long before anyone had ever heard of NGOs. Moreover, Alice Hamilton was intimately associated with the league from its inception. It was founded by Florence Kelley, with whom Hamilton was linked through her connections to Hull House and Jane Addams. One the league's founding missions was the promotion of occupational safety and health through statutory regulation, with a special emphasis on women's work. The league had a national agenda, but carried out much of its work through a network of local affiliates. Alice Hamilton was involved with the Consumers' League in a series of focused efforts, agreeing early in 1924 to serve as chair of a national committee for the organization "appointed to collect and disseminate the available information on industrial poisons."[22] In that period, she was particularly active with the New Jersey affiliate as well as with the national leadership of the league, working on behalf of a group of women workers made profoundly ill from painting radioactive radium on glow-in-the-dark watch dials. These advocacy efforts culminated in a high-profile conference in 1928 held in Washington, D.C., for which Hamilton was the leadoff speaker.[23] Other work with the league touched on the rayon industry directly: in *Exploring the Dangerous Trades,* Hamilton identifies the Consumers' League in Pennsylvania as a source of information on cases of carbon disulfide poisoning.[24]

Petricha Manchester was a long-term local operative with the Delaware branch of the league. In 1927, in her role as executive secretary of that group, Manchester had gone so far as to write to Pierre S. du Pont (eldest brother of Iréneé) seeking, without success, an increase in his $25 annual contribution to the league. Perhaps not surprisingly, du Pont was leery, particularly in regard to the league's agenda on child labor, which had the audacity to call for a raise in the minimum legal working age from twelve to fourteen years in the canning industry. He wrote back to Manchester, "As to Child Labor, I believe it would be a great mistake to establish a Federal Bureau. We have too many already and it is impossible for any set of regulations to be promoted to cover the whole United States."[25]

Manchester even had a national presence in the same circles as Hamilton. For example, Manchester served on the Committee on Hours of Labor and the Committee on Child Labor Standards at the Third National Conference on Labor Legislation in 1936, whose findings were published by the Division of Labor Standards, for which Hamilton then served as a consultant.[26] Thus, it strains credulity that Hamilton was unable to obtain any follow-up directly from Manchester or indirectly through her many ties to the Consumers' League.[27]

Hamilton may have mentioned no more than she did in her memoir because she was intent on guaranteeing the anonymity of the rayon factory in question, given the disparaging nature of the reminiscence. She does not even hint at the manufacturer's geographic location, beyond recounting that she met the factory's owner, by chance, at a dinner in Washington, D.C. Had she given away the location of the state from which the telegram originated, it would have been a de facto naming of the manufacturer, since it was the only rayon producer with a factory in Delaware. This was not DuPont, whose viscose production facilities were all out of state. A brief entry scribbled in Hamilton's pocket diary for Thursday, 3 March 1938, ultimately provides the key missing detail: "Dinner Mayflower . . . Supt. Delaware Rayon Co."[28]

The Delaware Rayon Company was one of a handful of smaller viscose rayon manufacturers (all on the Eastern seaboard) independent of the American Viscose Corporation and DuPont, although in sufficient collusion with them to be included among those charged by the Federal Trade Commission in its price-fixing case. This case already had closed out in 1937, the year previous to Dr. Hamilton's dinner encounter.[29] Presuming that the man at the Mayflower was indeed the president of the Delaware Rayon Company, he would have been John Pilling Wright, who held that position until his death in 1946 at age sixty-six.[30] He had come to viscose with two decades of relevant experience in making vulcanized fiber (a cellulose-based product), a business that he had established in 1906.[31] Going from strength to strength, soon after the flagship Delaware operation was up and running, Wright took advantage of the expanding rayon market by setting up a new subsidiary in Massachusetts, New Bedford Rayon, Inc.[32]

The Delaware Rayon Company factory, located in New Castle, Delaware, was founded in 1926 at the site formerly housing a Bethlehem Steel munitions plant from World War I.[33] Indeed, the previous owner established something of a local tradition of occupational disease. The Federal Writers' Project guidebook of 1938 for Delaware offers the curious visitor, as one of its recommended options, a sightseeing itinerary along the New Castle's River Road ("Tour 10"). This is coupled with an anecdote about the prior Bethlehem employees: "The workers, including many women, were called 'powder monkeys'—their skins stained a deep yellow from picric acid." The guide goes on to cite a local informant recounting that the building where some of these female workers lived was known locally as "the canary house."[34]

Picric acid not only stains the skin, but also causes serious internal toxicity. Ironically, Alice Hamilton had studied this poison at another facility when the Bethlehem plant was in operation. She paints a picture of the workforce in a large New Jersey operation she had visited in similar terms, describing "men so stained with yellow picric acid that they were dubbed canaries."[35] There is no record that, as part of her munitions investigation, Hamilton ever went to New Castle or inspected the plant where Bethlehem was packing shells with picric acid. Even had she done so, it is unlikely she could have made the connection to the site's prior use, since she was never allowed to visit Delaware Rayon.

During J. Pilling Wright's tenure as president of Delaware Rayon, aside from deflecting Alice Hamilton's interest in the plant, he oversaw his company's defense against a potentially costly negligence suit brought by one of his former employees.[36] Her name was Emily Bowing. She had been employed at the company for three years, a period that subsumed the 1933 date of Petricha Manchester's telegram to Alice Hamilton. But the case hinged on a seven-week period in January and February 1934, during which time Bowing later stated she was employed in the plant's "reeling room," where the rayon fibers were wound after spinning. The company claimed that she worked in that department for just eleven days. In the suit, Bowing complained of a range of neurological and psychiatric symptoms, including irrationality, unconscious spells, lack of self-control, nightmares, crying spells, headaches, weakened eyesight, and a loss of libido.

Preliminary court motions were heard on whether the employer, the defendant in the suit, could require that Emily Bowing be examined by not one, but five physicians of the company's choosing, not only a neurologist and a psychiatrist, but also a general practitioner, a gynecologist, and a "genitourinary" specialist.[37] The court decided for the defendant (Delaware Rayon) in this matter, also stipulating that Bowing was not allowed access to the experts' reports, only to their court testimony when the time came. In due course, the experts opined for the record that the patient suffered from endocervicitis and a retroverted uterus with adhesions; her "mental sufferings" were due to her marital, economic, and physiological conditions (the last seemingly meant to imply a uterine etiology, that is, old-fashioned hysteria).[38]

Central arguments revolved around whether the reeling room was inherently hazardous because of carbon disulfide exposure, and if so, whether Delaware Rayon thus had neglected its duty to inform Emily Bowing of such a risk (their legal duty to mitigate risk was not relevant under the "master-servant" precedent that applied). Delaware Rayon did not contest that there was some modicum of carbon disulfide exposure within the reeling room. It contended, however, that it simply was not present to such a hazardous degree that they were obliged to give warning, master to servant. To support this contention, the company argued that, after all, no one else there was ill. A desperate telegram to the contrary was not entered as evidence.

One piece of testimony in this record is chilling to read, even three-quarters of a century later. A former employee, who had been a foreman in the reeling room, testified that "drops of moisture falling from the ceiling gave at first a burning followed by a cooling sensation."[39] What the foreman seems to be describing is a room so laden with carbon disulfide vapors that they condensed in the rafters and rained down on the hapless employees of the Delaware Rayon Company, even though such heavy exposure would have been very unusual at the reeling stage of rayon manufacturing. Emily Bowing's case, which under the circumstances she could not have won, also is noteworthy because it at least gives a name to the person who had been affected and, even if tempered by legal constraints, serves to tell a version of the story from her point of view.

Other first-person accounts of life and work on the shop floor of a rayon factory in the early decades of the industry are also important. One of the most intriguing is contained in an unpublished memoir by Elinore Morehouse

Herrick. Herrick was executive secretary of the New York Consumers' League in the early 1930s and then went on to become a major figure in labor mediation later in the decade, playing a prominent role in the newly formed National Labor Relations Board. Before any of that, in the 1920s, she was employed as a rayon spooler in the then recently founded DuPont factory in Buffalo.[40] Her largely pleasant memories of her entry-level job are tempered by the sobering realities of the manufacturing machinery, including her fear of the twisting machine and the cuts that it made on her palms.[41]

Herrick's move up was also a move out. She was transferred to an even newer DuPont plant in Old Hickory, Tennessee (just outside Nashville), and promoted to a junior management position. In "I Break a Strike," Herrick recounts that she pointed out the "ring leaders" of a brewing strike, who were then promptly fired. She later saw to it that at least some were rehired, but looking back years later and with a world of experience behind her, she continued to "suffer a deep and burning shame" over her behavior in the episode.[42]

Another piece of the story is preserved in the archives of the National Women's Trade Union League of America (WTUL) in an account titled "Two Weeks at the Industrial Rayon Corporation in February 1930," written by a woman named Victoria Enos.[43] Her seven-page report on what it was like to work for nine and a half hours nightly (4:30 P.M. to 2 A.M.), plus five hours on Saturday (11 A.M. to 4 P.M.), on the "coning" line (where, rather than "reeling" fiber into skeins, thread was wound for knitting applications) emphasizes the grueling physical demands of the work: "I took especially good care of my feet and still I suffered very much from standing on them so long. They hurt so badly that I forgot it was 'feet' and I became sick all over. If there had only been a chair with a back and low enough so the feet could get a little relaxation."[44] As with Herrick's memoir, Enos leaves out any mention of chemical hazards in manufacturing arising from viscose work (she does note that only men worked in the steps of early fiber production).

An even more fragmentary report from the same WTUL archive likewise concerns Industrial Rayon, recounting what happened to a woman when she sat down under a tree outside its manufacturing plant. She was pondering her next move after an unsuccessful attempt to find work at the company when two women who were driving away from the factory stopped. They asked whether she was sick and needed help getting into town, explaining, "At the plant many girls get sick and have to go home during the day. We have picked them up before. We thought maybe you was sick, too."[45]

This unsigned report had been forwarded to the WTUL with a cover letter from the industrial secretary of the Young Women's Christian Association's West Side Branch in Cleveland, Ohio. The report had been brought back from a visit to the Industrial Rayon plant in Covington, Virginia (Cleveland was its other production site at the time). The intent was to get the information published in the *Life and Labor Bulletin,* the official organ of the WTUL, but the YWCA's Frieda Schwenkmeyer cautioned: "If you use this story will you again refrain from using the names, the cities, and the company, and the organization with which we are affiliated, so that we can continue to study more about these plants."[46]

The post of "industrial secretary" at a local YWCA may seem odd today, but in fact the YWCA, besides having an interest in the health of female workers, which it shared with the WTUL, in certain cities had a particular interest in rayon workers, consistent with the character of local industry. The YMCA was even willing to take on DuPont. A 1929 handwritten note from Brownie Lee Jones, industrial secretary of the YWCA in Richmond, Virginia, to the Trades Union Congress in Britain asks for its help in this regard (DuPont had opened a rayon plant in Richmond in 1927):

> We are studying the health hazards of Rayon work and wish any materials that may be available.
> Do you have any records of disabilities caused by Rayon in any reports that might be secured by us. [no question mark in original][47]

Through the collective "we," the voice of organized labor provides an additional perspective on those who worked in rayon production. Thus, seven years after the unnamed writer sat down under the tree, Job-like, in Virginia, a major union action took place at Covington in 1937, primarily over abysmally low wages rather than compromised worker safety and health. After a bitter struggle, a sit-down strike by Local No. 2214 of the Synthetic Yarn Federation of America (an affiliate of the CIO's Textile Workers Organizing Committee) was effectively broken. In its aftermath, the Commonwealth of Virginia invoked what had been intended as an antilynching law, enacted in 1928, to prosecute, convict, and imprison some of the union leaders.[48]

Elinore Herrick had referred to another labor action—the potential strike brewing at the rayon plant in Old Hickory. But the most tumultuous labor unrest in the rayon industry in that era occurred in Elizabethton, Tennessee, at another artificial silk–manufacturing center. The strike began with a series of walkouts in the spring of 1929 at two distinct but interrelated rayon producers, both outposts of European concerns: American Bemberg and American Glanzstoff.[49] The former made cuprammonium artificial silk, and the latter, viscose rayon. The two factories' proximity was far from coincidental. Bemberg in Germany had been taken over by Glanzstoff, and the facilities in Tennessee were jointly managed by one man, an organic chemist named Dr. Arthur Franz Felix Mothwurf.

Eilzabethton is situated in an area referred to locally as Happy Valley. For example, Happy Valley High School ("Our Motto: Expect More; Achieve More") was given its name in 1926, the year that American Bemberg started production.[50] Arthur Mothwurf came to Happy Valley twenty years before the eerily parallel fictional character Franz Kindler settled in small-town America. Kindler was the German agent on the lam in Orson Welles's film *The Stranger*; Mothwurf had more than Central Europe in common with Kindler. In 1918, while working in New York state as a Bayer chemical executive in aniline manufacturing, Mothwurf was arrested for foreign espionage.[51] He spent the remainder of World War I, and then some, incarcerated. In fact, as New Yorkers were celebrating the Armistice, Mothwurf was leaving Penn Station, bound for Fort Oglethorpe, Georgia, to join a group of interned Germans.[52] By 1921, Mothwurf had resurfaced in New Jersey as head of research for the Garfield Aniline Works.[53] His job in Elizabethton, under the aegis of Glanzstoff, represented a shift from his core expertise, but was not too great a stretch, since both rayon and synthetic-dye manufacturing are chemically intensive industries. By 1928, the two rayon plants under Mothwurf's supervision were up and running when Herbert Hoover inspected the Elizabethton facilities before addressing a huge presidential campaign rally on the outskirts of town; Hoover praised the prosperity such industry represented.[54]

Handling labor relations, rather than controlling chemical reactions, proved to be Mothwurf's undoing. The 1929 strike started first at the American Glanzstoff plant, initiated by a group of seventeen, among several hundred, workers in its all-female inspection department.[55] The day following their initial walkout, they returned to the plant long enough to lead the rest of the factory in a massive walkout. Five days later, the workers at American Bemberg joined in. The strike was largely over wages and paternalistic work rules, especially for its female employees. The first worker to walk out, Margaret Bowen, addressed the annual convention of the American Federation of Labor (AFL) later that year, recalling the lame response of her boss to her demand for a raise, "I said I could not live on what I was getting and he said I ought to have a bank account. I said, 'A bank account on $10.08 a week?'"[56]

Mothwurf and his German upper-management team contributed to the initial crisis; Mothwurf was inarguably the chief protagonist in the chain of events as the situation unraveled afterward. Wages were key, but health and safety issues were not ignored. Bessie Edens, another one of the activists, begins a description of her experience at Bemberg by emphasizing the chemical nature of the industry (and tying this back to the German plant management):

> The silk is made from cotton waste. They make this waste into a liquid form by using different kinds of chemicals, some very strong acids are used. No one knows the secret of making the silk but a few Germans.... It is very dangerous for the men to work in the chemical department. I have known several to lose their eyes by getting something in them in there. And two that I know of lost their lives. And it is very unhealthy for the men in the spinning room. Anyone not used to it can hardly get their breath when they first go in, the ammonia is so strong.[57]

Arthur Hoffman, who was the chief organizer in Elizabethton for the United Textile Workers (at the time, an AFL affiliate), also refers to unspecified "dangerous chemical combinations" used in making rayon. In describing the working conditions, he similarly focuses on acid exposures in the Bemberg process, from which "hundreds of men can be seen with skin diseases and facial disfigurement caused by acid and acid burns."[58] There does not seem to have been any contemporary recognition of carbon disulfide as an identifiable hazard specific to the Glanzstoff viscose operation. That is far from surprising, since no one *except the Germans* seemed to be privy to the secrets of the trade. Yet even if it could not be named, the workers knew that something was in the air. Christie Gallaher, an employee in the Glanzstoff inspection department, where the strike started, noted workers overcome by fumes where she worked: "About half a dozen girls faint everyday [*sic*]. I know several girls that have fainted and fallen on the concrete floor and hurt themselves very badly."[59] A co-worker from the same department corroborates her report: "We have very good sanitary conditions but there is something in the air that makes the girls faint. I have known as many as twenty-seven to faint in one day."[60]

Elizabethton garnered a lot of attention. The WTUL quickly dispatched its representative, Matilda Lindsay, there, and she reported back in the pages of the *Life and Labor Bulletin*. Its April 1929 issue tells the story of the initial walkout, providing background on the labor force, especially their homegrown authenticity: "Working in these two plants are approximately 5000 natives, 100 per cent Americans, the majority of whom are women and girls."[61]

Lindsay's first report is rather upbeat, highlighting an initial agreement reached in part with support from a local sheriff and General Boyd of the Tennessee National Guard. At one point the agreement nearly falls apart, but the hiccup is attributed to a miscommunication from "the president of the company who is a German [and] does not have clear and complete understanding of English."[62]

The July issue of the *Life and Labor Bulletin* paints a far less rosy picture, telling how Mothwurf reneged on the initial agreement, the strike resumed, the National Guard was called in (setting up machine guns at strategic locations at the plants), and mass arrests ensued. *Life and Labor* notes that the second strike was eventually settled, and the article retains a guarded optimism: "The day will come which will give to Elizabethton workers the right to speak of their community as 'Happy Valley.'"[63]

All this was captured on film, a remarkable photographic record of the second 1929 strike.[64] The images are taken from the perspective of Bemberg and Glanzstoff, quite literally. There is footage in which the camera looks down from the rooftop of the factory. Key shots are taken over the shoulders of the National Guard machine gunners, facing down the lane leading from the factory gates, running perpendicular to the main road. On that road, within easy firing range, a long procession of striking workers marches past, and in the background are the hills that demarcate the far side of the Happy Valley. Later, the camera captures the procession close up, scanning from the side of the road itself. The men first, followed by the women, march in separate contingents, perhaps organized by department. Many, indeed most, are in their Sunday-go-to-meeting clothes. Quite a few of the men are in neckties, others wear overalls, but over ironed white shirts. The women leading their contingent are draped in American flags. Only a single marcher among the entire group holds a picket sign aloft. It displays two roosters, the one on top labeled "Union Rooster," and below that the placard states, "Red Rooster Peppermint Delicious Gum." Red Rooster was the product of a local Elizabethton chewing gum factory that, apparently, saw the opportunity both to support the striking workers and do some advertising at the same time.[65]

Nor was reporting limited to the official organ of the WTUL. The Elizabethton strike made national news. There was repeated coverage in the *New York Times*, including a prominent report that Herbert H. Lehman, at that point acting governor of New York, had resigned in protest from his position as a member of the board of directors of American Glanzstoff.[66] Sherwood Anderson, visiting Elizabethton and comparing it to his fictional Winesburg, Ohio, found that its rayon mills "had that combination of the terrible with the magnificent that is so disconcerting."[67]

Back in Germany, Glanzstoff was displeased enough to move Mothwurf aside, dispatching a man named Konsul W. C. Kummer to serve as acting president of its American operations. Even when a settlement finally came after a protracted struggle, the Elizabethton situation continued to make news. Kummer had been credited with improving industrial relations; a press photo from the time shows him, with suit, cravat, and pocket handkerchief in place, sitting at a desk so polished that his elegant watch and French cuffs reflect back up sharply.[68] But appearances were not what they seemed. Two months after his arrival, Kummer wrote a terse note to his superior back in Germany, Dr. Fritz Blüthgen, "I cannot go any further with rayon," and then slit his wrists; in the aftermath, Mothwurf stayed on—he did not officially resign his position until early the following year.[69]

Shortly after the second Elizabethton walkout ended, a group of Bemberg and Glanzstoff workers attended the Southern Summer School for Women Workers in Industry at Burnsville, North Carolina, whose program, along with personal narratives by the participants, is preserved in its yearbook typescript, *Scraps of Work and Play*.[70] The Elizabethton rayon workers at the school that summer included Bessie Edens, Christine Gallaher, Elizabeth Hardin, Ida Heaton, and Hazel Jones. Frances Mullens was also there, a rayon worker from Roanoke, Virginia (where a plant was operated by the Viscose Corporation of Virginia, linked by the same holding company to the Viscose Company). The summer school's program director, Louise Leonard, was formerly an industrial secretary with the YWCA.[71] *Scraps of Work and Play* includes a "Class Prophecy" section that makes absurdly humorous predictions of the future. It ribs one of the participants while simultaneously mocking Mothwurf in a Dickensian name twist, foreseeing a day when "Bessie Edens had divorced her husband and had married Dr. Mockqurth, president of the Glanzstoff Rayon Mill."[72] Among the novels on the school's reading list was Sherwood Anderson's *Winesburg, Ohio*. Of the Elizabethton strikers, Anderson had written, "'At least,' I thought, 'these working men and women have got, out of this business of organizing, of standing thus even for the moment, shoulder to shoulder, a new dignity. They have got a realization of each other. They have got for the moment a kind of religion of brotherhood and that is something.'"[73]

Rayon labor had found its voice, allied with forces in the Consumers' League, the Women's Trade Union League, and the YWCA. The years that followed the Depression brought the progressive forces of the New Deal into play, promoting new initiatives in worker protection. And yet Alice Hamilton remained stymied in her efforts to pursue further the hazards of viscose. In 1934, Hamilton returned to the topic of carbon disulfide in a small section of *Industrial Toxicology*, a new general handbook she wrote. In *Industrial Toxicology*, she updates her previous discussion of the artificial silk industry and even goes so far as to emphasize viscose, rather than rubber vulcanization, as the principal source of occupational exposure to carbon disulfide in modern industry. Of particular note, she cites multiple medical reports of neurological diseases, including psychosis and Parkinsonism, that had been coming out of Italy, although she acknowledges that the industry also had been plagued with problems in France, Germany, and Japan.[74]

Yet in contradistinction to her earlier text, in the *Industrial Toxicology* handbook Hamilton does not refer to the Buffalo cases that she had seen personally, nor does she cite the earlier American case reports of severe illnesses from the nascent viscose industry, as she did when writing before about carbon disulfide. She does not refer to the telegram that she had received about an outbreak of disease in an American rayon factory, even though that message had come to her before publication of the handbook. In fact, Hamilton gives the impression that her prior concerns about the domestic viscose industry have been allayed, going so far as to add a footnote to the text that seems to let the U.S. industry off the hook altogether: "It is evident from a perusal of the foreign literature that the making of viscose-process artificial silk is attended with much more danger of poisoning abroad than here where it has been very largely mechanized."[75]

Such an off-base conclusion on Hamilton's part is atypical; she generally voices a healthy skepticism of a blanket clean bill of health for any industry. It may be explained, at least in part, by the fact that in 1934 she still was relatively unfamiliar with viscose rayon operations from firsthand inspections, and as a consequence, she did not seem to grasp fully the industrial processes involved. This may also account for Hamilton's misguided reassurance to Petricha Manchester that "there is only one place where there is normally any danger of carbon disulphide fumes and that is in the churn room."[76] In fact, carbon disulfide overexposure was a particularly vexing problem in rayon-fiber spinning as well, the step that occurs after churning and aging the viscose solution. Exposure also could persist even further downstream in the process, as was evident in the case of the Delaware Rayon Company spinner.

It did not take long for Hamilton to correct her mistake. In the latter part of 1935, following her age-forced retirement from Harvard, Hamilton took a position as a part-time consultant to the Division of Labor Standards of the U.S. Department of Labor.[77] In June 1936, she delivered a major address to the annual meeting of the Massachusetts Medical Society (published three months later in the pages of the *New England Journal of Medicine*).[78] In that presentation, "Some New and Unfamiliar Industrial Poisons," Hamilton covers dangerous chemical solvents then gaining popularity in industrial applications—in particular, carbon tetrachloride, tricholorethylene, and benzene. But she gives her most extensive consideration to the question of carbon disulfide. In so doing, she completely reverses her *Handbook* position: "In view of the wide distribution of rayon manufacture in this country and the general ignorance concerning its dangers, we must conclude that here is an important subject for study by the toxicologist, the psychiatrist and the industrial physician."[79]

A few months later, in September 1936, Hamilton, as an official delegate of the U.S. Division of Labor Standards, addressed the Convention of the International Association of Industrial Accident Boards and Commissions, held in Topeka, Kansas.[80] Her presentation, "The Making of Artificial Silk in the United States and Some of the Dangers Attending It," was her most detailed assessment yet of the industry and of carbon disulfide exposure. Hamilton took this opportunity to further correct her handbook, first with a caveat early in her talk: "I may say that I myself visited a few plants, but I know very little about what goes on in them. When I want to know more about artificial silk I go to the foreign authorities."[81] Later in the presentation, invoking those authorities, she specifically rectified her previous misconception that problems occurred only in the churn room: "The danger from carbon disulphide is not only in the churn room, where the cellulose is treated with it. In fact it is easier to control the escape of this gas there than in later processes which are open, the spinning, aging, washing, and filtering."[82]

When Hamilton moved from the industrial hygiene aspects of viscose production to the adverse health effects it caused, once again she summarized recent European medical reports, the only ones available. And she told the story of the telegram that came to her "not from a physician but from a registered nurse" (she had also referred to this in her Massachusetts Medical Society address). She concluded her presentation with the following words:

> Long ago I had to give up the comforting feeling that things that happen in other countries do not happen in ours, that our workmen are so much better paid and live so much better, and our plants are so much more hygienic that we do not suffer poisoning as the other countries do. I know that is not true, and I know that American workmen are made of the same flesh and blood and nerve fibers as foreign employees are, and the same things will injure both classes of workmen. I hope very much that this ignorance of ours would not be lasting, and we shall soon find out what is going on in our industry.[83]

Hamilton's hope was an implied call for an investigation, just as in her Massachusetts address she essentially laid out her field study "dream team" for an industry-wide investigation of rayon manufacturing. She was about to will that dream into reality through one of the most concerted studies of a workplace health hazard that has ever taken place in the United States, before or since. The study began incrementally, first through a localized pilot effort, which was extended to a statewide study, and then expanded to a multistate survey. Ground zero was located in that "great industrial State ... Pennsylvania," which let its workers "suffer and die of poisons against which they were helpless to protect themselves," the geographic target of Hamilton's critique in *Harper's* in 1929.[84]

By the mid-1930s, U.S. rayon manufacturing was flourishing. The factories were largely concentrated in a narrow geographic band that swept east and then north in a fertile crescent of industrial viscose production: from Old Hickory on the outskirts of Nashville across to Elizabethton in Tennessee, over to Parkersburg, West Virginia, next to Covington, Roanoke, and Richmond, Virginia, and then up in a final swing into Pennsylvania. It was there that two major rayon plants stood, both American Viscose Corporation facilities: one in Lewistown, near Harrisburg, and at the far eastern end, the granddaddy of them all, the founding factory at Marcus Hook.

Hamilton began her rayon work in Pennsylvania not simply because of the prominence of the industry in that state. Beginning in 1920, Hamilton was affiliated with Bryn Mawr College in Pennsylvania (in addition to Harvard), teaching as a special lecturer in industrial poisons at the Carola Woerishoffer Graduate Department of Social Economy and Social Research.[85] Its founding chair, Dr. Susan Kingsbury, led the department from 1915 until 1936. From the start, she was committed to research applied to questions of employment and health, especially among women (a class student project in 1915–16, for example, used the University of Pennsylvania Hospital as a base from which to study occupational disease).[86] Indeed, Bryn Mawr's Department of Social Economy and Social Research served as a nexus for a number of progressive, women-led forces striving to improve working conditions. The Women's Trade Union League, the YMCA's industrial sections, the National Consumers' League, and the Summer Schools for Labor all intersected, one way or another, at Bryn Mawr. It was the right place, and it was the right time to take on the hazards of viscose.

The plan was to systematically interview viscose rayon workers, a study approach consistent with the direct data-gathering methods common to many of the research projects done by students in Kingsbury's unit. To this was added another, very special investigatory component. Directly growing out of Alice Hamilton's suspicion that neurological disease (even insanity) due to carbon disulfide might never be recognized as work related, the cases committed to state mental institutions would be tracked. Although a pilot study, this was an ambitious, labor-intensive initiative. Born in 1869 and already retired at age sixty-five from Harvard, Hamilton was well suited to oversee this project, but not to do the fieldwork.

In February 1935, Susan Kingsbury wrote to the dean of Bryn Mawr, providing the academic background and bona fides of a new, temporary, three-month research assistant appointment for Adele Cohn, M.D.[87] She was to work with Kingsbury and her "Seminary in Social and Industrial Research" in order to carry out "an investigation of industrial disease in eastern Pennsylvania." Cohn's salary was set at $100 a month, which would be spent out of a $500 grant given anonymously to the Graduate Department in the summer of 1934 specifically for this purpose. In her appointment letter, Kingsbury outlines Cohn's previous training, including her medical studies and internship. She also notes that Cohn had benefited from three months of additional training at the New York Medical Center in the fall of 1934, with a focus on occupational disease, at the time a highly unusual clinical externship.

Over three months in the spring of 1935, Cohn carried out her work with the assistance of three of Kingsbury's graduate students. Most of their time and effort was devoted to a survey of lead poisoning, an impressive study in which they identified cases in large part by going through the records of twenty-three Philadelphia hospitals.[88] More remarkable still was what the team was able to accomplish in investigating carbon disulfide, because all the work was done in seven days.

In her weeklong study, Cohn focused on a single Pennsylvania town "where the worst rayon factory was said to exist"; cases were identified through union referrals, "since there were no known cases of carbon bisulfide poisoning in hospital records and the disease is not usually recognized by general physicians." But Cohn did find one other important source of information: "A visit to the prothonotary's [county clerk's] office, however, proved interesting, as there was found a list of commitments to institutions for insanity since the rayon factory had been established in the town. The clerks in the prothonotary's office had commented to the investigator, that for some unknown reason there had been much more insanity in the county since the rayon factory had come there."[89]

Although brief, Cohn's investigation included a visit inside the factory, which allowed her to provide a detailed firsthand description of the work process. Cohn's report emphasizes carbon disulfide exposure, consistent with the focus of her investigation. In the churn rooms, where the cellulose was mixed with carbon disulfide, the powerful solvent strength of the latter tended to break down the seals between pipes, allowing the chemical to escape into the workroom air. More than that, the workers were required to enter the churn tanks to clean them (thirteen tanks during each six-hour shift), where additional carbon disulfide overexposure occurred: "Theoretically, fresh air is supposed to be drawn into each churn before it is emptied and cleaned, but the workers interviewed, said that the time schedule never permitted this."[90]

Cohn considered a variety of other occupational factors likely to negatively affect the health of workers in the factory: a grueling rotating shift schedule, eye troubles from the acid that was ubiquitous in the spinning room (at their own expense, workers went through three pairs of gloves a week and one pair of shoes a month, all eaten away by the acid), and skin eruptions farther down the process line where the women were employed preparing the finished rayon thread.

In addition to visiting the factory, Cohn spoke at a meeting of the rayon workers' union, gathering information afterward on co-workers who had been made ill, according to the employees with whom she spoke.[91] Cohn includes in her report ten pages of brief narrative summaries of thirty-nine cases of work-related disease among the rayon workers, largely based on face-to-face interviews with the affected workers or survivors. Among these, she identifies twenty cases involving a "mental breakdown" or frank insanity (two occurred in family members of her contacts who also had worked in the factory), of which at least ten were reportedly committed to mental hospitals. Some of her summaries are quite detailed:

> J.L. is a white male, 45 years of age, born in the U.S. In May, 1928, he began to work in the churn room. For two months he continued to work there, losing weight, and one day he went "out of his mind completely." He said he didn't know what he was doing. At the end of June of the same year, J.L. and a co-worker in the churn room, R.R., were committed to a mental hospital where they remained for three months, at the end of which time they were discharged as recovered. Since then J.L. has worked in the wash room of the rayon factory and is free from any signs of insanity. He is, however, at present not entirely well mentally. He shows a timidity in speech and a difficulty in co-ordinating his thoughts and speech, and speaks in a slow monotone.[92]

Nonetheless, J.L. was one of the luckier ones. D.K., Cohn reports, based on her interview with his surviving sister, was only twenty-three years old when, after one year of churn room work, he became deranged, attacking people on the street. Since he was "an unusually husky young man," it took four policemen to restrain him, and he suffered a punctured lung from a rib fracture in the process; four weeks later he died of trauma-related infection in an insane asylum.[93] Other churn room workers also became violently insane. These are reported tersely by Cohn, almost in staccato: one tried to commit suicide by stuffing a stocking down his own throat; another first shot his wife, then shot off his own arm. Nor were the cases of mental illness limited to the churn room—Cohn also documents three among spinning room employees.[94]

The document that Dr. Cohn produced is really quite remarkable, all the more so since it appeared as part of an official governmental report to Pennsylvania governor George H. Earle from his secretary of labor and industry, Ralph Bashore. The "Bryn Mawr Study of Occupational Disease in Pennsylvania," case histories included, supported the main thrust of that report—a proposed new statute establishing a system for workers' compensation in Pennsylvania for occupational disease, one that would parallel existing mechanisms covering physical injuries on the job. The report from Cohn's 1935 investigation, however, was not officially transmitted until March 1937.[95] In the interim, on 20 November 1936, "Some Girls from the Reeling Room" at the American Viscose plant in Lewistown had sent a letter to the White House: "The acid or whatever is in the cakes eat our fingers so terrible sometimes they look like a piece of raw meat. . . . Please President Roosevelt won't you see the proper ones will look into this as see if something can't be done that our poor fingers won't get so sore—we can't tell you the agony we work in."[96]

Striking as the letter is, the rapidity of its follow-up is also noteworthy. By 14 December 1936, Verne Zimmer (Alice Hamilton's boss at the Division of Labor Standards), to whom the appeal had been referred, wrote to Ralph Bashore in Pennsylvania suggesting that he investigate the matter; three days later Bashore's unit wrote back to Zimmer to assure him that the plant (and the viscose plant in Marcus Hook) was being inspected and that a further visit to Lewistown would be made in light of the these complaints.[97]

In mid-1937, things began to move quickly. New legislation recognizing occupational disease as suitable for compensation was adopted in Pennsylvania on July 29 (unanimously passing the lower chamber of the legislature).[98] The law went into effect on 1 January 1938; only days later, a joint federal-state investigation of the rayon industry began in the commonwealth.[99] By the end of August, Ralph Bashore was in a position to provide a second occupational illness report to Governor Earle, this one devoted entirely to the question of rayon workers' health. In his cover letter for that report, Bashore especially acknowledges the help of "Dr. Alice Hamilton, medical consultant of the Division of Labor Standards of the U.S. Department of

Labor" and praises the "staff of distinguished medical specialists from the University of Pennsylvania" that carried out the study, "the first of its kind in either America or foreign countries."[100]

Bashore's verbiage was not empty hyperbole—the study was indeed a unique undertaking, one that could not have happened without Hamilton's initiative. It was she who secured the commitment of support from the U.S. Department of Labor for a multistate assessment of the viscose industry, albeit with the proviso that she first had to get Pennsylvania on board for a state-based study. With Bashore behind her, Hamilton assembled an impressive research team. She relied on an experienced governmental industrial hygienist and occupational health expert, Lillian Erskine, to oversee day-to-day study operations. Erskine, a special agent for the Division of Labor Standards, was lent to Pennsylvania for this purpose.[101] Erskine had previously been chief of the Bureau of Industrial Statistics for the State of New Jersey; Hamilton had worked with Erskine on occupational health matters during World War I.[102]

Hamilton recruited eight physicians to carry out the clinical component of the study. These physicians represented the disciplines of neurology, psychiatry, hematology, ophthalmology, internal medicine, and pathology. Moreover, the team's scientific leader was a world-class biomedical researcher, Dr. Friedrich Heinrich Lewy. His career in Germany had been cut short by the rise of Hitler, leading to Lewy's dismissal in 1933 from his post as the founding director of the Neurological Research Institute and Clinic in Berlin; his earlier pioneering work in the pathology of brain disease is to this day associated with his name as "Lewy body dementia."[103] After emigrating from Germany, he sojourned briefly in England, where he studied the neurological manifestations of industrial lead poisoning in a follow-up to a previous interest in occupationally caused neurological injury.[104] By 1934, Lewy was associated with the University of Pennsylvania School of Medicine (later adopting Frederic Henry Lewey as his Americanized name).[105]

The Pennsylvania study, though medically ambitious, had to grapple with practical considerations on the ground. Chief among these was a lack of direct, on-site access to the rayon workers through their employers and, in fact, overt antagonism to the project from that quarter. Participants were recruited by word of mouth and interviewed with a questionnaire designed by Hamilton (similar to what Cohn had done in her original project several years before). Then, extending greatly the reach of the earlier study, participants were brought to off-site locations so that they could be systematically examined by the medical team. Because the workers feared being fired if the employer found out about their cooperation with the investigation, transportation had to be done in unmarked state cars supplied by Bashore's office.[106]

Ultimately, 159 workers were interviewed, and 120 directly examined. Drawn from the only two rayon factories in Pennsylvania, they were all currently employed at the time of the study (that is, not disabled), and none was specifically recruited based on his or her symptoms. In that light, the prevalence of ill health documented among those studied was staggering. Mood or personality changes were reported by 75 percent, hallucinations by 30 percent, loss of libido by a similar proportion, and ideas of persecution by almost 10 percent. When it came to direct physical examinations, three-quarters were found to have evidence of peripheral nerve damage, nearly the same proportion displayed "psychic" signs (that is, abnormal findings on a psychiatric assessment), and one in seven showed evidence of Parkinsonism. Overall, one in four of the workers was determined to have "severe" carbon disulfide–related intoxication (in the most heavily exposed group, the churn room workers, half of those studied were afflicted this way).[107]

Verne Zimmer, in his role as director of the Division of Labor Standards, gave Dr. Hamilton the green light to expand her study to the rest of the industry. She was even provided personal entrée to the American Viscose Corporation's facilities in Virginia and West Virginia, in addition to the ones in Pennsylvania. Hamilton made these visits with the approval of the "president of the Corporation" and as the guest of AVC's founding corporate medical director, Dr. J. Alfred Calhoun, who had only recently been appointed. Calhoun, who hailed from South Carolina, was a collateral descendent of the nineteenth-century stalwart supporter of slavery John Caldwell Calhoun.[108] Calhoun recruited Philip Drinker, an industrial hygiene engineer and a Harvard colleague of Hamilton's, as a paid consultant to AVC.[109] Hamilton was careful to confirm that this collaboration met with Zimmer's approval: "It seems to me worth doing, if only to encourage them [Calhoun and Drinker] in their work of reform. You would wish me to, would you not? It should not delay much the completion of my report."[110]

Zimmer signed off, and Hamilton completed her report by the late summer, although its publication was pushed back to 1940.[111] The final document is similar in length to the state report; its core is a summary of that earlier document's findings among the Pennsylvania workers, although it also contains important new data on exposure levels of carbon disulfide, which had been measured in a number of the factories in the survey. The most dramatic new material in Hamilton's report is in a fourteen-page appendix authored by her assistant, Lillian Erskine, separately titled "A Study of Psychosis Among Viscose Rayon Workers."[112] Following up on Cohn's work (although she is not mentioned), Erskine extracted mental hospital files of rayon factory workers, summarizing twenty-seven of them. Some are clearly the same cases that Cohn had uncovered, including the twenty-three-year-old who had succumbed to injuries received from physical restraint. The cold details that Erskine adds in her brief synopsis of the medical record confirm and amplify Cohn's description: "Committed from jail. Manacled; beaten up, strapped. Stupor from drugs. Eyes closed. Six gashes on scalp . . . Violent, maniacal, forcible feeding. Talk irrelevant and disconnected . . . Diagnosis: 'Manic depressive psychosis.'"[113]

In addition to her notes extracted from hospital records, Erskine received nearly complete transcripts (names were redacted) of the case files for seven of those who were hospitalized. One of them was the same fatal case, complete with an autopsy report documenting the wound-related infection and its spread to his heart valves (there was no microscopic examination of the brain). This transcript also records, apparently verbatim, the notes of the psychiatric diagnostic assessment conference on that case, antemortem:

> Dr. a. Presents Case II for diagnosis: Manic depressive psychosis, manic type is suggested.
>
> Dr. b. "I saw this man several times and he was very uncooperative, resistive, and noisy all the time. I would think of manic-depressive psychosis, manic type."
>
> Dr. c. "I think most of the symptoms fit into that class. We had quite a few who have had the same symptoms with this delirious confused state. We should consider psychosis due to exogenous toxins."
>
> Dr. e. "Apparently he hadn't cleared up since he left the factory. It does seem to me that the gas had something to do with it. He had been in the factory for quite some time and had to leave on account of the mental and physical symptoms and has never been well since. I would be inclined to make it psychosis due to drugs or other exogenous toxins."
>
> Dr. f. "I think it is hard to say what influence his occupation would have. This man has had attacks during which he was violent, then he would clear up and get along comfortably until the next attack. It seems to me that if it was due to drugs it would be more or less continuous. The history of being dazed and confused doesn't sound so much like manic-depressive psychosis, but I think probably we can diagnose it as that."[114]

This sixth discussant, Dr. "f," had the last word—this case was not officially attributed to an "exogenous" toxic exposure in the workplace.

Other cases that Erskine summarizes in her appendix to the Hamilton report had not come to Cohn's earlier attention, including two sisters, ages twenty-three and twenty-four, reeling room workers hospitalized for two and six months, respectively, each with a violent psychosis; a thirty-three-year-old churn room worker brought in assaultive and handcuffed, with the parenthetic comment that two cases of "acute exogenous toxic psychosis" from the same churn room had been brought in the same week; and perhaps the only vignette with a relatively happy ending, the case of a twenty-nine-year-old spinning room worker: "Melancholy; hid from wife and children; unable to concentrate; irresponsible . . . Gradual adjustment. Escaped from hospital. Returned. Second escape final. Reported by letter from farm in West. In good health. Returned to home after 6 months. Reemployed."[115]

Even with some delays, Hamilton's surveys had been completed fairly rapidly. But this did not mean that it had all been smooth sailing. In Hamilton's report, for example, she summarizes pathological findings from an autopsy performed on a rayon worker done as part of that Pennsylvania statewide study, emphasizing the importance of its findings of chronic brain damage. What Hamilton did not reveal, either in the report or in her later memoir, was the attempt to sabotage this key undertaking. She finally disclosed the episode in an interview that she gave twenty-five years after the fact:

> I thought it would be really valuable if we could get a record of an autopsy. Well, it was very difficult. . . . But I did get word on one who had gone insane and then landed in the public hospital. I didn't do it myself, but I got the pathologist from the University medical school to ask for an autopsy and after some fuss and trouble he was given permission to, but when he got to the corpse he noticed there was a fairly newly made scar of an appendicitis operation and he said to the attendants, why this man that I'm looking for never had an operation for appendicitis. Oh, they said, they're [sic] must have been some mixup.[116]

In the end, the correct body was located, and the case of the purloined corpse proved to be only a minor glitch. The entire study, however, was almost derailed by an evolving political battle in Pennsylvania. In one of its skirmishes, the preliminary data on the health of rayon workers, then just in the process of being gathered by Hamilton's team, became a major bone of contention. The central figure in this fight was George H. Earle III.

Governor Earle had been swept into office in the elections of 1934, the first Democratic governor in Pennsylvania since 1890. To underscore the significance of this turnaround, it is worth noting that Roosevelt had lost the state in the 1932 presidential elections. Earle was, by background and breeding, the perfect Republican: a Mayflower-descended blueblood deeply rooted in Pennsylvania (including close familial associations with William Penn); a navy veteran who skippered his own yacht and was commissioned in World War I as a submarine chaser; a national polo champion; and a holder of strong capitalist credentials as a sugar mini-baron. But by conviction and perhaps by temperament, too, Earle was a Roosevelt man all the way. After all, other ancestors of Earle's included an English judge who had tried King Charles I. His lineage also claimed links to Lucretia Mott, the American social reformer and co-organizer of the women's rights convention in Seneca Falls. Having been a strong Roosevelt supporter in the 1932 campaign, Earle was appointed minister plenipotentiary (that is, ambassador) to the Republic of Austria.[117] He served in that role until his run for governor, and through that experience, he became something of an early ("premature" would emerge later in Red-baiting parlance) antifascist.[118] On assuming the governorship, Earle immediately began an aggressive agenda of long-overdue reform, which came to be characterized as "Pennsylvania's Little New Deal." It prominently included labor reforms, of which the new occupational disease statute and the related rayon workers' study were a part.[119]

Early in 1938, just as the Pennsylvania state rayon study was getting into full swing, Earle was gearing up for a run for the U.S. Senate (he was limited by statute to a single gubernatorial term). It was in this context that Ralph Bashore asked Lillian Erskine to prepare a preliminary report on the rayon workers' project, which she transmitted to him on 21 March. Erskine's seven-page report starts with a discussion of labor force supply problems, focusing on Mifflin County, where the Lewistown viscose rayon plant was located. She raises the possibility that "the large labor-pool maintained on call by the Viscose Company is increasingly necessitated by the serious health conditions among its workers." She then emphasizes the number of commitments for insanity from Mifflin County, arguing that there is "convincing proof that industrial exposure to carbon disulfide—especially in the case of the churn room of the Viscose Company—has been responsible for direct expense (for institutional care) to the taxpayers of Pennsylvania."[120]

Ralph Bashore was not only secretary of Pennsylvania's Department of Labor under Earle, but also Earle's close political ally, serving as secretary of the Democratic Party's State Committee. He would have been well aware of the political implications of what Erskine and Hamilton were uncovering. Two weeks after Erskine delivered her report, the American Viscose Corporation, as the Viscose Company had officially become, publicly attacked the Earle administration, arguing that high taxes were causing it to relocate its business to Virginia, where it was constructing a major new plant; as a *New York Times* article title put it: "Earl Taxes Assailed; Business Men Assert Industry Is Being Driven from the State."[121] Earle, armed with the information in Erskine's report, went on the counteroffensive. In a Senate campaign radio speech, Earle discussed the rayon health study and the commonwealth's new Occupational Disease Act, and then named names: "The Viscose Company was given the choice between installing safety equipment and going elsewhere . . . If that is considered 'driving industry out of the State' then I gladly plead guilty, and extend my heartfelt sympathy to the people of Virginia. . . . Viscose workers, in all too many instances, end up in insane asylums because the process brings mental as well as physical collapse."[122]

The Lewistown Chamber of Commerce characterized Earle's speech as a "dastardly attack," and the *Lewistown Sentinel* proclaimed in a front-page editorial that "our fellow-citizens among the Viscose employees enjoy as much, if not more, of the average good health, zest for living, keenness of mind than will be found anywhere in the United Sates."[123] The American Viscose Corporation was somewhat cagier, saying only that the company had dealt with no compensation case of any kind for "occupational disease," presumably referring to the three-month period since enactment of the new law that made any such compensation claim even possible.[124]

Three days after the story broke, Lillian Erskine wrote to Verne Zimmer in Washington expressing dismay that her report "had been used as political ammunition," but adding, "So far as my report on the conditions in Lewistown is concerned I have nothing to retract." Despite that, Erskine believed that the main survey, not yet in the field, had been compromised and at a minimum would have to be postponed. She ended by offering to fall on her sword: "Certainly I am not responsible for Gov. Earle's broadcast or his personal political fights. But since I am involved in the situation I desire that I fight it through without involving you or Dr. Hamilton further. I therefore tender my resignation."[125]

Erskine's resignation was not accepted; she went on to finish the study with little delay. Earle continued to counterattack on the "flight of industry" front—but he was willing to call a truce with viscose. In a follow-up address delivered in October 1938 as the senatorial election approached, Earl extended an olive branch:

> Republican spokesmen have shouted at great length about the supposed flight of industry, but have been most reluctant to name the industries that are fleeing. One firm that they have seized upon as an example is the Viscose Company, operating plants in Lewistown and Marcus Hook. No one was more bitterly critical than I of the Viscose Company when it did not provide proper safeguards for the health of its employees. Now that this Company is cooperating 100% with the Departments of Labor and Industry and of Health in protecting the health of its employees, the Company will find my administration most sympathetic in its progress.[126]

Earle was fighting a losing battle, and even Bashore was in trouble. The state attorney general at the time, Charles J. Margiotte, a former Republican who had been angling for the nomination for governor and was aligned with anti-Earle forces within the Democratic Party, succeeded in obtaining a grand jury indictment for bribery against key members of Earle's cabinet, including Bashore. Eventually all were acquitted, but not until two years later, by which time Earle had failed in his race for the U.S. Senate, a Republican was back in the Pennsylvania statehouse, and the "Little New Deal" was over.[127]

Despite this reversal, Pennsylvania's occupational-disease compensation law remained in force, putting the Commonwealth ahead of many other states, including some with major rayon manufacturing within their borders. West Virginia was one such state. On 6 June 1938, Fred G. Stewart sent a handwritten letter to Frances Perkins, the U.S. secretary of labor, which reads in part:

> Dear Miss Perkins,
>
> Just a line to you in regards to my case. I went to work for the *Viscose Co. Parkersburg* W. Va 1927 and worked until the acid ate my stomach very near up. The last day I worked was June 1st 1936 have been off ever since. My labor union took my case up but W. Va. has no occupational disease laws so I cannot get any compensation. My stomach is all ragged and full of *gastric ulcers* . . . I was O.K. when I went to work for Viscose Co. but now I am not able to work. My insurance pays me $29.33 per month but it runs out in June 1939. I have my wife and seven children six at home and can not get any relief of any kind.[128]

The letter was sent on to Verne Zimmer, whose answer told Mr. Stewart what he already knew, that in West Virginia there were no workers' compensation benefits for any occupational disease save one: silicosis, that is, silica-dust-caused lung disease.[129]

In Pennsylvania, legal compensation for disease not only allowed for the possibility of monetary damages, but also affected disease recognition in a broader sense. Even before the Erskine-Hamilton study was released in the summer of 1938, the *Journal of the American Medical Association* on 7 May 1938 carried a report by two Philadelphia-based scientists, Samuel Gordy (a physician) and Max Trumper (a Ph.D. toxicologist), on neurological disease due to carbon disulfide exposure in six rayon workers.[130] The paper never makes clear how the authors came to evaluate these individuals, but it is likely that it was through the workers' compensation process. Gordy and Trumper were active forensically, providing medical-legal testimony on behalf of injured rayon workers; they were associated in this work with a Philadelphia attorney, Henry Temin.[131] Gordy and Trumper's most publicized case was named John Nichols, a long-term American Viscose Corporation employee at its Marcus Hook facility. Nichols's illness was officially recognized by the court, but he was ultimately denied monetary compensation. It turns out that after nineteen years of employment, Nichols had ceased to be an employee of AVC (because of his illness) on 31 December 1937. It was the day before Pennsylvania's occupational disease law went into effect, nullifying his eligibility for any benefits.[132]

Gordy and Trumper's *JAMA* case series presents a far less systematic and insightful analysis than Lewy's contemporaneous and more rigorous study. The two data sets, even though from the same factories, were never analyzed together, nor does there ever appear to have been any direct collaboration between Alice Hamilton and either Gordy or Trumper.[133] Whatever the scientific caliber of their work, it was impactful, if only because it appeared in such a high-profile medical journal. Their paper made a significant contribution by asking an important policy question of the kind typically not addressed in the pages of *JAMA*: what airborne level of workplace exposure to carbon disulfide should be allowed? By 1938, such chemical-specific target exposure levels were just beginning to be set. In their *JAMA* paper, Gordy and Trumper called for an exposure limit of no more than 10 parts carbon disulfide per million parts of air (10 ppm).[134]

Philip Drinker, Hamilton's Harvard colleague and AVC's consultant, did not agree. In fact, he held that reeling and other downstream rayon-manufacturing steps carried nothing more than a negligible exposure risk for workers (even though several of Gordy and Trumper's cases were, in fact, reelers). Drinker was willing to concede that in spinning, as well as in churning, exposure to carbon disulfide was present (as was exposure to the irritant gas hydrogen sulfide, a by-product of spinning). Giving a talk to a Safe Practice Conference jointly sponsored by the Pennsylvania Department of Labor and Industry and the State College of Pennsylvania (held at its Nittany Lion Inn), Drinker made a point of downplaying poisoning problems in the rayon industry: "Of course, some carbon disulfide and hydrogen sulfide adhere to spun rayon, so that both may occur (in very slight concentration to be sure) in the various drying, washing, and chemical cleaning processes subsequent to spinning. In the American Viscose Corporation this source of carbon disulfide and hydrogen sulfide is of no consequence at all. The churn and spinning rooms are, therefore, the only ones with which we are particularly concerned."[135]

Workplace chemical-exposure limits of the sort that Gordy and Trumper called for could be promulgated only on a state-by-state basis—no mechanism for legally binding national U.S. limits existed in 1939 (nor, indeed, would they for another three decades, until the creation of the Occupational Safety and Health Administration). Nonetheless, quasi-official targets were being recommended at the time through a consensus process involving industry and scientific experts, and carried out under the aegis of an organization called the American Standards Association.[136] In his Penn State talk, Drinker alluded to then-ongoing discussions of a committee of that body on which he sat, along with Alice Hamilton, and which was charged with recommending a carbon disulfide exposure limit. Drinker called for a target value of 30 ppm, invoking Hamilton as agreeing that 10 ppm was "needlessly stringent."[137]

Hamilton addresses the same question at the end of her federal report, but she is less cavalier than Drinker.[138] Hamilton notes that DuPont, based on its own animal-testing data, had adopted an exposure limit of 10 ppm (Gordy and Trump's recommendation as well). Although most European standards of the time were thirty to forty times as large as that value, Hamilton also acknowledges that the official Soviet target limit was by then 3.2 ppm. Hamilton reports that at the 30 June 1939 meeting of its Sectional Committee on Standard Allowable Concentrations of Toxic Dusts and Gases, the American Standards Association agreed to a carbon disulfide target of 20 ppm, doubling the limit favored by Gordy and Trump and DuPont, but at least not watering down protection to the extent that Drinker had proposed.

In a June 1940 meeting of the American Medical Association, Friedrich Lewy presented an overview of nervous system damage from carbon disulfide, based on experimental animal research along with the limited human pathological data that were available (including the one autopsy from the Pennsylvania study that Hamilton had emphasized in her report).[139] When the paper was opened for discussion, both Gordy and Drinker jumped in. Gordy astutely homed in on the combined nervous system effects of peripheral nerve damage, central nervous system effects on the basal ganglia (the area of the brain that is key to Parkinsonism), and disease involving the arteries of the brain, cerebral atherosclerosis. This last effect was an avenue of carbon disulfide toxic insult that would come to be appreciated only years later. Drinker, whose comments also were extensive, made the point that industry was well on its way to achieving reductions of exposure to 4 ppm (he did not volunteer that he was pushing for a less stringent target). Another participant, Dr. Noble R. Chalmers from Syracuse, New York, was brief and to the point: "I should like to ask what are the principal sources of carbon disulfide poisoning in industry (?)."[140]

Dr. Lewy, responding to this query, covered carbon disulfide's past uses in rubber vulcanizing but pushed the viscose rayon industry to the forefront.

He allowed that there were a few other minor applications, going so far as to include repairing golf balls as a potential scenario for carbon disulfide exposure. What Lewy did not mention was the manufacturing of cellophane.

Similarly, Hamilton, Drinker, and even Gordy and Trumper, despite their attention to rayon, all failed to take into consideration the potential hazards in viscose processes that used regenerated cellulose to manufacture film and related products (one of which was viscose bottle closures, known as cellons). No one was talking about cellophane by that or any other name. In the biomedical literature throughout this period, apparently no formal study addressed cellophane workers' health. This does not mean that all was well with cellophane. In 1934, a doctor in Wisconsin wrote in to *JAMA*'s "Queries and Minor Notes" question-and-answer column (a kind of Dear Abby for clinical conundrums); the doctor was concerned over an alarming situation that he had encountered.

> A local concern, which manufactures caps for bottles made of cellulose, the process being similar to that used in manufacture of rayon, is having trouble with the men who work in the "viscose" department of the plant.... The workmen complain of becoming dizzy, their heads seem to swell, a numbness and tingling sensation spreads to the hands and down the legs to the feet, and they feel as though they were drunk. The sensation soon passes off if they can get out into the open air but it has created a marked mental reaction in three of the men, in that they have a perfect horror of returning to the job.... I was talking to the wife of one of these men who has been in the department for about two months, and she said that her husband had become increasingly irritable and had lost his appetite and did not sleep soundly at night. This particular man stated that the main reaction was his dread of returning to work for his shift, stating that he would do almost anything if he did not have to go back to that smell again. I went out to the plant and inhaled enough of the gas to experience the same sensation.[141]

JAMA's answer to the query was unequivocal in pinpointing the culprit and recommending workplace controls, although, remarkably, it categorized the cases as being only "mild carbon disulphide intoxication."[142] The manufacturer itself goes unnamed, but undoubtedly was the Cellon Company of Madison, Wisconsin, a manufacturer of viscose bottle caps and cellulose film active in this period.[143]

Beyond this isolated medical advice column, there was a single systematic review published on the health hazards arising from the manufacture of cellophane and related materials. Appearing in the trade serial *Industry Report* in 1932, "Transparent Wrapping Materials" discusses the acute explosive potential of carbon disulfide, as well as the potential chronic ill effects to workers from exposure, including tremor, disturbed vision, and chronic dementia.[144] The corporate publisher of the review had every reason to be concerned over both explosive conflagration and longer-term liabilities from cellophane manufacturing; the moral hazard associated with carbon disulfide was right up its alley. The Retail Credit Company, in Atlanta, Georgia, was in the business of advising insurance companies on the risks they assumed by carrying particular policyholders. The Retail Credit Company is still in business today, currently operating as Equifax.

A few months after Hamilton's federal report appeared in 1940, Gordy and Trumper published another study of carbon disulfide poisonings, twenty-one additional cases this time.[145] All were males, Gordy and Trumper having restricted the group to churn room and spinning workers. The authors emphasize neurological symptoms, but also summarize other illnesses; for example, gastrointestinal problems were present in nearly three-quarters of the group, including one with multiple stomach ulcers. Once again, Gordy and Trumper do not provide any information on how the cases came to them, but in the "Comment" section of the paper (where a biomedical journal article would typically contextualize its findings), they state that the cases of carbon disulfide poisoning being reported, together with those from the previous *JAMA* article, included workers from factories in Pennsylvania, Virginia, West Virginia, and Delaware.

In this disclosure, Gordy and Trumper implicitly "outed" the Delaware Rayon Company, but they went further to explicitly address the telegram that Alice Hamilton had cited more than once as being emblematic of the hazards of carbon disulfide in the rayon industry. Gordy and Trumper state that in 1934–35 they personally examined six cases of carbon disulfide intoxication from among forty women affected at the plant (which they locate in Delaware) whose epidemic of disease had led to Hamilton's telegram. All the cases worked in the factory's viscose fiber reeling room. Gordy and Trumper were puzzled at first by the extent of illness linked to this late viscose process step, especially since the fresh-air intake of that unit had been increased once trouble started, albeit to no avail. They found the answer in contamination of the supposed "fresh" air intake: "We discovered that the concentrated carbon disulfide vapors were regularly exhausted from the churn room at a point less than 100 feet from the large and powerful fresh air intake of the reeling room.... It happened that the prevailing winds made certain that some of the carbon disulfide would be carried in front of the reeling room intake."[146]

Ultimately, Alice Hamilton succeeded in accomplishing what she had set out to do: call attention to unrecognized hazards of carbon disulfide in the rayon industry and bring about improved conditions for its workers. In her autobiographical reminiscence, Hamilton ends her chapter "Viscose Rayon" on an optimistic, even Panglossian note: "The control of this dangerous trade was slow in coming, but when it came it was astonishingly rapid and complete. It is safe to say now that no large viscose rayon works is a dangerous place to work in and probably few of the smallest ones."[147]

Hamilton published *Exploring the Dangerous Trades* in 1943. World War II, then at its apogee, was to prove how the tragedy of viscose was only just beginning to unfold.

5

Rayon Goes to War

July 14, 1942

To all Guests and Employees of
American Viscose Corporation
Marcus Hook, Pennsylvania

This booklet, "Rayon Goes to War" has been prepared to show some of the more important contributions made by "Crown" rayon to the war effort.

Not all of the various types of rayon required for the war effort are made here at the Marcus Hook Plant. The six other plants of the Company are doing their part.

We have felt it to be most fitting, however, that the first distribution of this book be made here at Marcus Hook on the day when the men and women of this plant are being honored with the Army-Navy "E" for their excellent contribution to the war effort.

Sincerely,
F. Farwell Long[1]

This letter, inserted loosely into a profusely illustrated pamphlet with an eagle along with red, white, and blue stars and stripes on the cover, is meant to set the stage for the panorama that follows. The banner on the cover declares, "Rayon Goes to War."² Image one shows a near-naked dark-skinned native crouching in the undergrowth, watching a warplane coming in to land on a jungle runway: "For example, when the laboratories of American Viscose were collaborating with the manufacturers in the development of a new and stronger cord for the tires of *your* car . . . the average person would never have dreamed that the resulting basic improvement would solve a vital problem in making the tires of our biggest bombers *safer*."³

Other pages illustrate sailors on a ship deck in their rayon uniforms; a rayon-tired half-track "combat car" taking hills, boulders, and shell holes in stride; electric lights directing an incoming night landing, thanks to rayon extension cords; and rayon parachutes facilitating drops of relief supplies.

DuPont did not want to be left out of the advertising action. Its full-page multicolored print advertisement from around the same time, titled "The silent enemy in the steaming jungle," shows a GI about to bite into a ration biscuit. Its ad copy reads:

> War correspondents' reports from jungle battlefronts at Guadalcanal and New Guinea describe the astounding destruction of equipment and supplies caused by the moist, steamy climate. . . .
> The U.S. Army Quartermaster Corps has to find the answer to protect foods in such a climate. And that's why they use Du Pont Cellophane for the vital protection of many Army field rations.⁴

Viscose may have been coming into its own in World War II, but the military roots of the rayon industry go much farther back than that. In fact, in the 1920s a recurring critique of the rapidly expanding artificial silk industry, and especially its multinational corporate interconnections, was rayon's potential use as a platform for rapid conversion to munitions manufacturing. This concern, not without foundation, was driven in large part by the close chemical and manufacturing links between artificial silk made through the nitrocellulose process and the production of explosives. In the period after World War I, however, the trepidation was that the rayon industry as a whole could be exploited to undermine disarmament. These fears were made explicit by Grace Hutchins, who, writing in 1929 in *Labor and Silk* (under the heading "Munitions and Rayon"), had exclaimed: "The secret of this big jump forward in rayon production since the war is connected with preparation for the next war!"⁵ She elaborates further:

> Both rayon (artificial silk) and dynamite can be made from nitro-cellulose. The nitro-cellulose process of making rayon in an artificial silk factory can be changed overnight into the production of dynamite. Under the innocent

name of artificial silk factories, munitions plants are extended and maintained. It is probable that equipment in all rayon plants, not only those using the nitro-cellulose process, can be adapted for explosives.

DuPont, largest munitions corporation in the world, and Nobel, dynamite maker and donor of the "Peace Prize," are now making large additional profits from artificial silk. Tubize, an international explosives trust and artificial silk corporation, also connects the rayon industry with the chemical industry.[6]

Hutchins had it partly right. DuPont, of course, did enter into viscose (rayon fiber and cellophane) in a big way, but did not become a commercial producer of nitrocellulose-based rayon. Nonetheless, one of DuPont's major viscose facilities in the 1920s was located in Old Hickory, Tennessee, where a huge munitions plant manufacturing "gun cotton" (that is, explosive nitrocellulose) previously stood. That munitions factory had been constructed and operated by DuPont during World War I under U.S. government contract (with foresight, though perhaps not outright foreknowledge, DuPont had purchased the land before U.S. entry into the war), transferred to the U.S. government control after the war, closed down and abandoned by government, and then purchased back by DuPont.[7] At that point, DuPont established its newly built viscose manufacturing facilities on the site. But this was not a direct munitions-to-rayon conversion, any more than DuPont's initial rayon factory, in Buffalo, converted from a rubber reclamation facility whose construction was still uncompleted by the end of World War I, represented a direct war-industry spin-off.[8] Still, DuPont was certainly linked to munitions, and in 1934 it was subjected to a congressional investigation into possible war profiteering, saliently targeted to Old Hickory contracts.[9] Remarkably, the official U.S. Senate's website refers to these as the "Merchants of Death" hearings.[10]

Despite Hutchins's claims and Alfred Nobel's personal scientific interest in nitrocellulose fiber, the industrial interests he established were not directly involved in rayon manufacturing in this period, neither during World War I nor in the interwar period that followed. In contradistinction, Hutchins omits mention of Courtaulds or its U.S. subsidiary, the Viscose Company. Viscose itself did not have any direct connection with armaments manufacturing, although it was such a major economic force in the British chemical industry that in the midst of World War I, Charles Cross was elected to membership in the Royal Society as "the founder of a great industry . . . [and for] the discovery of 'viscose'-cellulose."[11] The closest Courtaulds came to the arms business was the Viscose Company's purchase of the other major guncotton facility owned by the U.S. government at the end of World War I. The Viscose Company maintained the plant, located in Nitro, West Virginia, as a facility for manufacturing viscose precursor cellulose from "cotton linters."[12] This feedstock is a form of cotton consisting of the short fibers teased off seeds after ginning, not fit for weaving but well suited to making viscose or nitrocellulose. In the mid-1930s, as the Viscose Company was transforming itself into the American Viscose Corporation, it shut down the linters operation at Nitro and opened its first rayon staple plant there.[13]

Of the corporations that Hutchins invoked, it was Tubize that could claim the greatest continuity with munitions manufacture. Fabrique de Soie Artificielle de Tubize (whose major factory was in Tubize, Belgium, only eight miles distant from Waterloo, a pre-guncotton battleground) manufactured rayon by the nitrocellulose process. Its American offspring was the Tubize Artificial Silk Company of Hopewell, Virginia. This American outpost, established in 1920, was allowed to use the Belgian manufacturing patents in exchange for a major stock interest.[14] Tubize was intent on putting an American veneer on all this. As the *New York Times* put it: "In the effort to thoroughly Americanize the Belgian artificial silk industry," Tubize sent a group of young women from the United States on an all-expense-paid trip to Belgium ("under proper chaperonage") to be trained in its manufacturing process.[15] The choice of Hopewell, too, linked Tubize with the U.S. armaments trade because that was where DuPont had had another huge munitions operation; also abandoned after the armistice, it reemerged (under the DuPont aegis) as the Hopewell Industrial District. Tubize became one of its important new tenants.[16]

Well before coming to the United States, Tubize had looked to the East. In 1910, Tubize took over a nitrocellulose plant in Sarvar (Vas), Hungary, that had opened there in 1904. More precisely, Tubize took over the second such plant in Sarvar. The shared chemistry of nitrocellulose guncotton and nitrocellulose artificial silk meant not only that this product was highly flammable, but also that this manufacturing process was prone to explode. On 15 October 1905, the year after the plant opened, a blast ripped through Sarvar, killing six workers outright, blinding five others, and leaving more than two score injured. The factory was rebuilt, over public opposition (the factory had come to Hungary in the first place when siting in Italy was turned down because of the danger of conflagration).[17]

In 1911, Tubize was linked with a new enterprise in Tomaszów Mazowiecki, Poland (then under Russian control). That operation was founded by a Pole named Feliks Wiślicki, who had resigned his position as engineer with Tubize's Belgian facility to start this operation in his homeland.[18] Production was interrupted by the war, but the Tomaszów factory resumed its production of nitrocellulose silk thereafter, and was even reported to be supplying munitions to the newly independent Poland.[19] By 1930, the factory had abandoned nitrocellulose and changed over to viscose production; the Sarvar operation had shut down in 1927.

The U.S. Tubize plant in Hopewell stayed in operation until mid-1934. That was when its workers, represented by Local 2170 of the United Textile Workers of America, went out on strike. Decades later, the events that transpired in 1934 continue to engender counterreaction, at least as evidenced by the none-too-subtle phraseology of the *Hopewell News* in a nostalgic look back in 1976, headlined "Since Tubize, Hopewell has Never Been the Same": "Summer, 1934. Hitler was grabbing power in Germany. Labor was shaking the Depression-sick economy with strikes."[20]

At the end of June 1934, striking workers brought nitrocellulose production at the plant to an abrupt halt. Claiming that the nitrocellulose drying out in the pipes was likely to explode, the company used state police to ferry in a group of nonunion employees to clean out the dangerous material.[21] Several weeks later, Tubize announced that it would permanently shut down production in Hopewell.[22] The next year, in June 1935, Tubize announced that the idle equipment from Hopewell was being shipped to Brazil as part of a new joint venture, Companhia Nitro Química Brasileira, a small but landmark event in the history of exporting hazardous manufacturing to developing countries.[23]

Tubize could maneuver as it did in part because several years before, in 1930, it had carried out a highly strategic merger with the Italian-controlled, U.S.-based American Chatillon Corporation of Rome, Georgia.[24] The merger created the Tubize Chatillon Corporation and diversified Tubize's production, adding to its nitrocellulose line both viscose and acetate rayon, which were manufactured by the Georgia plant. Tubize held out its Georgia operation as an odd counterexample to any labor agitation, claiming in a public notice that "working conditions in our Hopewell plant are equally good . . . [as] those at Rome."[25] Equivalent conditions could hardly have made for a strong talking point: years later, Rome locals could remember the acid eating workers' pants and shoes and darkening coins in their pockets. They saw neighbors led home by plant security at the end of a shift at the rayon plant after being blinded by the workroom atmosphere.[26]

All these reports are indicative of excessive acid mist, sulfur fumes, and other production by-products released by a poorly controlled viscose operation. They were not an exaggeration. A professional look back at the first ten years of Tubize Chatillon's operations in Rome, written in 1939 by the plant's chief chemist, H. L. Barthelemy, confirms and even amplifies the assessment that overexposure at Rome was a fact of daily life.[27] He voices misplaced pride in the fact that there were only three cases of severe carbon disulfide nervous system disease, of which only one had to be committed to a mental hospital, and the other two recovered after a mere four months. On the other hand, eye problems were admittedly more pervasive, although Barthelemy, just as his British counterparts had done a decade before, downplays the epidemic as an outbreak of simple conjunctivitis, rather than the more significant medical condition of keratitis. He certainly does not allow that optic neuritis might also have been present among the afflicted. Choosing to highlight the statistics for only one month of each year of the multiyear outbreak, Barthelemy tallied up 332 cases of eye disease in December 1933, 85 in December 1934, none the next December, and then 71 in December 1936. But Barthelemy discounts these alarming yuletide figures as nonrepresentative, maybe even outright humbug, because December always manifested the highest prevalence. The Christmas season, it seems, was when the plant's windows and doors were typically kept shut against the cold.

Insanity and blindness underscore that this part of Tubize Chatillon's operations in Rome had much in common with the wider viscose industry. It was its sister cellulose acetate operation, however, that tied Tubize in Rome to the armaments industry, much as nitrocellulose had anchored Hopewell to war materiel. Technically, cellulose acetate is a modified cellulose product, which differentiates it from nitrocellulose, viscose, or cuprammonium artificial silk. Manufacturing processes for the last three start with cellulose, which is chemically manipulated but ultimately "regenerated" as the final cellulose product. The cellulose in cellulose acetate, however, is permanently altered, and its chemistry allows it to be manipulated in new ways. Most importantly, it can be dissolved into a liquid form and spread out as a synthetic coating material. As a commercial artificial silk fiber, cellulose acetate was a marginal enterprise in the twentieth century; it was World War I that kick-started the business as a coating material for military applications.

Air warfare of the new age demanded novel supporting technologies. Airplane doping was one of these. The word "dope" was an Americanism that, long before meaning an illicit substance, referred to a coating of pitch applied to the bottom of shoes to facilitate gliding across sun-warmed snow.[28] It was a logical extension of this meaning to use it for the coating applied to a cloth-covered airplane, a coating that allowed it to better glide across the sky. An ideal coating, or "proofing," was one that provided both waterproofing and air-proofing. In addition, such a coating could guarantee that the cloth covering stretched across a frame remained taut. Doping had been tried first with nitrocellulose, which could also be dissolved into solution. But this had a major disadvantage, especially in combat: the nitro in nitrocellulose not only made it explosive during manufacturing, but also highly inflammable even when dried on an airplane wing. To circumvent this drawback, the British fixed upon cellulose acetate, even before World War I, as the preferred airplane doping material; the practice soon was adopted elsewhere.[29]

A key challenge to using cellulose acetate dope was to find just the right chemical dissolving agent to turn it into a spreadable, fast-drying coating. Luckily—or unluckily, depending on whether you were a pilot or an airplane fabricator—a new family of solvents was coming into commercial use just as demand for airplane dope was picking up. Many of these solvents, all chlorinated hydrocarbons, are still on the market today, for example, as paint strippers and dry-cleaning agents. All the chemicals in this family of toxic substances can cause health problems of various sorts. But the chlorinated solvent used in cellulose acetate doping was one of the most deadly ever introduced into commerce.

Called tetrachlorethane, this chemical caused severe and often rapidly fatal liver disease in the workers who inhaled its highly volatile fumes or who got substantial quantities on their skin, through which it was easily absorbed. World War I started in August 1914. By September and October of that year, cases of liver damage and jaundice were being reported out of London's suburban Hendon Aerodrome, and the first fatality had been documented by the end of November.[30] Even before the outbreak of hostilities, a tetrachlorethane case fatality already had been reported in a doping operation at the Johannisthal Airfield in Berlin.[31]

Britain came to mandate better ventilation and initiate a search for substitutes for tetrachlorethane, although these actions may have been driven primarily by cellulose acetate shortages. By the end of May 1917, when the king, queen, and Princess Mary went to visit the Grahame-White Aviation Company at Hendon, they were assured that its dope room had been made safe, even if not idyllic:

> The hundreds of workers in all parts of the factory gave the Royal Party a rousing reception, and as the King and Queen proceeded through the different bays the workmen and workwomen formed a cheering guard of honor. . . .
>
> They went through the fitting shop, the seaplane department, and the "dope" room. In this latter place some hundreds of girls were engaged in "doping" the fabric which covers the planes, rudders, and ailerons. The mixture with which these are covered gives off very poisonous fumes, and for that reason powerful fans draw off the bad air and pumps supply fresh, so that there is no danger to any of these workers, although the smell from the "dope" is anything but pleasant.[32]

Britain sought to alleviate its cellulose acetate shortage by building a domestic industry that could replace the imported material. By the time of the royal visit to Hendon, cellulose acetate production was coming online at British Cellulose in Spondon, England, thanks to a sweetheart deal between the British government and the Cellonite Company of Switzerland. This agreement provided the company a virtual domestic monopoly on the product and allowed it to reap substantial war profits. Once the war ended, there was far less need for dope, but the same military manufacturing apparatus could be adapted to civilian synthetic textile production. British Cellulose morphed into British Cellulose and Chemical Products, whose major brand of cellulose acetate, Celanese, hit the market in 1921.[33]

The United States entered the war after the initial European epidemic of toxic jaundice from tetrachlorethane, and that chemical was not widely used in the American aircraft industry of the time. But as in Britain, a postwar renaissance in cellulose acetate textiles meant that tetrachlorethane toxicity could still be a threat. The one homegrown U.S. cellulose acetate plant operating in the immediate postwar period was called the Lustron Company, located in South Boston. Its Boston base reflected close ties to a local man (an MIT alumnus) who made Lustron possible, Arthur D. Little.[34] This was an outgrowth of Little's original expertise in the chemistry of the paper industry. He held an early patent for a carbon light filament based on cellulose acetate[35] and, by 1902, a further acetate silk patent jointly with two other chemists, Harry S. Mork and William H. Walker, which was assigned to an entity called the Chemical Products Company.[36] The Lustron Company was the eventual corporate outgrowth of this work.

Although Little (by then at the helm of ADL) was not active directly in this new cellulose acetate business, his nephew, Royal Little, was both an employee of and an investor in Lustron.[37] He was a Harvard dropout who had joined the American Expeditionary Forces (his division, the 42nd of the AEF, experienced thousands of war gas casualties).[38] A young veteran when he started out as a salesman at Lustron, Royal Little worked his way up the ladder. Unfortunately, not only was the company ailing financially, but its workers were threatened physically as well.

Lustron had staked its manufacturing process on the continued use of the toxic solvent tetrachlorethane. The company called in experts from the Harvard-MIT School of Health Officers (later the Harvard School of Public Health) to consult on worker health. The academic consulting team published two scientific papers based on its studies of Lustron workers without ever revealing the name of the company involved or even its general location.[39] They did document, however, that more than 25 percent of the production workers at Lustron had become ill over a five-month period beginning in January 1920. And yet even though all but three of twenty-one cases were overtly jaundiced to varying degrees, the findings of liver injury were downplayed; nonetheless, the authors of the reports had no trouble singing the praises of the enlightened owners of "a plant well adapted as to management, personnel, and control of poisoning to illustrate the possibilities open for the use of tetrachlorethane in peace time."[40]

The clearest advice to Lustron came from a visiting foreign expert taken to tour the company's plant in south Boston, accompanied by Dr. David Edsell of Harvard (soon to be founding dean of the Harvard School of Public Health). This visitor was none other than Thomas Legge, then still His Majesty's senior medical inspector of factories. Legge was all too familiar with the British doping experience: he had visited the Hendon Aeroplane Factory in early December 1914, shortly after the first cases had been reported (the visit was made together with William Willcox, who two decades later was to attack Legge for being antibusiness in pushing for carbon disulfide controls).[41] Seeing the continued use of the unnecessarily dangerous material, Legge recommended that Lustron either find another solvent to substitute for tetrachlorethane or shut the plant down.[42] Lustron did neither. Indeed, as late as 1925, Harry S. Mork and his aptly named coinventor, Charles F. Coffin, filed a new cellulose acetate patent for Lustron, still basing the process on the use of tetrachlorethane.[43]

Ultimately, the workers at Lustron were spared not through the substitution of a less hazardous solvent, but by the business being bought out and closed down by American Cellulose & Chemical Company, a subsidiary of the British airplane-dope-cum-artificial-silk producer. American Cellulose, not keen on competition, already had set up its own U.S.-based Celanese operation in Cumberland, Maryland, at the end of World War I.[44]

Royal Little left Lustron before it was closed down, but he had not sold his stock, which became worthless at the rock-bottom price that American Cellulose paid for the assets. Undeterred, Royal started his own textile chemical research firm in South Boston; it was underwritten by his former Lustron boss, Eliot Farley, and employed its chief chemist, Mork.[45] In a series of mergers and acquisitions that later made Royal Little a business icon and a very wealthy man, his Special Yarns Company merged with Franklin Rayon Dyeing, based in Rhode Island, in 1928, morphed into an expanded Atlantic Rayon Corporation in 1932, absorbed New Hampshire's Suncook Mills, and then announced in 1944 its name change to Textron, Inc. This culminated a period of exponential growth, spurred on largely by the manufacture of rayon parachutes.

The Royal Little Story, a testimonial pamphlet put out by Harvard University's Graduate School of Business Administration in 1966 to commemorate the establishment of an endowed Royal Little Professorship, takes pains to underscore that he was no mere war profiteer. It quotes one of his friends, who emphasizes, "I can assure you that, as a government contractor, he was primarily interested in making more goods better and faster, more economically and more efficiently for the troops. He was *not* there as a manufacturer interested solely in making a lot of money from the war."[46] But then again, *The Royal Little Story* points out, "While turning out the best possible war supplies he knew how to make, his builder's instinct kept active, and he proceeded to acquire several marginal New England mills that were viable wartime investments or brought brand names, new products, cash or tax advantages to his enterprise."[47]

Royal Little, going from doughboy on the gas-riddled western front to salesman for an airplane doping spin-off to rayon mill magnate producing wartime parachutes, amply represents the ties that bind artificial fibers to the military-industrial complex. But this relationship, spanning the entire period of the world wars, is even better personified by another figure, a German applied scientist named Ernst Berl. Berl was a quintessential citizen of the world. Born in the Austrian Empire in an area that later became part of Czechoslovakia (Bruntál, originally Freudenthal), Berl first studied chemical engineering at the Technische Hochschule in Vienna before obtaining a doctorate at the University of Zurich in 1901 under Alfred Werner (who later became a Nobel laureate in chemistry). Berl then moved over to the Eidgenössische Polytechnik (now the Swiss Federal Institute of Technology), before leaving to take an industrial chemistry position in Belgium.[48] From 1910 until the outbreak of World War I, Berl was the chief chemist at Soie Artificielle de Tubize.

Writing in 1939, Ernst Berl looked back on 1914. He recalled how he was "called to the cause on the first day of the war," that cause being, of course, his Austro-Hungarian homeland.[49] Berl was immediately assigned to a governmental explosives plant near Vienna. He rapidly rose to become "Chief Chemist of the Austro-Hungarian War Ministry," in charge of its wartime munitions industry. Being a chemical engineer well versed in nitrocellulose, he was ideally positioned for his new role. It was more of a segue than a switch, as he noted: "There was no preparation for war. Rayon and other plants had to be converted into munitions factories."[50]

For nitrocellulose rayon feedstock, the conversion to explosives was not much of a reach, technically or geographically. One of the key sites for the changeover to munitions manufacturing was in Vienna's own backyard: the Tubize nitrocellulose plant at Sarvar in Hungary. After all, it had exploded even during its civilian use for making rayon. Berl was familiar with the risks. In a further reminiscence, Berl recalled a conflagration in Belgium (presumably under his Tubize aegis) that occurred when tubing carrying liquid nitrocellulose under high pressure broke, shooting out a spray that hit an open flame, ignited, and jumped back to the rayon equipment, quickly destroying the entire spinning room.[51]

After World War I, Berl took a prestigious position back in academia: professor of chemistry, electro-chemistry, and technology, and director of the Chemical-Technical Institute, at the Technische Hochschule in Dortmund, Germany. He continued to be interested in cellulose chemistry, publishing papers on the subject as well as obtaining new patents.[52] Berl remained in Dortmund from 1919 until he resigned in 1933. By the time of his 1939 reminiscence, Berl's hometown of Bruntál had been annexed to the Reich (along with the rest of the Sudetenland). In the meantime, Berl had resettled in America, secure in a position as research professor at the Carnegie Institute of Technology in Pittsburgh. He worked hard there on research that could be applied to U.S. preparedness for the looming war. Although this research included munitions, Berl's most significant contribution was made in the chemistry of coal and petroleum substitutes. It is work for which Berl today is considered something of the progenitor of the synthetic biofuels movement. Even at the time there was considerable attention to his work in the popular press, including a piece in *Time* magazine from September 1940, "Recipe for Fuel": "To the American Chemical Society convened in Detroit last week, Professor Ernst Berl of Pittsburgh's Carnegie Institute of Technology made an astonishing announcement. He said he had made, experimentally but successfully, oil, coal, coke and asphalt from grass, leaves, seaweed, sawdust, scrap lumber, corn, cornstalks, cotton."[53]

Another émigré whose career in rayon arched across the interwar years, but in an entirely different trajectory, was a German textile engineer turned industrial magnate. He was born Oscar von Kohorn, but by 1918, thanks to his economic service to the war effort, he was raised to the minor nobility as Freiherr (Baron) Oscar von Kohorn zu Kornegg.[54] Both a brilliant innovator and a highly successful entrepreneur, von Kohorn had started off in the carpet-weaving business. In 1913, he patented a production method for broadloom carpets, and during the war, he had adapted to wool shortages by using alternative fibers in the manufacturing process.

This led von Kohorn to rayon. He had the foresight to realize that given the rapid growth in the industry, it would be highly lucrative to manufacture and supply equipment meeting the specifications of his own patents. Moreover, von Kohorn could supply a team of engineers to do the start-up and train local operatives to take over (along with, where suitable, owning a piece of the action). He began by establishing two rayon factories in Czechoslovakia after

World War I, then equipped others across the Continent, expanded across Europe and the Soviet Union, and then extended even farther east. By 1934, von Kohorn had so much business in Japan that he established an office there.

Based in Chemnitz, Germany, von Kohorn was living the lifestyle of the rich and famous, 1920s mode. He and his wife were great patrons of the arts. Richard Strauss, a frequent guest at Villa Kohorn, in 1925 composed and dedicated to the family "Hymne auf das Haus Kohorn" (a cappella for two tenors and two basses). Critically, the baron was farsighted in more than just business. As conditions in Germany deteriorated, he sent his older son to England for higher education and then on to Japan to work in the business there. By the end of 1936, von Kohorn, his wife, and his younger son (still of secondary-school age) were all safely in Japan as well, initially ensconced in a suite at the Imperial Hotel, Tokyo. The rayon business was booming there; von Kohorn set up fully equipped factories and then made them operational at multiple sites for Japanese firms. But by the spring of 1938, it was time to move on again. Baron von Kohorn, wife, and youngest son joined the oldest boy in America, who by then was already set up in New York, supplying U.S. rayon manufacturers with patented equipment and processes.

Wilhelm Hueper was yet another German émigré linked to viscose, particularly in the 1930s, but his was a far different biography from that of either Berl or von Kohorn. Hueper had come to America a decade before the Nazis' rise to power. In fact, in the fall of 1933 he had written (on the University of Pennsylvania stationery of his then employer, Dr. Ellice McDonald of the Cancer Research unit of the University of Pennsylvania) seeking an academic position back in the new Germany.[55] This plan to return to the fatherland did not come to pass, but another tack Hueper had taken even earlier did yield results. In 1930, Hueper had drafted a memorandum proposing that DuPont create a new research laboratory to assess the biological hazards of toxic agents they were using. The specific impetus, aside from potentially finding a job, was an outbreak of occupational cancer that had emerged at a DuPont dye factory that Hueper had visited under McDonald's aegis (her laboratory was supported in part by DuPont funds). McDonald passed Hueper's memorandum on to Irénée du Pont himself, but the proposal was squashed internally by the argument that the DuPont Dyestuffs Department could handle the problem without additional help. Irénée du Pont was informed that current business conditions were not favorable to the proposed expense.[56]

As the tally of bladder cancer cases grew, it became clear that the question of toxicity had to be reassessed. The turning point in establishing the toxicology laboratory was a "Proposal for Scientific Medical Research" put forward by

Dr. George H. Gehrmann, DuPont's medical director, in November 1933.[57] In his memorandum, Gehrmann included the bladder cancer outbreak as part of the rationale for a new, toxicity-focused research department, but he put viscose manufacturing, rather than dye works, at the forefront, starting with cellophane: "Recently there have arisen many problems relative to the moisture-proof cellophane that we make. This material is used very extensively in wrapping foodstuffs and must necessarily be kept free of any possibility of even traces of substances which may cause health disturbances. Two years ago we were suddenly faced with the necessity of changing the chemical make-up of the plasticizing material and eliminating Tri-Cresyl-Phosphate because government research had definitely proven Tri-Cresyl-Phosphate to be highly toxic."[58]

That government research did not happen by chance. In 1930, a massive outbreak of paralysis due to poisoning struck the United States, affecting thousands. Many were crippled permanently. The epidemic was quickly tied to an extract of Jamaica ginger and, in fact, to a single manufacturer. The elixir, known as "ginger jake," was being ingested for its alcohol content as a way around Prohibition restrictions. But the precise culprit turned out to be a chemical adulterant that had been added to the product: tricresyl phosphate (TCP).[59] It was the same chemical that Gehrmann was concerned about.

Sophisticated experiments carried out by scientists at the National Institutes of Health, which only recently had been established, identified one specific form of three structural variants of TCP, triortho-cresyl phosphate, as the active agent that accounted for ginger jake's potent toxicity to the nervous system.[60] Unfortunately, commercial-grade TCP was a mixture of several forms (known as isomers) that included the ortho- culprit.

The makers of contaminated ginger jake had acquired their multiple-isomer TCP, trade-named Lindol, from the Celluloid Corporation of Newark, New Jersey.[61] Celluloid also supplied Lindol to DuPont as well as to another company, Riegel Paper, which was attempting to break into the cellophane market with a waterproof product of its own.[62] Riegel had set its sights on the lucrative market for tobacco wrappers. Celluloid played down any potential problem to Lorillard (the tobacco giant) when it was considering the Riegel product and inquired about problems with Lindol, in light of the ginger jake epidemic. The superintendent of Celluloid's chemical specialties department wrote reassuringly about Lindol: "Externally this material is non-toxic, and in our plant where the material is made our operators come into contact with it, getting it on the hands and other parts of their body and they have suffered no ill effects. We know also, from tests that have been made, that fruit juices, fats, and oils do not dissolve appreciable amounts from a coating material such as is used on Cellophane. Only the faintest traces of phosphate were found."[63]

Lindol was never publicly identified as the brand name of the poison in the paralysis epidemic, nor was the Celluloid Corporation named as its source; Lindol's use in cellophane was never openly disclosed either. As the definitive history of technology at DuPont notes, the company had reformulated its waterproof cellophane soon after TCP's toxicity became known (as Gehrmann's memorandum makes clear), but "still, had this episode been made public, it might have caused a serious setback for the company's fastest growing and most profitable product."[64]

Gehrmann's proposal did not stop with the TCP example. Another argument he made for establishing a toxicity research group was based on the ongoing problems that DuPont was having with eye complaints among its viscose rayon spinners. This was happening at more or less the same time that Tubize Chatillon was facing the same syndrome in Rome, Georgia. Gehrmann, however, came to understand from one of DuPont's major European competitors and sometime ally that rayon spinner's disease was no mere nuisance conjunctivitis, as it was incorrectly viewed at Tubize. Of the eye condition, Gehrmann noted: "The damage is much deeper and more extensive. It involves the outer layer of the cornea and is an actual destruction of the tissue. I learned these facts from the Medical Scientific Research Laboratories of the I.G. Corporation during my recent visit to Germany."[65]

Gehrmann's memo helped carry the day: DuPont established the Haskell Laboratory of Industrial Toxicology in early 1935, and Wilhelm Hueper, the originator of the idea of the laboratory in the first place, was hired as one of its core scientific staff members.[66] Viscose was high on the Haskell Laboratory's research agenda. In September 1936, Hueper published a paper on an outbreak of a novel skin condition among DuPont's viscose rayon workers.[67] It was marked by unusual fluid-filled blisters on the tips of fingers, which were linked with the introduction of a new method of handling the spun rayon called "crew doffing," in which a team of workers rapidly removed completed "cakes" of the fiber. Because there was so much skin exposure to dripping viscose solution, laden with acid and carbon disulfide, the crew wore protective gloves. But despite that, carbon disulfide came in contact with skin. Hueper figured out that the gloves actually trapped the carbon disulfide, which, forming a vapor chamber around the fingers, caused the blistering.

At the same time that Hueper was tackling the problem of skin blistering, he and his Haskell Laboratory colleagues were working on the far more serious problem of brain damage from carbon disulfide. By carrying out their animal experiments over five months, a far longer exposure period than had been studied before, they could show that airborne levels of carbon disulfide previously touted as safe were far from it; their work formed the basis for a new internal DuPont exposure limit of 10 ppm.[68] Moreover, they finally took on, in print, the four-decade-old data of the German toxicologist Karl Lehmann that were so often invoked to argue for permissive levels as high as 300 ppm, albeit their refutation was made in polite scientific tones: "The results reported by Lehmann which are frequently quoted in the literature, are evidently not conclusive because they refer to short exposures."[69]

Other scientific publications came out of the Haskell Laboratory in those early years. Despite this successful scientific record, however, DuPont changed its policies, disallowing most open scientific publication of its scientists' work. By 1939, all of the original Haskell team, including Hueper, had left.[70] Hueper had come to America to experience a forced resignation of his own.

A key motivating factor in the DuPont lockdown on knowledge sharing was its fiercely competitive stance on research and development, in particular vis-à-vis the "I. G. Company," whose Medical Scientific Research Laboratories Gehrmann had visited and where they had educated him about eye disease in rayon workers. The I. G. Corporation is better known as I. G. Farbenindustrie. DuPont and I. G. Farben negotiated on and off for several years to arrive at mutually beneficial joint licensing agreements. Although they never achieved a master arrangement broadly covering multiple patents, late in 1938 the two companies agreed to share data on DuPont's monovinyl-acetate process and I. G. Farben's work on styrene—both on the path to the holy grail of a workable industrial process of manufacturing synthetic rubber. Both DuPont and I. G. Farben were also keenly interested in an entirely new avenue of synthetic fiber production, one independent of cellulose—the novel fossil-fuel-based polymer nylon. DuPont had claims to nylon-6,6 (the standard form used today in clothing textiles), and I. G. Farben held the patents to another material, nylon-6 (trade name Perlon), which still has major applications in furnishings (especially carpets).[71]

DuPont was inarguably a major U.S. rayon producer (second only to the American Viscose Corporation) and the dominant force in the U.S. cellophane market. For I. G. Farben, in contrast, viscose was never a core business, although it was a growing activity for the combine. When I. G. Farben was created in 1925 through the merger of a group of major chemical manufacturers, one of these (Hoechst) brought with it the cellophane maker Kalle.[72] Farben's intentional expansion into viscose began in 1926 when it took over Koln-Rottweil AG, which was a German innovator in rayon staple (its *zellwolle* was branded as Vistra).[73]

Nonetheless, Glanzstoff, not I. G. Farben, was the German giant of rayon. Even as Farben consolidated, Glanzstoff had taken over one of its key domestic competitors, Bemberg, and its connections to Courtaulds positioned it more strongly in the international viscose business than any other German firm. Glanzstoff also created a rayon staple spin-off of its own, manufacturing it through its subsidiary Spinfaser AG (Kassel) and naming the product Flox.[74]

In the period just before World War II, two other rayon conglomerates that emerged in Germany came to assume major strategic roles in the manufacture of staple. In 1935, four producers joined together to form one "rayon staple working group" (Zellwolle-Arbeitsgemeinschaft). In April 1938, a new entity was created, a more tightly interlocking *arbeitsgemeinschaft,* which pulled away one of the original four from the first working group and added to it four others, creating the new conglomerate Phrix. In the summer of 1938, the first group remade itself too, replacing its member lost to Phrix with several new facilities and turning overall leadership to one of its member firms, Thüringischen Zellwolle.[75] This new corporate entity arrived just in time to pluck the jewel for its crown, a newly constructed viscose operation in post-Anschluss Austria established at the site of a long-standing paper mill in a town named Lenzing. This facility would incorporate the latest technological know-how in staple production, and there were plans to complement it with a workers' village for two hundred families, one that would be constructed according to clear ideological principles that triangulated life, soil, and blood ("work" being within the "blood" apex of the design).[76]

At nearly the same time, the Phrix group was constructing a major new facility at Wittenberge Elbe. The construction and initial operations of this plant are preserved on film in a short documentary from UFA (Universum Film AG), *A Phrix Factory Comes into Being* [Ein Phrix-Werk Entsteht].[77] By 1938, when this film was shot and edited, UFA was under close oversight by the regime, and so it was careful to produce movies appropriately aligned with the state's interest. This film, fully narrated in voice-over, begins by documenting the large scale of the building project at hand, including dredging and filling along the Elbe-side construction site, pile driving, caravans of concrete being poured into rebar flooring, and then brick being laid for a massive factory tower. The structural work was nearly complete by May 1938, marked in *A Phrix Factory Comes into Being* by a brief interlude showing the swastika waving in the wind, with a flowered maypole standing behind. After the equipment is installed, the narrator takes the viewer down the production line, with pedantic exposition explaining each and every step of the rayon-staple-making process. At one point, a worker is shown reaching into the precipitation bath to facilitate the gathering together of a tow of newly generated fibers. His left hand and forearm are gauntleted by an industrial rubber glove; unfortunately, he is performing the task in question with his right hand, unprotected.

One other aspect of Phrix's production process at Wittenberge—its very first step, in fact—is noteworthy. Whereas a typical viscose facility would have shown a mound of wood pulp as the starting material (or possibly a cotton linter's by-product), instead the viewer sees a pyramid of straw. This was meant to be Phrix Wittenberge's unique attribute. Wood pulp was a relatively limited resource for Germany in 1938 and required long transport. To spin staple out of straw, Rumpelstiltskin-like, was a propaganda coup for Phrix that UFA clearly captured on film.

There was no growth spurt for cellophane in Germany comparable to that in viscose rayon, although Kalle's 1928 introduction of a process to manufacture continuous "seamless" viscose film tubes was a real marketing plus in the sausage market niche.[78] But German cellophane did find itself playing a part in the Nazi propaganda apparatus. The vehicle was a product line of cellophane filmstrips intended for home projection. Called Ozaphan, it was a creation of Agfa, also part of I. G. Farben, working with Kalle.[79] Ozaphan began to be sold in the early 1930s, and the comarketing of an affordably priced home projector followed thereafter. The initial Ozaphan film library was weighted heavily to abridged documentaries (the very first was on herring fishing in the North Sea) and animated short subjects. By the end of the 1930s, however, the Ozaphan catalogue came to be dominated by a new collection of films, a number of which fell under the overall heading "Germany's Greatness in Past and Present." Although there was also a series called "Places of Work," it did not include an Ozaphan filmstrip about the Kalle cellophane factory itself.[80]

The dramatic industrial consolidation of German rayon manufacturing, and in particular rayon staple production, was in large measure a result of economic policies ranging from fiscal inducements to outright edicts, including specified percentages of rayon staple to be mixed into all wool or cotton blends.[81] This fit in with an overarching goal of economic autarky that was to be achieved in the 1936–39 period of the Four Year Plan.[82] As evidence of the attention the state gave to the synthetic fibers sector, the health of rayon staple workers even figured into an assessment titled "Industrial Hygiene and the Four-Year Plan," which was offered by the highest-ranking Nazi health official of the time, Dr. Prof. Hans (Karl Julius) Reiter, president of the Reichsgesundheitsamt (Reich's Health Administration).[83] Reiter characterized as an unqualified success the changes in industrial technology introduced to control fumes emanating from rayon precipitation baths, although he misstated the problem as being hydrogen sulfide rather than carbon disulfide.

Synthetic fiber production, whether rayon continuous-spun fiber (*spinnfaser,* generically known as *kunstseide* unless otherwise stated) or staple (*zellwolle*), was perfectly suited to National Socialist iconography. Rayon connoted modernity and technological sophistication, and above all, it held both real and symbolic importance, given the priority put by the regime on self-sufficiency from imported raw materials, among them natural fibers. In its role as a diva of the new economic order, rayon acquired star status and, like any 1930s celebrity, had a studio photographer at its disposal to make it look as glamorous as possible. That photographer's name was Paul Wolff, always keen to be known professionally as Dr. Paul Wolff, lest his previous medical training go unnoted.

Wolff's images appear in a 1938 book put out by Süddeutsche Zellewolle of Kelheim (one of the enterprises taken into the Thüringischen group) with an appropriately uplifting title, *Zellwolle: Ein Weg zur Freiheit* [Rayon Staple: The Way to Freedom].[84] The text is couched in the bombastic jingoism to be expected of the time and place, highlighting, for example, the "self-reconstruction" of the German people that has marked them "indelibly" with "gratitude to their leader!" The photographic images get the intended messages across: a young worker sitting in a small field of grass dotted with flowers, his hair cut short on the sides and longer on top in a fashion presaging a hipster look of our own time (the caption reading "Health and Creativity"); other health-conscious images include a doctor examining a worker, workers in the showers (discretely clothed in trunks); and, for safety, a factory firefighter.

In the same year when *Zellwolle: Ein Weg zur Freiheit* was published, Wolff photographically illustrated a similarly themed but more artistically ambitious project for the Flox subsidiary of Glanzstoff: *Zellwolle: Vom Wunder ihres Werdens* (loosely, Rayon Staple: A Miracle in the Making).[85] The book includes lots of Leni Riefenstahl-esque images of workers, a number of shots of manufacturing equipment, and even a rather stunning view, inspired by New Objectivity (Neue Sachlichkeit), of a factory building at night, lit internally from its windows and outside by a few bulbs over silhouetted doorways. Other than the implied hygienic message of the spotless machinery and the robust workers (one shirtless, lifting up a loose skein of spun fibers), there is nothing much in the "miracle" directly pertaining to health and safety. The viscose spinning baths, at least as shown in close-up, are open and without any apparent special ventilation.

Yet another 1938 appearance of Wolff's rayon images is even more telling. *Der deutsche Rohstoffwunder* [The German Raw Material Miracle], several hundred pages in length, meant to catalogue the myriad ways in which the new Germany was freeing itself from dependence on sources of raw material outside its control. Rayon features prominently, for example, in four plates among the limited number of photographs that illustrate the book.[86] Three of these are by Wolff, one being a reappearance of his shirtless *zellwolle* worker. As a testament to its popularity, *Der deutsche Rohstoffwunder* went through multiple printings: republished twice in 1938, then again in each of the next four years. With the fourth printing in 1939, the book's author, Ernst Lübke, added a supplement (one is tempted to translate this as an "appendage") enumerating the newly Germanized raw materials available in the Sudetenland, especially coal. That *Rohstoffwunder* was noticed at the highest levels of the regime is beyond question: Hitler's personal copy of the book, with a presentation inscription by Lübke, is in the holdings of the Library of Congress.[87]

These books were intended for public edification, and the rayon companies also put out regular internal magazines primarily for their employees. Glanzstoff's organ in this period was called *Wir von Glanzstoff=Courtaulds*, emphasizing the "we" (*wir*) in the equation of its links with the British manufacturer, even though only its Cologne facility, among its multiple factories, was a joint venture with the British concern. Each issue of *Wir von Glanzstoff=Courtaulds* was thematic. The January 1939 publication, for example, featured the Glanzstoff facility in St. Pölten, which, like the Lenzing plant, was formerly in Austria but was now a part of the Reich.[88] The September 1938 issue focused on injury prevention, including accidents outside the workplace (one cartoon illustrates the dangers of riding a bicycle while holding onto the back of a moving truck). Workers are admonished to follow a variety of regulations, but chemical exposure hazards conveniently go unmentioned. An article toward the back of the issue about the operation of a spinning bath appears to be there purely for educational purposes; it is no more concerned with safety than the chess column that precedes it or the gardening and woodworking articles that come after.[89]

The magazine of the Phrix conglomerate carried a name that, to modern sensibilities informed by Bernard Malamud's novel of blood libel, has a wholly unintended Yiddish flavor: *Der Phrixer*. This publication was remarkable for its high production values, putting Glanzstoff's in-house organ to shame. Even the May 1940 issue of *Der Phrixer*, coming out nearly a year after the start of the war (and thus at a time of ever more limited civilian resources), was multicolored, printed on heavy paper stock, and illustrated with small swatches of real rayon cloth representing Phrix's output. It also included a tiny glassine packet of sugar. This was inserted onto the page of an article, "Zellwolle und Zucker?," concerning by-product extraction from sulfite cellulose waste liquor, touting this not only as a source of sugar and alcohol, but also as a base for "biological protein synthesis" [*biologische Eiweißsynthese*] using cultured yeast to yield nutritional supplementation that could enrich the German diet.[90]

Enmeshed in the German state apparatus as the viscose industry was, south of the Brenner Pass, *viscosa*, if anything, was even more integral to the Italian Fascist milieu. One commentator has placed the prominent role of viscose in Italian Fascism in the context of a "long-standing tendency to conflate aesthetic and political manifestations of power and selfhood," a fusion that seemed specific to Italy: "In no other country could a *fashion mobilization* built around the cult of national fabrics have been undertaken with such apparent urgency" (emphasis in the original).[91] For much of this period in Italy, viscose mobilization was led by a single industrial concern, SNIA Viscosa. During the 1920s, SNIA had witnessed exceptional economic growth, beginning with a round of acquisitions of many of the other Italian rayon producers of the day; its manufacturing base was concentrated around Turin, Milan, and Pavia, in northern Italy.[92] This initial growth spurt was completed by the end of 1922, just as Mussolini was marching into Rome.

The man behind all this was SNIA's founder, Riccardo Gualino. He was an industrialist, a patron of the arts, and a modern grandee. A 1922 portrait by the Italian modernist Felice Casorati (known for his compositions evocative of Piero della Francesca) shows Gualino at his zenith: dressed in black neck tie, dark suit with wide lapels, narrow collared white shirt; neatly combed, short silver hair; magisterially seated before an open folio that displays large early printed devices, leather or perhaps vellum bound.[93]

For the remainder of the 1920s, SNIA continued to expand, thanks in large part to a booming export base (spurred on by low wages, a weak lira, and foreign inflation) and underpinned by an aggressive and arguably unsustainable program of other acquisitions. Overextension led to fiscal vulnerability, which Courtaulds and Glanzstoff seized upon, jointly gaining a major stake in SNIA. Eventually, Gualino was forced out of SNIA's management, and then, following his involvement in a bank scandal, late in 1930 he was singled out by Mussolini as the very personification of business corruption.[94] A few months later, Gualino was arrested, tried, and exiled to the Aeolian island of Lipari.[95]

SNIA went on under new leadership more to the liking of the Fascist regime, retaining market dominance and solidifying its iconic position in Italy as the 1930s progressed. A major domestic competitor in Italian viscose manufacturing did emerge in this period: Compagnia Industriale Società Anonima (CISA) Viscosa, a consortium of producers geographically concentrated closer to Rome. CISA and SNIA joined forces to market rayon to the Italian public in a massive campaign that included a roving truck caravan touring the country; the effort was furthered by a promotional "rayon page" appearing regularly in many daily papers.[96] By 1939, SNIA had taken over CISA, solidifying its hegemony over Italian rayon.[97]

In the meantime, SNIA had continued to innovate. It entered the rayon staple trade with its own product, Sniafiocco (SNIA marketed continuous fiber as Sniafil).[98] In addition to staple, it had expanded into a noncellulose synthetic named Lanital. This new fiber, known generically as "artificial wool," was derived from casein (milk protein) by a process for which SNIA held valuable patents through its inventor, Antonio Ferretti.[99] The production process for Lanital had much in common with that used to make viscose, both chemically and mechanically, including the formation of an alkaline solution pushed through the same spinnerets and precipitated in a bath of acid. In fact, carbon disulfide could be added to the alkaline solution used in making Lanital; viscose and casein-based solutions could even be mixed together to produce a hybrid synthetic. Such a novel product was worthy of special promotion. In the fall of 1937, accompanied by Mrs. Harrison Williams (well known at the time from her standing on best-dressed lists), Princess Caetani (of the Florence Antinoris) sashayed into New York's St. Regis Hotel as Lanital's "social representative."[100] At the same time, in an initiative paralleling the Phrix rayon-from-straw enterprise, SNIA was developing a process to make viscose out of reed grass. This source of cellulose was abundant in parts of Italy, making it an attractive alternative to imported pulp. All this fit in well with Mussolini's post-Ethiopia boycott-responsive autarky initiative.

SNIA had also embarked on a major new building initiative. Following an open competition, it hired a talented up-and-coming Italian architect, Alessandro Rimini, to design its new corporate headquarters in Milan. The sixty-meter multistory tower (Torre SNIA Viscosa), constructed from 1935 to 1937, became Milan's tallest building. It was an instant landmark, popularly known as the "Cloud Stealer," rather than by the literal Italian translation for skyscraper (*grattacielo*).[101]

Contemporaneously, SNIA was embarking on a grandiose industrial construction project in the northeastern corner of Italy. Close to reed-bearing marshes, a huge new factory and satellite city came into existence, solely to produce viscose from domestically sourced cellulose. The long courtship between SNIA and the Fascist state culminated in a symbolic marriage in September 1938 when the new factory and city, Torviscosa, was inaugurated in the presence of Il Duce himself.[102] A dedicatory ode for the event was composed by the major Futurist poet Filippo Tommaso Marinetti. Marinetti had written the original manifesto of Futurism in 1909 and then, ten years later, co-wrote a founding manifesto of the Italian Fascist movement.

Marinetti had established a relationship with SNIA Viscosa as something of its poet in residence through his 1937 paean to Lanital, "Il poema del vestito di latte" [The Poem of the Milk Dress].[103] His epithalamium to the union of Fascist ideology and technical advancement represented by Torviscosa was titled "Gli aeropoeti futuristi dedicano al Duce il poema di Torre Viscosa: Parole in libertà futuriste" (briefly, The Poem of the Viscose Tower). "Aeropoetry" was a contrivance of Marinetti's in which the poet would imbue his writing with the aesthetics of flight. The poem's concluding line says it all: "In alto viaggiare viaggiare senza fine la nuova constellazione le cui stella formano la parola AUTARCHIA" [On high, traveling, traveling without end, the new constellation whose stars spell out the word AUTARCHY].[104]

Marinetti's "Torre Viscosa," whose title expands the contracted place-name Torviscosa but also echoes the corporate headquarters in Milan, is full of imagery invoking the reeds destined to be chemically and mechanically transformed into cellulose feedstock.[105] But the piece has very little to do with the viscose process at Torviscosa beyond those initial steps. Marinetti made up for this in his next foray into Futurist synthetic-textile-focused poetry. In it, he engages in an ambitious poetical reinterpretation of the chemical process of viscose synthesis itself. This was followed by a poem on fashion, the concluding component of Marinetti's SNIA tetralogy.

The chemistry-focused "Poesia simultanea della luce tessuta" [Simultaneous Poem of Woven Light] juxtaposes an anthropomorphized sequence of cellulose's transformation into rayon fiber or staple with an idealized factory in which humans are absent from the scene (and thus, implicitly, unexposed to its toxic chemicals). This dichotomy is clearly elucidated in a piece of prose text that Marinetti wrote in preparation for the poem:

> Step by step the cellulose is decanted purified pressed milled with no intrusion of human hands.
>
> It is alive, no doubt: indeed, it requires a mysterious period of response. I sense it transforming itself in enormous horizontal tanks whose slow rotations make them into the stomachs of giant silkworms.
>
> The monstrous patience genius of legions of chemists long ago discovered the formula for achieving solubility: so let alkaline cellulose wed the carbon

bisulfide to which it was betrothed to liquefy in a xanthate orange- and rust-colored like a bloody weapon amidst the festive drumbeat of belts that fearlessly stretch between lower and upper wheels.

Left to sit the solution turns purple in an acid bath that causes it to precipitate into the thread's base components solid flocks or floss.

Not a human in sight.[106]

Marinetti may not have seen them, or may not have wanted to see them, but there were humans in the factory. One source of information about just how bad the working conditions were in SNIA's plants is unimpeachable—the reports of Fascist secret police spies who were meant to infiltrate the regular rank-and-file shop floor workers. Sent in to identify and denounce a suspected anti-Fascist cell among the Turin SNIA workers, the agents sent back to their handlers reports of the deplorable working conditions that they encountered. One reported on the irritant fumes in the bleaching operation, which were so potent that after only three days of exposure, and despite the correspondent's self-described iron constitution, the would-be spy suffered from an incessant cough. Another report told of a worker making Lanital who complained that the manufacturing line was malfunctioning and was likely to catch fire; the foreman told the man to keep the equipment going at all costs. A few minutes later the worker was enveloped in flames and threw himself into a nearby vat of acid to douse the fire; he expired two days later.[107]

Bruna Bianchi, the historian who ferreted out these reports in the Italian State Central Archives, also researched the clinical case records of the psychiatric hospital of Padua. It, like Turin, was another center of SNIA production. Of the multiple hospitalizations for mental illness linked to SNIA employment, one of the most notable case files contained a letter from the SNIA plant director, a man named Piergallini. He had been notified by the treating physician that one of his employees, Bartolomeo Brigo, had been confined and that the doctor believed Brigo's insanity was due to carbon disulfide intoxication. Piergallini's response was a categorical denial that carbon disulfide was to blame, claiming that there was no exposure in the man's department (fiber washing, where residual carbon disulfide, in fact, can be off-gassed) and that, anyway, the worker was an alcoholic and had relatives with mental illness.[108]

There is no reason to believe that things were any better at CISA in Rome than at SNIA in the North. Another historian, Alice Sotgia, has analyzed the archives of a mental asylum, Santa Maria della Pietà, in Rome. Sotgia was able to extract case records for intoxicated rayon workers stretching from 1927 through 1940. The physicians there, however, seem to have been particularly unfamiliar with the rayon industry. A case admitted on 18 August

1940 is the only one recorded in which the medical chart identifies carbon disulfide as the cause of the patient's insanity. The case record also specifies that the patient, who happened to be Jewish, experienced delusions of religiosity, including the hallucinatory experience that God, in personified form, took him by the hand and told him, "Son, I will accompany you to the Ministry of War."[109] This case history, in fact, was reported in an article that appeared in 1941 in an Italian psychiatric journal. That publication, in which this case was one of six reported, but the only one for which the "race" of the patient was specified, provides additional details for the twenty-four-year-old "Mario F.": his initial presentation was characterized by manifestations of "an unusual enthusiasm for the Fatherland, the Regime, and the War" ("un insolito entusiasmo verso la Patria, il Regime e la Guerra"). Later, he expressed delusions of persecution, believing that the nurses and other patients were all acting as surveillance agents of the party and the government.[110]

Openly disseminating information about the case of this poisoned worker in a medical journal appearing in year 19 of the Fascist era may seem to add yet another layer of absurdity to an episode already replete with tragicomedy. But in fact, biomedical investigations and reports on the ongoing hazards of the Italian viscose industry appeared consistently throughout the Fascist era. Indeed, the very first scientific papers on the subject in the mid-1920s appeared after the March on Rome that installed Mussolini in 1922. In 1929, when the Italian Society of Occupational Medicine formally established itself, consistent within the Fascist system of workers' "syndicates," the key early leaders in the study of carbon disulfide toxicity were all signatories of its charter.[111] In June 1934, the Istituto Nazionale Fascista per l'Assicurazione contro gli Infortuni sul Lavoro (INFAIL, Italy's industrial accident insurance fund, today known as INAIL), though its official organ, the journal *Rassegna della Previdenza Sociale,* published a sixty-three-page, exhaustive review of carbon disulfide poisoning written by the physician Gustavo Quarelli, who had been central to documenting Parkinsonism, among other problems, in the industry.[112] In October of the same year, when the 11th National Congress of Occupational Medicine convened in Turin, the cost of the meeting was underwritten in part by INFAIL as well as by the Confederazione Fascista degli Industriali. The official program included an afternoon site visit to the local SNIA factory, followed by a scientific session devoted to carbon disulfide; published proceedings on the subject ran to several hundred pages.[113]

In 1939, Quarelli saw fit to publish *L'impotenza sessuale nel solfocarbonismo professionale e la sua grande importanza nel problema razziale* [Sexual Impotence in Work-Related Carbon Disulfide Illness and Its Great Importance to the Racial Problem].[114] The journal *La Medicina del Lavoro* continued to come out in Milan, albeit on a somewhat curtailed basis from its former monthly schedule: its combined January–March 1944 issue carried a piece on muscle dysfunction (myopathy) caused by carbon disulfide. This report by Enrico Vigliani (who was then becoming prominent in the field) and his associates, centered on the case of "Francesco C." Francesco, a thirty-one-year-old SNIA rayon staple worker, first developed symptoms in 1940. The article is rather erudite, including the results of a muscle biopsy diagnostic of the myopathy. A postscript note acknowledges that in the fall of 1943 another case of carbon disulfide–caused myopathy, authored by the ubiquitous Quarelli, had appeared in a different journal just as Vigliani's paper was going to press.[115] —

For two decades a sort of tarantella, classically characterized by low movements and a lot of footwork, had been danced by the *viscosa* industry and the Italian Fascist state. A fitting homage to this partnership as it neared its end can be found in the pages of a pseudo-memoir of a self-purported spy going only by the initials "S.K." His book, *Agent in Italy*, was published in the United States in 1942.[116] It is the story of an anti-Fascist German who, in 1939, is forced to get out of the Reich quickly, and so he goes to Italy, the only place he can enter without a visa. For the next two years, S.K. ingratiates himself into Italian Fascist and German circles of the rich and powerful, garnering information where he can and passing it on to an American contact. The intent of the book is to paint a picture of a thin veneer of true-believer Fascists controlling the mass of Italians, who were sympathetic to a growing underground and chafing under the growing domination of a de facto German occupation (this was before the collapse of Mussolini and outright control by German forces). Tellingly, one of S.K.'s main protectors is the man he had worked for in Germany, where he had acted as the foreign sales agent for a major Italian artificial silk manufacturer. This upper-echelon businessman is well connected with the regime, but privately predisposed to S.K.'s clandestine activities. In *Agent in Italy*, the artificial silk company is named "Seta," and its director, the man who helps S.K. at every turn, is named Luigi Venturi. The novel's *Luigi* Venturi is clearly an invocation of *Lionello* Venturi, the name of the art historian who worked so closely with Riccardo Gualino, the founder of SNIA (for which Seta is *Agent in Italy*'s stand-in).

Rayon held a prominent place, economically but symbolically too, in the Fascist cosmologies of both Germany and Italy, though it played itself out in different ways. It would be hard to imagine, for example, a German equivalent of Marinetti's Futurist odes to SNIA that would not have triggered a critique of their degeneracy. But it also is important to note that rayon was also of service to other states as an instrument of propaganda. In the United States as well as in Great Britain and the Commonwealth, rayon became the preferred vehicle for carrying patriotic images and messages of war sloganeering. This was particularly manifest in wartime rayon scarves. In the United States, for example, Echo Scarves, circa 1940–41, produced such items as the "Pledge of Allegiance," "Gettysburg Address," and "Fighting Words" rayon scarves (the last decorated with "Fire When You're Ready, Gridley," among other admonitions). Britain produced its own rayon paraphernalia, sometimes quite bellicose, such as an "On to Berlin" printed scarf whose streams of red suggested rivers of blood. But more typically, rayon was a vehicle for more benign home-front encouragement and even extended to larger pieces of clothing, such as a "Dig for Victory" rayon house dress, whose images supported the British victory garden program, or a woman's rayon blouse with a tropical flair but also emblazoned with motifs of the Royal Australian Air Force and Navy.[117]

In the United States, the war propaganda value of rayon may have hit its high mark on 30 October 1942, the day that the Brooklyn Museum's *Inventions for Victory* exhibition opened to the public. It included a range of synthetic products, some not yet on the market, intended "primarily as a forecast of what civilians may expect after Victory." Rayon was the fashion star of the show. DuPont was particularly well represented: a coat designed by Vera Maxwell with a brushed knit rayon lining, displayed on a wall pedestal; a futuristic rayon evening gown designed by Kiviette using DuPont rayon usually purposed as tire cord, prominent on a pillar platform; and three Sally Victor hats, one of spun rayon and two made with a new DuPont product called Bubbifil, a "continuous stream of cellophane air-filled bubbles."[118]

Propaganda textiles reached their zenith not in the West but rather in the Japanese designs of this period. Although these were predominantly printed silks, rayon had its day here too. Particularly striking were examples of children's kimonos made of rayon with propaganda motifs. From the 1938–40 period, one kimono shows planes, trains, and Mount Fuji, along with the flags of the growing empire: the Japanese flag is flanked by those of occupied China and the puppet state of Manchukuo (occupied Manchuria). Another example was decorated with a design of searchlights, planes, and falling bombs.[119]

For Germany and Italy, rayon meant textile independence. In Japan, silk played this role. The key for Japan was in rayon exports, which became a pillar of foreign trade before the war. By 1939, Japan led the world in rayon staple exports; its exports of rayon yarn were second only to Italy's.[120] This export power meant potential influence in other ways. In the summer of

1941, months away from Pearl Harbor, Japan attempted to broker a deal with Mexico to trade rayon for mercury, a vital strategic resource, an effort that took place at the ministerial level but was ultimately unsuccessful.[121]

As domestic rayon consumption in Japan increased steeply, paralleling the rise in exports and escalating further in the war years, the consumer market spread from standard rayon continuous-fiber textiles to rayon-staple-based blends. Unlike Germany's embrace of *zellwolle* and Italy's affair with all things SNIA, including Sniafiocco, the response of the Japanese public to rayon-staple consumer products was decidedly negative. The major product, a rayon staple blend with wool or cotton, was called Sufu. This product purportedly tore easily and fell apart after a few washings, becoming synonymous with poor quality and, by extension, other defects. In Japan in the early 1940s, to call someone a "Sufu head" was to call him or her stupid. That did not stop Sufu propaganda textiles from being made, although—not surprisingly, given the textile's inherent flaws—few examples of the product have survived. One of the rare extant pieces is a boy's kimono with a "running soldiers" design (also decorated with rows of planes and dirigibles).[122]

As the war went on, and well before American Viscose put out *Rayon Goes to War*, viscose became a prominent item in the modernized inventory of materiel for an emerging global conflagration. As early as the Spanish Civil War, the Mussolini regime had touted its autarky campaign as stretching to rayon-blended army uniforms, which extended limited stocks of wool and cotton.[123] Rayon production in Spain, which began even before World War I, had been concentrated in Catalonia and was not tied to Italian interests.[124] Following Franco's Mussolini-assisted victory, however, Spanish-Italian cooperation on rayon materialized in the form of the Sociedad Nacional de Industrias Aplicaciones Celulosa Española. The new concern's initials just happened to spell out SNIACE, homage to SNIA, its 25 percent co-owner. Ground laying in Torrelavega (Santander) took place in October 1941, although the factory was not even partially operational until 1943.[125] These SNIA machinations did not go unnoticed. As if taking a page from *Agent in Italy*, the U.S. Office of Strategic Services (forerunner of the CIA) sent around a memorandum in late 1943:

> It has been reported that the rayon manufacturing business of the Italian SNIA Viscosa monopoly of Turin and Milan will be transferred to Spain.
> SNIACE ... has arranged for the Saimo Transportation Company to take care of the shipment of machinery which is expected to take place any day.
> The blueprints of the SNIA viscosa will be deposited with the Spanish Consulate in Milan to be transferred later to Madrid, and complete legal

and financial arrangements have been made with affiliates in Berlin and Lugano, Switzerland.

It appears that control will be vested in Berlin and dummy control may be in Switzerland with the Society Nautilus at Lugano.[126]

No such wholesale transfer of SNIA property ever took place, although it might very well have been part of the Nazis' contingency planning following their occupation of Italy.

There was indeed a huge shift in rayon assets during the war. But this was the outcome of a struggle that took place far from any theater of military engagement. As D. C. Coleman notes from a British perspective in his authoritative history of Courtaulds: "In the summer of 1940 Britain and Germany fought alone in the Second World War. Behind the battle of men and guns and aircraft lay the trial of economic resources. In this conflict, Britain's ability to buy war materiel in the U.S.A. was of vital import; by the time that the Battle of Britain had been won the financial resources to do so were dwindling fast."[127]

As lend-lease legislation to support the war effort was being put forward by the Roosevelt administration in the winter of 1940–41, the promised sale of privately owned British assets in the United States emerged as a major political deal sweetener for passage by Congress of the enabling legislation and, even more important, for approval of the vital appropriation bill to make the program operational.[128] Never publicly traded (there were only estimates of what its likely huge worth might be), Courtaulds' American Viscose Corporation was referred to as "the Madame X of U.S. Corporations" by *Time* magazine.[129] Whether meant by *Time* to be a metaphor for something high class and desirable but unattainable or, conversely, to imply an element of moral lassitude, the immediate sale of AVC by Courtaulds emerged as the U.S. quid pro quo for lend-lease.

Key players in what became an American-British showdown over AVC's sale included the U.S. secretary of the treasury, Henry Morgenthau, and his opposite in the negotiations, Sir Edward Peacock, a director of the Bank of England. The tentative AVC sale arrangement that emerged was not to Britain's liking. John Maynard Keynes was dispatched to Washington to see whether he could do better or, as Morgenthau put it, simply "sabotage the viscose deal."[130] An opponent whom Keynes did not count on was someone who was already an expert on the machinations of AVC, Morgenthau's lieutenant, Harry Dexter White. White, after all, had cut his economic teeth on the study of protective tariffs for rayon. Another U.S. economist, a self-described "fly on the wall" for the lend-lease negotiations, recalled, "I saw Harry White and saw something of his unpleasantness when he forced the British to sell American viscose, if you recall that episode."[131]

AVC was sold to a consortium in March 1941, and then the stock was resold on the open market a little over two months later. The investment bankers who managed the sale, led by Morgan Stanley, made a bit under eight million dollars on the deal.[132] The heady mix of U.S. wartime politics and economics influenced rayon in several ways. The Berry Amendment, for example, first passed by Congress in 1941, restricted Department of Defense expenditures to home-manufactured (but not necessarily homegrown) textiles, including U.S.-made rayon.[133] More salient still was the dramatic 1942 political confrontation between the administration and Congress over the use of rayon to replace cotton as a reinforcing fiber in rubber tires. William M. Jeffers, head of the Rubber Production Board, took on a group of southern senators, going so far as to growl at Senator "Cotton Ed" Smith (Democrat, South Carolina). A news account noted, "It was the first time in years that a Government official had 'talked back' in such strong language to a committee that had called him on the carpet," and a political cartoonist portrayed Jeffers as a latter-day Daniel in the lions' den.[134]

Although the U.S. government clearly was willing to take a stand on who owned the rayon and who bought and used it, it did not invest much political capital in championing wartime worker protection in the industry. About as far as it went at the Department of Labor was a six-page pamphlet put out by its Children's Bureau in December 1942 as part of its *Which Jobs for Young Workers?* series. The mission statement on the title page, "The young worker of today is the skilled worker of tomorrow—Protect him from injury," sets the tone, and the pamphlet, *Advisory Standards for Employment Involving Exposure to Carbon Disulfide*, assures its readers that "as a result of advances in the methods of control, there is no longer the degree of hazard from carbon disulfide in industry that at one time existed."[135] Still, just to be safe, it is recommended that those younger than eighteen not work in churning or spinning rooms or in handling fresh-made fibers, although any kind of work with finished yarn, or in inspection or even laboratory work, was deemed fine for someone age sixteen to seventeen.

By this time, the United States had entered the war. Meanwhile, "neutral" Spain was not the only noncombatant nation still getting in on the rayon action. The Swiss concern Viscose Emmenbrücke began to go into rayon staple in a big way beginning in 1940.[136] Coincident with ramped-up wartime production, the number of officially recognized illnesses related to carbon disulfide increased dramatically in the Swiss workers' compensation system.[137] Sweden, seeing war on its borders and wanting to expand its own domestic rayon industry, too, with governmental involvement began in 1942 to construct a major new facility in 1942 and had Svenska Rayon online by 1943.[138]

Across the Baltic, up the Gulf of Finland, and squarely in a theater of action, one major rayon factory became coveted war booty early on in the fighting. Originally a Finnish manufacturer, Kuitu Ltd. (Kuitu Oy) was a rayon staple facility located in Jääski on the Karelian isthmus. After the Red Army pushed into this territory in 1939, the factory's reconstruction became a matter of considerable priority. From the spring through the fall of 1940, a series of memoranda flew back and forth within the Council of People's Commissars.[139] These documents, some marked "urgent," slated the Gulag of the NKVD (secret police) as a major source of equipment for the effort and as the responsible administrator for rebuilding and then running the factory. Underscoring the priority placed on this high-profile operation, the Economic Council of the Council of People's Commissars decreed that multiple civilian economic and industrial councils were obliged to cooperate with the NKVD on the matter. At this time, the Council of People's Commissars was under the leadership of Nikolai Alexandrovich Bulganin, who in later years rose to become premier of the Soviet Union.

The Jääski effort was interrupted by the renewed hostilities with Finland that began in conjunction with the German invasion of the USSR in June 1941. After the front shifted, the rayon staple factory reopened, but back under Finnish control. This may or may not have been a break for Gulag laborers, but it spelled trouble for the Finnish workers in the plant. Their situation might never have come to light except for the unlikely scenario of an occupational medicine physician, Leo Noro, having been in a frontline Finnish panzer tank unit in Karelia. Two factory engineers whom he encountered told him about poisoned workers down the road at Jääski, and Noro investigated.[140] His survey of twenty-four workers poisoned with carbon disulfide documented many of the classic symptoms of intoxication (including depression, irritability-nervousness, loss of libido, and changes in vision).[141] Noro's work is most notable not for these observations per se, but because he was able to publish his findings in a Swedish medical journal in 1944 (this was just before Karelia changed hands yet again), his being a rare medical investigation of a rayon factory population to come to light during World War II rather than after.

Across occupied Europe, medical data on rayon workers' health were being collected, and limited airborne exposure measurements were taken, but this information was not disclosed at the time. The stark reality of conditions came to light only in medical reports that appeared afterward. For example, not until later was there documentation of the conditions experienced by

Polish workers in one of the rayon staple plants there, an operation known as Widzewska Manufaktura and run in Łódź. The viscose operation had been set up in a former cotton mill, a major prewar textile factory. The conversion to rayon was carelessly engineered at best. Carbon disulfide overexposure was rampant, made even worse by a grueling work schedule that did not allow for "voluntary" lapses. We know about the extent of carbon disulfide exposure only because air levels of the chemical, several hundred parts per million (ppm) and more, were precisely quantified by a technician for the plant's management. Only in retrospect was it possible to partially reconstruct the extent of illness among the employees, including many victims of insanity who were sent off to a local mental asylum and never heard from again.[142]

In Belgium, a major prewar center for viscose manufacturing, the "honorary inspector general" of the governmental Occupational Health Service later detailed a series of cases of carbon disulfide intoxication that occurred at multiple rayon factories operating under the occupation. Rather pedantically, Dr. Langelez reported:

> During the war, due to the shortage of other textile materials, the manufacture of synthetic products intensified; the work was carried out in three shifts resulting in continuous occupation of the premises and non-stop operations, compromising the exchange of air in the workshops. This situation was aggravated further by the necessity of blackouts, forcing the workers to close all openings to the outside at night, reducing even normal ventilation in workplace. Finally churn cleaning, a dangerous operation as said, became more difficult because of the use of beech wood pulp, necessitated by circumstances, in place of fir pulp that was usually employed in fabrication.[143]

It seems that the beech pulp made for stickier viscose, thus intensifying exposure during cleanup. The same point was made by an expert reporting the French experience, although in this case aspen as well as beech pulp was blamed, and it was made clear that this poorer-quality raw material came from Germany. As in Belgium, the collection of work illness statistics in France continued apace throughout the period, and the toll was staggering: more than 500 "registered" cases of carbon disulfide poisoning from 1942 through 1945, even down to the detail of lost workdays (8,648 such days in 1942, for example).[144]

In France under the occupation, the rayon industry was transformed, growing in prominence even as conditions worsened for its workers. French viscose took on the trappings of Italian and German autarky, but distinctly restyled itself as haute couture. A counterpart to German centralized synthetic fiber manufacture, the business entity France Rayonne was established.

This new combine brought together what had previously been independent producers of filament and staple, the latter in France going by the name Fibranne. Fibranne had been a relatively small component of domestic French production and consumption before the war, but that changed fast, in keeping with German priorities. To further promote rayon staple applications, a number of fashion houses were strategically recruited to the cause. In July 1941, an exhibition solely devoted to the new fabrics opened in Paris at the Petit Palais; other, regional exhibitions followed, also promoting a new synthetic high style. French fashion journals of the time loudly trumpeted new innovations in wool and specialty silk substitutes, variations on Fibranne but each with its own trade name, all of which have long since fallen into obscurity: Astraline, Triklidou, Jolibab, Doussalba, Cote Bérénice. Had there been a costume designer for the documentary footage in the film *The Sorrow and the Pity* (1972, directed by Marcel Ophüls), Fibranne and its offspring would have constituted the fabrics of choice.[145]

Rayon profiteers collaborated, but the workers in the French factories, who bore the brunt of the deteriorated conditions the material was produced under, contributed actively to the Resistance. This has been documented best in the case of La Société Nationale de la Viscose à Grenoble, a major rayon factory located in the French Alps. Many of its employees were active in the underground, and they were generally supported not only by the overall labor force of the plant but also by its sympathetic factory director, a man named Pierre Fries.[146]

Nonetheless, the French Resistance fighter most closely associated with rayon was not a factory worker at all before her arrest, but rather a museum curator. Agnès Humbert worked at the Musée des Arts et Traditions Populaires, which was closely linked to the Musée de l'Homme, both located at the Palais de Chaillot, in Paris.[147] At age forty-three, she found herself in occupied France, and along with colleagues at the museum, she formed one of the first active cells of the Resistance. Humbert and her colleagues were arrested in April 1941; after a high-profile trial early in 1942, her male comrades were executed and she was sentenced to five years' prison labor. Soon thereafter, Agnès Humbert was deported to Germany. Initially in a prison fortress called Anrath, Humbert (along with other prisoners, most of whom were German women, many convicted of minor wartime crimes such as black-market trading) was "outsourced" to do factory work. The factory in question, on the outskirts of a town called Krefeld and near the Adolf-Hitler-Rheinbrücke, was Rheinische Kunstseide Aktiengesellschaft, also known as "Rheika," one of the Phrix conglomerate's viscose manufacturing sites.

Humbert's account of her wartime experiences was set down just after her liberation and first published in 1946, when her recollections were those of recent and searing events. She styles the memoir as that of diarist, organized by place and time (often by the month when an event occurred, but sometimes tethered to a specific and particularly memorable date). The first sight of the Phrix factory, after a march from the prisoner workers' barracks in town, together with her detail (*kommando*), Humbert dates to 11 April, 1942:

> The factory complex consists of several large red-brick buildings, all of them very modern and harmonious in design. The first courtyard is embellished with flowers and a lawn. The largest building has an immensely tall tower that reminds me, though on a much smaller scale, of the tower of the Palaccio Vecchio in Florence....
>
> ... We find ourselves in an enormous hall; it has a metal structure, painted green throughout. The floor is of beautifully polished wood. And everywhere, all around, are spools of rayon in every shape and size, heaped in gleaming white piles, reminding me of brides in their thousands. The hall is a symphony in white and green.[148]

Humbert's first job was in this rayon-winding department. Even though she and her fellow prisoners worked alongside German civilians, their conditions were quite different. The work was dry and dusty, but the prison laborers were not allowed to take any water, even though a hygienic drinking fountain was near at hand. The hours were long, toilet breaks were strictly limited, and the sadistic warden came out from the Anrath fortress for surprise inspections and the punishments that followed. Yet this was an idyll compared to what followed.

The spinning department had been manned by foreign nonprisoner workers. Humbert saw them in the factory dining hall where she also received her limited rations:

> In the Rheika restaurant we see more and more groups of workers—Dutch we hear—in a pitiful state. They are "free" workers whose job is to make rayon. Their prison overalls are in tatters, eaten away by the acid; their hands are bandaged, and they appear to be suffering terribly with their eyes, to the point where they often can't manage by themselves. A fellow worker will hold them by the arm, sit them down, put their spoon in their hand. They appear to be racked by excruciating pains. What sort of work can this be that causes such torment? I had no idea that the manufacture of rayon was such an agonizingly painful process.[149]

Agnès Humbert was soon to learn.

In the face of the terrible working conditions, most of these "free" personnel went AWOL. In July 1942, Humbert and her cohort were drafted into the viscose-spinning department. Over the many months that followed, Humbert and those she worked with suffered the terrible burns and blinding fumes that she had seen the effects of in the Dutch workers but whose cause she had not understood. At one point, remembering the days before the war and how her mother once had been pleased with herself over an investment in SNIA shares because they gave such a good return, Humbert wrote:

> It gives a good return! Poor Maman, so good, so sensitive, how could she ever have imagined that this "good return" was founded on human suffering? True civilian workers in the artificial silk industries of civilized countries are treated very differently from the prisoners in the Phrix factory.... When the eyes of civilian workers start to hurt they are given immediate treatment, whereas we prisoners are forced to stay at our machines until we are blind. To protect them from burns caused by the treacherous viscose, civilian workers are equipped with rubber fingerstalls and gloves; we prisoners work with our bare hands.[150]

She was to continue working there until 22 August 1943, a date Humbert could pinpoint because that was the night the factory was put out of commission by an Allied bombing raid. With the walls crumbling around her, Humbert was sent back into the building to recover the precious rayon spinnerets from the machines, lest their valuable gold and platinum be plundered from the wreckage.

Humbert never knew that in addition to the visible burns and bloated eyes, the insidious hazard of carbon disulfide vapors, everywhere but unseen, would also have been working their damage. At the time, she did not even know that a chemical by this name was a key part of the viscose process nor that it might account for the ill health of a co-worker whose "nervous system is completely shattered," or of another who attempted suicide, "drugged and fuddled by the acid vapors," or of a woman who "has just thrown herself out of the clothing store window."[151] Suicide attempts were common enough but attributed to the overall plight they all faced. Neither Humbert nor any of her fellow prison laborers seemed to be aware that neurological insult was a classic manifestation of viscose intoxication, or that viscose even contained a chemical called carbon disulfide.[152]

Approximately fifty-five miles from Krefeld, another factory of the Phrix rayon chain, Rheinische Zellwolle AG in Siegburg, also relied on "forced labor." This is a term that has come to be used in translation from the German *zwangsarbeiter*, but is arguably little more than a euphemism for what in practice was slave labor.[153] The situation in Siegburg fit this picture. The prison there, officially the Zuchthaus und Strafgefängnis, meaning both a prison for hard labor and for routine confinement, maintained a number of satellite camps to serve local employers, including Rheinische Zellwolle.[154] But this was only one source of laborers for Phrix in Siegburg: another was forced laborers from the East (Poles and Russians), and yet another, prisoners of war (also from Poland and the Soviet Union, as well as France, Belgium, and Holland). The vast majority of forced laborers in Siegburg served Phrix; in a tally from June 1943 there were 1,326 of them.[155]

We know something of the plight of the Phrix workers in Siegburg through a contemporary report contained in clandestine prison letters written by Erich Sander and smuggled out of the Zuchthaus. Erich was the son of August Sander, broadly recognized as one of the greatest German photographers of the twentieth century. Politically active on the left, Erich Sander appears together with three fellow activists in an iconic photograph taken by his father in 1926, *Werkstudenten* [Working Students], one of the key images in Sander's 1929 landmark series *Antlitz der Zeit* [Face of Our Time]. In the photograph, Erich and his young comrades sit closely together, all looking straight at the camera, intense but by no means dour.[156]

Erich remained politically committed. In 1934 he was arrested and sentenced to a lengthy prison term. At Siegburg, his principal work was in the prison infirmary, although, amazingly, he also managed to gain access to photographic equipment and supplies intended by the prison authorities for documentation geared to their purposes. In an image far less well known than *Werkstudenten*, Sander can be seen in Siegburg in 1943. Still bespectacled, but now wearing an orderly's apron rather than the jacket and tie of his student days, he sits hunched over a desk at work in the prison infirmary.[157]

Because he took care of the sickest prisoners as they passed through the infirmary, Sander saw firsthand the ravages of work at Phrix. Moreover, even though trained in economics and certainly not in industrial chemistry, he nonetheless was well informed about the nature of materials used in making *zellwolle*. In a letter dated 13 June 1942, he wrote: "The worst conditions in the Zellwolle-Factory, where the most severe damage to health occurs, come through the poisonous vapors. There have already been 8–10 men who have gone insane from carbon disulfide intoxication and some put in an asylum."[158] Two weeks later, when he wrote about the factory again, infirmary admissions had gotten even worse: "We still have, nearly day after day, people hospitalized from Zellwolle with severe symptoms of poisoning [carbon disulfide poisoning], which is manifested in mental derangement. It is terrible when you see young guys who are simply mental ruins. One screamed 'Tempo! Tempo! Tempo!' ceaselessly; another, 'Jump, jump, jump!' ... The death toll here is rising alarmingly, during the night Thursday to Friday three of them died suddenly."[159]

Unlike Agnès Humbert, Erich Sander did not survive his imprisonment. Yet the record that he left, even if fragmentary, is just as moving. Another victim of the Phrix factories, although we have nothing in his own words, may be most well known of all. His name was Johann Wilhelm Trollmann, but he was called Rukeli. Rukeli Trollmann was a boxer, and a good one.[160] He was famous for his dancing style in the ring; his record included 31 wins, with 11 knockouts. In 1933, he became, briefly, the German light middleweight national champion. He outperformed his opponent, but was initially denied the win when the judges, under pressure, called what was a clear victory a draw instead. When the crowd threatened to riot, the judges reversed themselves. Shortly thereafter, he was stripped of his title anyway: Rukeli Trollmann was of Sinti heritage (a branch of the Romani), and it was intolerable to the National Socialists that a non-Aryan could be dominant. In a forced rematch, set up to go against him because he was not to be allowed to move his feet, Rukeli floured his body white and bleached his hair. The publicity for a film from 2013 on his life, *Gibsy*, refers to him as the "Muhammad Ali of the 20's and 30's," which may be a bit of hyperbole but nonetheless resonates.[161]

The story that followed was no less dramatic. Trollmann was not allowed to box, but was suitable for conscription into the Wehrmacht. Wounded on the eastern front 1941 and sent back to Germany, he was then rounded up and deported to the Neuengamme concentration camp. Neuengamme, located outside Hamburg, was established as an independent camp in 1940, initially for Germans and then for other nationalities (in particular foreign prisoners of war), with only a subset of Jewish prisoners. Neuengamme came to function as a hub for scores of satellite subcamps, some near at hand in the Hamburg area, and others far afield.[162]

The Phrix works at Wittenberge Elbe, the building of which in 1938 had been the stuff of the slick UFA propaganda film, established a satellite concentration camp whose prisoners were supplied by Neuengamme. Rukeli Trollmann became a Wittenberge inmate. His transfer there did not happen by chance, but through the help of fellow prisoners who saw that factory as the lesser of evils; Rukeli had become dangerously weakened at Neuengamme.

At Phrix, Rukeli was not assigned to viscose chemical work but rather to the slave-laborer work gangs that faced the grueling task of hauling huge bales of straw for the process. This was the ultimate reality of the raw-material miracle meant to be the hallmark of Phrix Wittenberge. And even so, Rukeli Trollmann might have survived the pharaonic conditions to which he was subjected except for a sadistic capo who recognized the boxer. He forced Rukeli to fight him. Rukeli won. Not long after, the capo took revenge, using a stick to beat him to death in the straw field.

There also were approximately one thousand "eastern" laborers at Phrix for whom the Neuengamme prisoners, who began arriving in August 1942 and whose numbers peaked at 500 souls, were a supplemental workforce.[163] Nor were they the first slave laborers on the premises. In the spring of 1942, a group of Jewish slave laborers from Poland was sent to Phrix, but not as part of a concentration camp annex. Their relocation occurred in response to direct requests from Phrix's corporate headquarters; when they complained that not as many workers had been delivered as promised, the SS intervened. By the time the Neuengamme detail arrived, any Jewish laborers who had managed to survive up to that point had been sent back to Poland to suffer their fate there.[164]

Phrix's bargaining chip in its call for more forced workers was a proposed yeast production operation, the same operation it had touted in the slick pages of *Der Phrixer*. This project, which required additional labor for construction and operation, planned to use the factory's sulfite wastewater from straw processing, grow yeast on this substrate, then exploit the yeast as a human food supplement.[165] There was nothing revolutionary in this: yeast production from growth on sulfite waste was already up and running at several German wood pulp factories (all using a species of yeast called *Torula utilis*). But Phrix promised big results, with production that would help close what the Germans referred to as the "protein gap" by providing vital nutrition for the Wehrmacht.[166] Phrix had the ear of the SS: the company also obtained slave labor for its staple plant in Küstrin on the Polish border (in Polish, Kostrzyn nad Odrą), where it similarly planned to have a yeast production sideline.[167] Another Phrix factory, the fifth in its combine of five, was located in Hirschberg (in Czech, Jelenia Gòra) near the Czechoslovakian border. It was serviced by inmates of the Gross-Rosen concentration camp. One of the work details there was responsible for unloading by brute force train cars full of heavy logs—if the wood was not thrown far enough from the tracks, the prisoners were beaten.[168]

Yeast-for-food production as a by-product of Nazi rayon manufacturing became a major factor at another rayon facility, Zellwolle Lenzing. Not tied to Phrix, this factory was part of the Thüringischen Zellwolle group. The SS became very interested in an alternative approach to growing yeast that Lenzing had adopted. Lenzing used a different yeast species, *Oidium lactis,* and employed innovative culturing and processing methods. The resulting ersatz sausage was named Biosyn-Vegetabil-Wurst.[169] The company had direct, high-level ties to the SS through Lenzing's general director, Dr. Walther Scheiber, himself an SS-Brigadeführer (general major).[170] The SS was rigorous. It wanted confirmation of the efficacy of Biosyn-Vegatibil-Wurst as a nutritional substitute. Scheiber set out to prove this, working closely with a physician named Dr. Ernst-Günther Schenck. Schenck, another prominent SS member, had expertise both in nutrition and in concentration camp affairs.[171] The human experimentation that Scheiber and Schenck orchestrated at the Mauthausen concentration camp led to the deaths of hundreds. Many of the victims suffered from severe adverse gastrointestinal effects associated with the artificial diet to which they were subjected.[172]

Whether Biosyn-Vegetabil-Wurst was a success or not, Scheiber still had a rayon business to run. Lenzing had long supplemented its operations with forced laborers "from the East," but as the war progressed even that source was insufficient to meet its need for workers. Mauthausen was relatively close. In fall of 1944, shipments of female concentration camp inmates began to arrive to serve Lenzing. They were based at a satellite camp under the aegis of Mauthausen, but located in the town of Pettighofen, adjacent to Lenzing.[173] One of the prisoners, no more than a girl, years later recalled her job at Lenzing, where she filled large sacks with rayon staple:

> I don't know how long the shifts were since we had no way of telling the time. All I can say is that a day shift would begin before dawn and end well after sunset. But at least it was warm in the factory and I did not know then that the heat was due to overworked, unoiled, overheating machinery releasing dangerous fumes. There were cold and warm water taps above a deep, square trough against the wall. We could go to the toilets two by two, escorted by soldiers, at given times when the machines were turned off for that purpose. Then back again to the mind-numbing routine of bag grabbing and button pressing while going round and round on the turntable. And hungry, always hungry.
>
> I was now four months from my thirteenth birthday.[174]

For much of these terrible times, Walther Scheiber thrived. He had risen in the ranks, reporting directly to Albert Speer (minister of armaments and war production) and taking on multiple projects. For example, because of Scheiber's expertise in synthetic fibers specifically and chemical engineering generally, with I. G. Farben experience under his belt as well, he was deeply involved in the development of chemical war agents.[175] But above all, Scheiber prospered as the de facto *zellwolle* czar within the Nazi apparatus, garnering power, influence, and personal wealth. In addition to directing Lenzing and other operations within the Thüringischen Zellwolle conglomerate and also having a stake in the Biosyn-Vegetabil-Wurst start-up, Schreiber oversaw yet another enterprise, Schwarza-Zellgarn AG. Zellgarn's corporate mission was to bring rayon manufacturing to Łódź, Poland (in the geopolitical atlas of the Nazi regime, Litzmannstadt in the Reich-annexed administrative region of Reichsgau Wartheland). This location presented a highly attractive business opportunity because it was a prewar textile center and, even more importantly, because of its abundance of forced laborers. It was Zellgarn that was responsible for the abysmal conditions at Widzewska Manufaktura, where so many workers had gone insane. Later in the war, Zellgarn used the methods it employed at Widzewska to convert a cotton thread factory in Łódź for viscose production as well.[176] By then, there were only Polish workers to be had: the Łódź Ghetto, the second largest in Europe after Warsaw, had been liquidated. One of the many industries that had been sited there was a *zellwolle* rag-recycling operation.[177]

Phrix and Thüringischen Zellwolle were prominent in their use of forced laborers, prisoners of war, political and nonpolitical hard-labor prisoners, and concentration camp inmates. But they were not alone in this. I. G. Farben, which became infamous for its large-scale participation in slave-labor war crimes, used concentration camp labor in its Vistra brand *zellwolle* plant in Wolfen and forced labor in its sister facility in Premnitz, the flagship production site for rayon staple in Germany.[178]

The Glanzstoff factory in Lobositz (the German name for Lovosice, in the annexed Sudetenland of Czechoslovakia) took advantage of the nearby location of Theresienstadt (Terezín) to exploit its inmates, who were transported for the thirty-minute distance in a factory truck and then worked as forced day laborers for construction as well as tasks more directly related to rayon production. There were relative advantages to being assigned to the Arbeitskommando Glanzstoff Lobositz: there was a ration of potatoes, albeit often rotten, and it was possible to make illicit contact with Czech workers, through whom messages could be clandestinely passed to the outside. One of the prisoners reportedly assigned to Glanzstoff Lobositz was Josef Beran, a future archbishop of Prague and later a cardinal of the Catholic Church.[179]

Across all of Glanzstoff's operations, by 1943 one in four workers was a forced laborer or a prisoner of war. At Cologne, Courtaulds-Glanzstoff was notable on several counts. Early on it had a contingent of German Jewish "workers," many of them local residents forced into servitude. The company maintained personnel cards on them; they were identifiable as part of the Jewish contingent because either "Israel" or "Sara" was added to their names.

The longest tenure was that of Moses (Israel) Aron, who lasted as a viscose worker until 1944—by then almost everyone else in his group of 200 had been deported.[180]

In the interim, Courtaulds-Glanzstoff's prisoner-of-war population grew after the German occupation of Italy, when an influx of over 300 Italian military officers was pressed into labor. These soldiers' writings, later anthologized, give voice to their experience within the "damp walls of Glanzstoff" (the image used by one of the prisoners in a poem written at the time). One recounts working on the viscose line until being temporarily blinded by the fumes. Another was given something to drink supposedly to prevent poisoning. It was passed off as wholesome milk, but the ersatz brew was diarrhea-inducing. He eventually became intoxicated by the fumes in any event.[181]

Glanzstoff and Phrix had stakes in Hungarian viscose concerns, too.[182] Slovakia was unusual in that its major viscose producer (rayon fiber and staple as well as cellophane) was wholly Czech owned before the war and then fell under German control. The acronym SVIT (Slovenské Vizkózové Továrne, or Slovak Viscose Works) designated not only the manufactory, but also the model factory town around it, built to the master plan of F. L. Gahura. Nestled at the foot of the Tatry Mountains, SVIT was established in 1934 as one of the many industrial projects of Jan Bat'a, the corporate king of Bata shoes. Viscose presented an opportunity for diversification.[183]

A period photograph of Svit shows an almost idyllic scene: rows of neat housing units facing a tree-planted lane, the nearby mountains as a backdrop. It is no wonder that in March 1942, when nineteen-year-old Alice Jakubovic was rounded up in Prešov, Slovakia, she was relieved to hear that she would be sent to Svit to work at Bata. It was a ruse. She was on her way to Auschwitz.[184]

When the war in Europe was finally drawing to an end, work had ceased at most German-controlled rayon fiber and staple production sites, east and west, because of either physical destruction or the transfer or dispersal of the labor force. Courtaulds-Glanzstoff was notable in that it did not suffer serious aerial bombardment, despite the destruction of most of Cologne's industrial base (Courtaulds' facilities in Coventry did not fare so well).[185] The Phrix leadership had preserved itself, relatively unscathed, at its Hamburg corporate headquarters. At Thüringischen Zellwolle, Walther Scheiber's star already had fallen: he had been purged from the Nazi Party for corruption and dismissed from service with Speer toward the end of 1944. Still, Scheiber was smart enough to end up in the western zone at the end of the war. In contrast, his collaborator in the Biosyn project, Dr. Ernst-Günther Schenck, found himself in Berlin at the end. Schenck was the physician who treated the injured in the Reich Chancellery and who attended Hitler's wedding reception at the nearby bunker, as portrayed in the film *Downfall* from 2004.[186]

By the war's end, much of the extensive Japanese rayon industry had been destroyed by aerial bombardment as well; other factories had been scrapped or repurposed for other types of war production, although some plants had stayed in operation, including one set up in Japanese-controlled Manchuria.[187] The von Kohorns, long since decamped, had known this industrial landscape firsthand (the youngest boy was being groomed for the business in 1938 before he left Japan). The junior von Kohorn, fluent in German and speaking passable Japanese, was obviously an intelligence asset. After America entered the war, he was called into an OSS bureau discreetly located in a Manhattan office building. Asked to look at reconnaissance photos of industrial sites, he was easily able to confirm their specifics, and later believed that many had been successfully targeted.[188]

In the United States, the viscose production boom of the World War II era was only one small part of a far larger mobilization effort. Unfortunately, there was no parallel wartime expansion in experimental research into the dangers of carbon disulfide. The last U.S. scientific paper to appear on the subject before Pearl Harbor was published in November 1941, on the response of cats' nervous systems to carbon disulfide inhalation. That research found an intriguing pattern of damage to the vascular system of the brain, an observation that would prove to be all too relevant to rayon workers in later decades.[189]

It was not until October 1945, after the war's end, that the next scientific progress in the field was reported. "Degeneration of the Basal Ganglia in Monkeys from Chronic Carbon Disulfide Poisoning" was the first research to establish unequivocally that carbon disulfide caused the kind of brain damage seen in Parkinson's disease. The paper represented the fruits of research and analysis spanning the entire period of World War II. Exposure of the first monkey began on 1 December 1938, two months after the Munich crisis. On 20 August 1939, twelve days before the invasion of Poland, the first changes in the test animal's appearance and behavior were observed. By 11 December of that year, the monkey could no longer move. Detailed pathological studies showed disease in precisely those parts of the brain in which the destructiveness of Parkinson's disease is most manifest in humans. Four more exposed monkeys all demonstrated similar findings. The last to go, after more than a year and a half of intermittent exposure, was killed in February 1942: "The animal had become somewhat less active and had shown some rapid action tremor, chiefly of the arms. The monkey was exposed to the gas and left in the gas chamber during the night of February 15, 1942. It was found dead the following morning."[190]

Archives have preserved the official memoranda and related materials documenting the heinous activities carried out by the German viscose industry in active collaboration with the Nazi state. A generation of modern scholars, largely German, has dedicated itself to exploring and exposing this history. Its victims have left us their testimonies. Added to this, there is a single object, preserved in the collection of the German Historical Museum (Deutsches Historisches Museum, Berlin), that speaks volumes on its own. It is a strip of yellow cloth, printed in black with eight rows of Jewish stars meant to be cut out of this mass-production item. This remaining fragment, although showing where multiple stars already had been cut out, still has more than fifty remaining. The yellow material, felt-like, is catalogued by the museum as "regenerated cellulose," that is, *zellwolle* by another name.[191]

Given the breadth and depth of the German viscose industry's involvement in the crimes that took place, it is striking that this history is not more widely known. But these activities were never completely hidden. In fact, there were those in America, presumably with access to the relevant intelligence, who seemed to be aware of what was going on. The American Viscose Corporation's promotional pamphlet *Rayon Goes to War* ends with the words of Lieutenant Colonel Howard Norris, taken from a speech he gave to two hundred executives at a textile industry luncheon in March 1943: "We will show the Axis nations that democracy can outfight, out-produce any efforts that they can achieve under slave production."[192]

6

The Heart of the Matter

The war was not yet over. In fact, it was just shy of two weeks after the Normandy landings when Samuel Courtauld, chairman of Courtaulds Ltd., sent off a letter. It was brief, little more than a note, but of no small import. Dated 19 June 1944, it was directed to Lord Woolton, a secretary in the War Cabinet. Courtauld began, "I understand that you would be interested in seeing a special display of samples which have been made from our rayon staple in conjunction with the Bedford Dyers' Association, and which have been prepared for post-war exports."[1] Courtauld suggested that Lord Woolton might come to lunch the week after, adding, "Should you care to bring someone else with you who is interested in textiles, we would of course be happy to see him too."

Frederick James Marquis, Lord Woolton, was both a university-trained scientist (quantitative and social) and a man of business. He was the director of a major department store chain before his appointment by Neville Chamberlain as minister of food in 1940. In that post he oversaw Britain's successful food-rationing program. Indeed, despite other political accomplishments during the war and in the years following, he is best remembered today as the eponymous popularizer of "Lord Woolton pie," a shortage-driven concoction of root vegetables combined with gravy and baked in a crust.[2] In 1943, Woolton was appointed secretary of the newly formed Ministry of Reconstruction. He was the only person to ever hold that position, since the ministry was dissolved at the war's close.

Lord Woolton did come to lunch, although one assumes that he did not dine on Woolton pie. The day after the luncheon, Courtauld wrote again, appreciative that "our little exhibition interested you."[3] He enclosed a copy of a five-page Courtaulds report that had already been sent to Hugh Dalton, MP, at the Board of Trade (he, too, had also seen the "little exhibition"). The memorandum enumerated eleven separate points focused on the same overriding concern: future exports and the balance of trade. Specifically, if nothing was done, postwar demand for rayon staple, even domestically, would outstrip Courtaulds' current production capacity. Samuel Courtauld spelled it out for the minister: "We feel it is not too early for us to approach you in regard to the granting of special priorities to enable us to proceed with the construction of new plants with no delay, as soon as conditions permit."[4]

This memorandum touts the superior qualities of Courtaulds' staple, brand-named Fibro, and, inter alia, highlights Fibro-based fabric modifications currently in development, along with new color-dyeing options. The latter (under point 5) includes "dope dyeing," a process in which rayon was colored in the viscose solution even before being extruded as thread in the spinning bath. The elephant in the room does not go unmentioned. Samuel Courtauld points out that the United States, based on his sources, will have its staple production up sevenfold from 1939 levels by war's end. Getting down to brass tacks, Courtauld asks for the official permission then required under reconstruction planning to build a new unit at one of his existing factories (he does not request financing). The plant in question, located in Greenfield, North Wales, would indeed later prove pivotal to the company's postwar standing, but in ways Samuel Courtauld could not have predicted.

At the war's close, late in his career and near the end of his life, Samuel Courtauld held preeminence in textiles and was the epitome of a British industrial magnate. The comic device in the postwar British film *The Man in the White Suit* is the invention of a synthetic textile that does not soil and never wears out, threatening a death sentence for any future sales. Whether or not the wizened Sir John Kierlaw in the film was meant to evoke Courtauld, *The Man in the White Suit* captures the capitalist angst of an industry in transition.[5] Courtauld was not the only one in the textile business jockeying for position in anticipation of peace dividends as well as potential new business rivals. Nearly as soon as the war was over, teams of specialists were dispatched, from Britain and the United States, to assess the rayon production capabilities of the defeated Axis powers and to give particular attention to any technological developments that might be of present or future value.

Japan's textile industry, especially in rayon, was of great interest to the United States in the aftermath of the war. In mid-January 1946, a delegation jointly sponsored by the U.S. State Department and the U.S. Army left Washington for Japan. The group included five Americans and an equal number of "foreign observers" (one representing Great Britain). Among the U.S. delegation was H. Wickliffe Rose of the American Viscose Corporation, which now really was American after its lend-lease-driven public sale. After a ten-week investigation, the group presented its summary recommendations to General Douglas MacArthur, but Rose separately drafted a more extensive report focused solely on synthetic textiles.[6] His summary findings, more than three hundred pages profusely illustrated with the author's own photographs and diagrams, was published in the fall of 1946 by the Textile Research Institute, an industry-supported technical body.[7] It is not entirely clear who was the intended audience for *The Rayon and Synthetic Fiber Industry of Japan*—possibly future American business travelers, if the odd, four-page Japanese-English glossary at the end of the report is any indicator. For example, under the letters *I* through *J*, one finds:

> ichi—one.
> jima, shima—island.
> jinken—artificial silk (from *jinzo*, artificial, and *kenshi*, silk); rayon
> jinzo—artificial[8]

Wickliffe Rose was encyclopedic in his approach. There were twenty distinct rayon business entities in Japan in 1946, with forty-seven separate factories. Rose lists them all, including their addresses. As a testament to the war's toll on the Japanese industrial base, of the original forty-seven, only sixteen were still in operation at the time of his assessment. All but four of these were personally visited by the team in the making of the report; four others were inspected even though currently inoperable.

One of the defunct operations, visited on 6 March 1946, was a former rayon staple producer, Kinka Spinning. Its manager, J. Hirose, was on hand for the inspection. Before the end of the war, the equipment from Kinka had been taken apart and the parts reused elsewhere, and any excess metal was scrapped. This deconstruction occurred despite the fact that the factory was relatively new, having been put in operation only in 1935. After removal of the equipment, the main factory edifice, which was made of reinforced concrete, had been taken over as a military dormitory. Rose had a special reason for seeing this gutted factory, located outside Hiroshima's city center:

> Since there was no rayon machinery in the plant, the interesting point was the effect of the atomic bomb.... This plant, on the edge of town, several

miles from the explosion, had every window smashed and all wooden roofs and beams smashed. All combustible buildings were crushed and burned. The wood sills and frames wherever exposed, were charred. The plant manager had just reached the plant at 8 o'clock in the morning when the bomb exploded. He was knocked down but not injured.[9]

Indirect as it is, the Hiroshima segment is virtually the only reference to human health in the entire text of *The Rayon and Synthetic Fiber Industry of Japan*, except for a critique of the crowded workers' dormitory system at certain factories, which served as effective vectors for tuberculosis transmission.

In fact, even though Rose did not allude to the health problems from carbon disulfide, the rapid increase in Japanese staple manufacturing indeed had led to an epidemic of disease, including classic outbreaks of insanity.[10] The Japanese industry-wide Research Committee on Occupational Health, established in 1938, had purportedly promoted the installation of glass barriers to keep carbon disulfide vapors contained in the spinning baths, after cases of eye problems and other ailments started to mount.[11] In 1939, in an experimental protocol evoking an image of the canary in a coal mine, a series of sophisticated experimental studies carried out in Japan investigated the effects of carbon disulfide inhalation on domesticated finches.[12] Despite all that, during the war and into the early 1950s, "the working environments in the rayon industry in Japan had deteriorated so much that various classical types of carbon disulfide poisoning were observed," with exposure levels averaging 40–50 parts per million (ppm) and peaking as high as 300 ppm (that is, back to the old, debunked "safe" levels proposed by Karl Lehmann).[13]

It was not the diseases of workers but rayon production capacity (actual and, even more salient, potential) that was of interest to Wickliffe Rose and his team. Possible competition for future markets in particular warranted a clear understanding of any technological innovations in synthetic fibers that might have taken place in Japan, especially as part of its war effort. One new process driven by Japanese war needs was the same sort of "dope dyeing" that Samuel Courtauld cited as a key innovation on the British postwar horizon. It was also known as "spun-dyed" staple, and multiple Japanese manufacturers had been using it for several years on a large scale, albeit with a color palette limited to navy blue or khaki for government-requisitioned materials.[14] In another area of new product development, the Japanese had experimented with blending in protein with viscose. It was akin to the hybrid of milk protein and viscose produced in Italy, but the Japanese did not have the dairy stocks to make casein protein polymers such as SNIA's Lanital. Control of Manchurian territory, however, provided access to soybean crops, which could serve as an industrial protein source. Although the hybrid textile was a promising innovation, production of it stopped in 1944.[15] Industry in Japan had also kept up in technologies to produce high-tenacity rayon suitable for tires and other applications. Even more importantly for future markets, the Japanese had moved forward in "crimped" rayon, a staple product so fine that it was compared to Merino and received the woolen accolade of being a fabric that possessed "loftiness."[16]

The British were more directly engaged in postwar assessments of the rayon industry in Germany than in Japan. They cooperated with the U.S. Field Information Agency, Technical (FIAT), and participated in joint U.S.-British Combined Intelligence Objectives Subcommittee (CIOS) operations. Later, they went solo as the British Intelligence Objectives Subcommittee (BIOS). Under the aegis of each of these, delegations repeatedly fanned out across Germany to assess and report on individual plants, selected technical processes, and more broadly defined industrial groupings.

One of the earliest efforts was an ambitious FIAT undertaking by the Synthetic Fibers Team, made up of five Americans, later joined by four Brits.[17] The Americans were led by LeRoy H. Smith, general manager of the American Viscose Corporation plant in Roanoke (DuPont was represented through its Nylon Division). The British members included representatives of Imperial Chemical Industries and Courtaulds. Smith ultimately integrated a number of FIAT reports into a single tome, *Synthetic Fiber Developments in Germany*. Running more than a thousand pages (along with a separate volume of technical figure illustrations), it, too, was published by the Textile Research Institute.

The Americans arrived in London on 15 June 1945, barely a month after the German surrender, and were on the ground in Leipzig less than two weeks later. They rushed to the East first in anticipation of the transfer of control of that sector to Russian forces. Over the weeks that followed, the intrepid team traveled widely, extending its reach beyond Germany (after Leipzig, it was restricted to the Western-controlled zone) and Austria to SNIA plants in Italy and to rayon facilities in France and the Netherlands. There was a bit of mission creep in all this, as acknowledged by Smith: "One of the purposes of the mission was to gather information which might be of help in the prosecution of the war with Japan but as most of the members felt the war would be over before we finished our work, the main efforts were directed towards securing information which might be of assistance to Allied industry."[18]

To that end, Smith and his team paid close attention to product development (for example, there were two plants in Germany manufacturing spun-dyed rayon staple, and several varieties of "crimped" rayon were being produced as well). As with the report on Japan that was to follow, *Synthetic Fiber Developments in Germany* has virtually nothing to say specifically about health hazards or even more generally about working conditions in the German wartime viscose industry. Almost offhandedly, Smith admits that under pressure to produce quantities of product, quality had been sacrificed and that some of this "could be blamed on the quality of workmanship as many plants were using Polish, Russian, French and Italian slave labor."[19] There is one other telling comment by Smith. Many of the plants visited were found to have suffered relatively little damage from aerial bombardment. One plant, however, was utterly destroyed, but through another route: "Some plants like the one at Wolfen entirely escaped damage until the slave labor was liberated by the Allied troops and then in revenge they burned everything in the plant that could be fired. The Germans were at a loss to understand such actions, but such revenge would be quite a natural reaction to years of enforced labor under concentration camp conditions."[20]

By mid-July 1945, the Synthetic Fibers Team's itinerary had taken it to Lenzing, Austria. When the delegation visited the Lenzinger Zellwolle und Papierfabrik, it met with the chief chemists and engineers, but also with the man who had been approved by the occupation forces to serve as the caretaker trustee-manager of the enterprise. It was none other than Herr Dr. Arthur F. F. Mothwurf, formerly of Elizabethton, Tennessee. Unlike the visiting textile experts, Mothwurf had not been brought in from the outside—he was an inside man. His precise whereabouts after leaving Tennessee are unclear, but in the interim he had managed to obtain U.S. citizenship (a status Mothwurf made sure to let his American visitors know about). Then, in the late 1930s, Mothwurf returned to Germany. He resurfaced there as a technical specialist and inventor under contract to the Thüringischen Zellwolle group (of which Lenzing was a part). The American team was particularly interested in Mothwurf's invention of a continuous process for spinning the rayon yarn, an approach previously applied only to staple. Mothwurf had filed patents for this process in both Germany and the United States.[21]

In his introduction to *Synthetic Fiber Developments in Germany*, Smith gratefully acknowledged the help of military government officers in making logistical arrangements and in "adding their 'persuasive powers' when necessary, to make some German executive talk or deliver the correct information desired."[22] Six months after Smith's book was published, a British team revisited many of the same sites in Germany (but did not go to Austria or further afield). This BIOS team produced its own report, *The Viscose Continuous and Rayon Staple Fibre Plants of the British, American, and French Occupa-*

158   *The Heart of the Matter*

tion Zones of Germany. Unlike Smith, they were rather charmed by their German "hosts":

> The team visited Germany with one primary object which was to assess the value of any suitable plant likely to be seized as Reparations, in accordance with the Potsdam Agreement.
>
> A very agreeable and secondary objective was the acquisition of any further technical knowledge.... It is pleasing to mention that not in any instance was any reluctance shown by any of the German technicians when information was called for. Indeed we found the reverse to be the case, and this was our good fortune.[23]

The first stop on this love-fest junket, 6–7 December 1945, was Rheinische Kunstseide in Krefeld, the Phrix plant where Agnès Humbert had been a slave worker two years before. They met with a presumably very cooperative Mr. Heim (managing director), Dr. Mueller (chief chemist), and Dr. Bergk (works manager). No account was made of the illness and injury that had occurred on the premises, but a careful inventory was taken of the plant's valuable gold-platinum and tantalum jet spinnerets (more than 25,000 in all). Noting that "this company can quite well manage with considerably fewer jets than the stock shown," the team recommended that a hefty subset be secured as war reparations.[24]

One week later, on 13 December 1945, the team arrived at the Courtaulds-Glanzstoff plant in Cologne, where the Italian prisoners of war and others had suffered (Dr. Paske, works manager and director, was on hand). The plant, minimally damaged, was already up and operating, and thus must have had a full complement of its precious-metal spinnerets. But for this partly Courtaulds-owned operation, the team recommended neither seizure of hard assets nor technology transfer. The next day, the team went on to Phrix Siegburg, whose slave laborers Erich Sander had cared for in the prison infirmary. They were hosted by Dr. Kaiser, works manager, and Mr. Vollrath, chief chemist. The team was very interested in an innovative carbon disulfide measuring device that the factory was using. The equipment measured with precision how much carbon disulfide went into the formulation, not how much of the poison was off-gassed into the workroom atmosphere.

Just after the New Year and nearly at the end of its tour, the group visited the only plant of the former Thüringischen Zellwolle combine in German territory not in the Soviet zone: Süddeutsche Zellwolle in Kelheim. Although virtually undamaged, the plant was on a production hiatus due to a lack of coal. There, the chief engineer, Mr. Heyme, was extremely forthcoming and knowledgeable, not only from his work at Kelheim but also through his close ties to Phrix (he had helped build the plant operations at Wittenberge, which, we are told, he considered to be the best in Germany).

As part of an investigation of the German war effort to grow yeast on sulfite waste (independent of questions specifically related to viscose fibers), a separate U.S. FIAT team paid a visit to the Phrix corporate headquarters in Hamburg in July 1945, holding meetings that included its general director, Richard Eugen Dorr. The FIAT report notes:

> The efforts of the Phrix Company in the production of wood sugars by pre-extraction of wood or straw and its subsequent conversion to yeast has been highly overpublicized in Germany. Their Wittenberge and Kustrin plants were in Russian territory and were not visited by any representatives of this team. It is known, however, that no significant production occurred at either of these plants.... During interrogation of the technical personnel of the company at the Hamburg offices, it was difficult to keep clear what they had actually accomplished and what was proposed. It was since learned that the people interviewed particularly the Director, Dr. Dorr, had a tendency to brighten up the truth.[25]

Dorr's "truth brightening" extended beyond his interviews with Allied interlocutors. After the war, it was business as usual for Phrix, although it retained only its Krefeld and Siegburg plant sites: the others were under Russian control. In 1948, Dorr accused another Phrix executive of embezzlement. Dr. Adolf Groms, the accused, happened to be an internal rival of Dorr's for control of the company. As soon as Groms was under criminal investigation and out of the picture (he was later exonerated), Dorr went ahead with a Groms-opposed business plan to resurrect Phrix's protein synthesis operation. Dorr also set up a Swiss-based shadow company, Orgatex, through which he funneled Phrix sales in a money-laundering scheme. Meanwhile, a horde of precious-metal spinnerets had gone missing from the Phrix vault. These were not the ones that the British had eyed at Krefeld, but ones "evacuated" from the Phrix Wittenberge and Kustrin plants ahead of the Red Army advance. Finally, in 1955, it was Dorr's turn to be arrested. He was tried and eventually convicted on charges arising from the Orgatex affair.[26]

Dorr may have been caught playing fast and loose with Phrix's finances, but he was never indicted for war crimes, despite his leadership role at Phrix throughout the war and thus his responsibility for Phrix's abuses at Kustrin, Wittenberge, Hirschberg, Krefeld, and Siegburg. Indeed, the only person ever tried for war crimes for rayon slave labor was the I. G. Farben plant leader overseeing, among other facilities, Premnitz and Wolfen (where the liberated slave laborers torched the factory). Fritz Gajewski ultimately was acquitted on the slave-labor indictment, as were most other I. G. Farben defendants except for those directly involved in the synthetic-rubber plant at Auschwitz.[27] The court narrowly found that only those defendants who could be shown to have actively procured slave labor (and not merely used them) were guilty, a position that came under a withering attack in a dissenting opinion crafted by one of the three presiding judges, Paul Macarius Hebert, dean of the Louisiana State University School of Law.[28] —

Unlike Dorr at Phrix, Walter Scheiber did not stay at the helm at Thüringischen Zellwolle (Mothwurf's employer). He had been deeply involved in the Biosyn-Vegetabil-Wurst enterprise and had overseen Lenzing (which maintained a satellite camp of Mauthausen). Yet Scheiber, too, was never indicted. Far from it, in fact. In 1948, Scheiber was recruited for the U.S. postwar secret chemical weapons effort known as Operation Paperclip. He had parlayed his chemical know-how in poison nerve agents (experience that he acquired at I. G. Farben before moving into rayon) into U.S. protection.[29]

The only prominent figure associated with the rayon industry that ever served any real prison time for his wartime role was the Biosyn-Vegetabil-Wurst human experimentalist physician, Dr. Ernst Günther Schenck, who had stayed to the bitter end in Berlin. Captured by the Russians in 1945, he was home in time for Christmas in 1955, after ten years in a Soviet prison.[30] He landed on his feet, though, getting a job in the 1960s with the German drug manufacturer Grünenthal, which had a pattern of employing former Nazis. Grünenthal, which is still in business, is best known for its innovative pharmaceutical program, although Schenck was not involved in one of its most high-profile products: he started to work there only after Grünenthal had put thalidomide on the market in the late 1950s.[31]

Two key figures of the older generation of German neurologists whose work had touched on carbon disulfide survived the war, but did not live long after. Dr. Lewy, who had left the Neurological Research Institute and Clinic in Berlin, died in the United States in 1950. Dr. Bonhoeffer, who had stayed in Berlin, emeritus from the Charité, died in 1948. Two of his sons and two of his sons-in-law were executed by the Nazi regime weeks before the war ended.[32]

Although its workforce was small, the German capacity for primary carbon disulfide manufacturing to supply its viscose industry was a key component of postwar industrial renewal for that sector. This was one of the few areas where an Allied assessment team seemed to have anything explicit to say about worker health under renewed production. A BIOS report from 1946, in describing a carbon disulfide manufacturing facility picturesquely located at the foot of the Bavarian Alps ("Except for the chimneys it could be taken for a large farm"), notes, under the heading of "Welfare": "The plant operators are provided with gas masks and good washing and dining accommodation. Regular tests of the plant atmosphere are performed."[33]

While the rayon industry was being "rehabilitated" in postwar Germany, a parallel process was taking place in the East. The majority of Phrix's German facilities, nearly all of the facilities of the former Thüringischen Zellwolle industrial combine, and the key I. G. Farben viscose plants at Premnitz and Wolfen were all situated in the Soviet-controlled zone. In the West, Phrix and Glanzstoff persisted, and Farben, de-conglomerated, continued in its constituent parts (for viscose, Hoechst was particularly relevant). But in East Germany, new state corporate entities emerged. The old Phrix flagship plant at Wittenberge became the VEB Zellstoff- und Zellwollewerke Wittenberge, "VEB" being an abbreviation for "Volkseigener Betrieb," meaning a publicly owned enterprise.[34] The former Farben factory plant at Premnitz, the originator of rayon staple, reemerged as the "Friedrich Engels" plant; Thüringischen Zellwolle at Schwarza, the "Wilhelm Pieck" factory; the Glanzstoff facility in Elsterberg, the "Clara Zetkin" plant; and the former Kuttner fiber plant in Pirna (one of the few prewar independent rayon producers), the "Siegfried Rädel." Pieck, Zetkin, and Rädel were all notable German communists, but Rädel (murdered by the Nazis in 1943) holds the distinction of actually having worked in the Pirna plant and organized its workers in the 1920s.[35]

East German viscose production quickly emerged as a powerhouse among the Eastern Bloc's COMECON (Council for Mutual Economic Assistance) member states: in 1950, its annual production surpassed that of the USSR by a factor of three. Indeed, the industry was even attractive to West German investors.[36] The "great catastrophe" of World War II had been put behind, and it was business as usual—all the more so for the viscose industry in East Germany. Protecting the health of the workers engaged in this revitalized enterprise, however, may have been given lip service on ideological grounds, but was of questionable success. A medical report on a serious outbreak of keratitis (the same eye problem that had plagued the industry for more than twenty years) puts the failure of effective controls in focus. It documents that from 1950 to 1951 at the newly reopened plant at Pirna (which had been shut since the war's end), there were scores of new cases of chemical eye injury each month, one time peaking at thirty cases in a single day.[37]

Little additional contemporaneous information on conditions in East Germany emerged. In 1972, a retrospective review from Dr. Herbert Zenk (at the Central Institute for Occupational Medicine in East Berlin) downplayed the extent of disease among viscose industry workers. Zenk based his assessment on the observation that there had been only a little over one hundred accepted workers' compensation claims for carbon disulfide poisoning over a decade (starting in the mid-1950s), a finding he attributed not only to improved hygiene, but also to certain inherent attributes of the East German workforce. As the author states in the idiosyncratic prose of his own English-language summary: "In our opinion the regression of the intoxication complaints is due to the individual disposition of the autoprotective and compensatory forces and to improved occupational hygienic conditions, including medical control examinations which should be carried through in a more narrow-mashed way in future."[38]

After the war, Poland became the other major rayon producer in COMECON. Indeed, it was neck-and-neck with the USSR in output in the early 1950s, although both were dwarfed by the DDR. For Poland, we have a far clearer picture of the compromised working environment in this period. An early and surprisingly frank medical report published in 1946 describes two plants that reopened shortly after the war's end (both are unnamed in the article, but the only two plants operating in Poland at the time were at Tomaszow and Chodakow). A joint medical–industrial hygiene team measured carbon disulfide in the air in the factories and linked it to an outbreak of severe poisoning. The presence of disease was not surprising, since documented airborne levels of carbon disulfide in one factory were as high as 150 ppm and peaked at the other plant at 300 ppm, levels long since recognized as dangerous.[39]

The Polish medical paper includes the details of the particularly dramatic case of a fifty-nine-year-old *rabotnik* (worker), G.A. He began working on the rayon staple line in January 1946, and after only a little more than a week's exposure, came back from his night shift manic, yelling, and seeking an ax to commit mayhem. G.A. was committed to an asylum and straitjacketed until he regained his senses, which took several weeks.

The authors end their report with a call for better control of toxicants in the workplace, but it is doubtful that much was done to get the problems under control. For example, another medical report from Poland more than a decade later extols the virtues of occupational health services for workers. Rather than advocating the curtailing of exposure to carbon disulfide, it describes a kind of modified "magic mountain" approach taken to mitigate its adverse effects on workers: "Part-time nursing homes, or 'night sanitoriums,' where workers spend all their free time at night hours under the best possible conditions for the treatment of their maladies, have been organized by various factories such as those producing synthetics fibers, in which there is the danger of exposure to carbon bisulphide."[40]

In Slovakia, the viscose plant in Svit came through the war physically intact. Ironically, in the late winter of 1945, one of the first members of the Czech Army to reach the newly liberated plant was an escaped Auschwitz survivor and fresh military recruit named Ivan Brod. The entire Baťá enterprise in Czechoslovakia was later nationalized; the Svit facility morphed into Chemsvit and went on to be a major air polluter through its carbon disulfide releases.[41]

In Italy, the aftermath of World War II touched the rayon industry in a variety of ways. At the war's end, Italian military officers who had been forced laborers at Glanzstoff-Courtaulds returned.[42] Alessandro Rimini came home, too. He was the architect who had created the Torre Viscosa skyscraper in Milan. After the war, he could work again under his own name. When the racial laws had come into effect in Italy in 1938, Rimini had been forced as a Jew to submit his architectural designs under others' names. Later betrayed by an informer and arrested, he was sent to a holding camp pending a final deportation, but managed to survive.[43]

Postwar renewal in Italy also had to address the extensive physical damage to the industrial infrastructure of rayon manufacturing. For example, SNIA's flagship Torviscosa facility, which had not been spared aerial bombardment, needed serious repairs in order to become functional once again.[44] This restoration was accomplished with impressive rapidity. At its opening, in 1938, SNIA had celebrated Torviscosa with Marinetti's Fascist verse. At its reopening, in 1949, it marked the successful revitalization of the Torviscosa operation with a film. *Sette Canne, un Vestito* [Seven Reeds, One Suit] is a ten-minute minidocumentary that follows the production of rayon at the Torviscosa facility: harvesting the cane, digesting the cellulose, churning the viscose (carbon disulfide is specifically mentioned by the narrator), spinning the fiber (the spinning batteries are not enclosed), and reeling thread on the filament line or baling cut material in the rayon staple unit. Many of the images are quite compelling. The opening sequences have an almost surrealistic quality, showing the reed harvesters hooded in cage-like coverings (apparently to protect them from flying debris). A later shot of a container of processed cellulose being pushed into an industrial elevator and then rising up out of view conveys an eerie disquiet. The eclectic musical score, at times almost like marching music, at other points jarringly discordant, adds to the atmospherics. Then, after nine and a half minutes of the proletarians keeping to their duties, *Sette Canne, un Vestito* ends with a fashion show of rayon evening wear that underscores who really stands to gain from the surplus value of this labor. Without the film's opening, a viewer might be left wondering how an industrial documentary could be so riveting. *Sette Canne, un Vestito*, however, credits

Michelangelo Antonioni as its director—the film was made one year before his first full-length feature, *Cronaca di un amore* [Story of a Love Affair].[45]

As the European viscose industry recovered, the manufacturers in the Netherlands, Belgium, and France were in the game as well. For France as a whole, however, rayon was not embraced with open arms. The high-fashion hype for the pitiful viscose substitutes of the occupation had been put behind; the haute couture New Look by Dior, no friend of rayon, had arrived; and the cheap rayon Dior knockoffs that came out in the U.S. market only heightened French disdain for the synthetic.[46] An early postliberation French ballad, "Rose Pâle," captures this view perfectly. It tells the story of a lonely flower girl selling her bouquets on rue Pigalle. Her pathos is matched by her outfit: the opening line of the song tells us that Rose is dressed in two bits' worth of rayon.[47]

Not to be left out, Courtaulds was extending its reach on the Continent. It still had a 50 percent stake in Glanzstoff's Cologne enterprise, owned outright a French-based viscose factory in Calais, and held a 20 percent voting capital share in SNIA. Courtaulds even sued for war damage claims, first against the French government, to help pay itself back for reconstruction costs at Calais, and then from Italy, for an amount proportional to its share of SNIA. Courtaulds then used that money to leverage a 50 percent share in a new SNIA acetate operation in Magenta, Italy. The company even toyed with the idea of negotiating the formation of a pan-European viscose multinational that would include German Glanzstoff (VGF), Dutch AKU (formerly Enka, and already partnering with VGF in the postwar period), and the French Comptoir des Textiles Artificiels (CTA). This grandiose scheme, using Shell Oil and Unilever as multinational models, never got beyond the memorandum stage.[48]

In contrast to Courtaulds' machinations on the Continent, rebuilding and reopening at home in the United Kingdom appear to have been a lower priority. Despite his powerful standing, Samuel Courtauld's pitch late in the war to get governmental aid with which to substantially expand its operation in Greenfield, Wales, came to naught. Courtaulds limited itself to only modest improvements there, most notably a carbon disulfide–producing unit with a 180-foot chimney tower that became known locally simply as Landmark.[49] When Samuel Courtauld died in December 1947, he left a major bequest of art to a landmark of a different sort, the self-named Courtauld Institute of Art, which he had helped found in 1933.[50] The *Times* obituary, in addition to noting Courtauld's "Huguenot rectitude, shrewdness, and tradition," took as a particularly outstanding personal characteristic his "sympathy and understanding towards all workers however employed": "He continually had in mind their welfare and their happiness."[51]

Across Europe, as the recovering viscose business was regaining momentum, it was critical that the well-being of its labor force not continue to be neglected or, at best, left to the paternalistic oversight of its industrial barons. This was all the more pressing in the aftermath of the terrible toll taken by the deplorable wartime working conditions so many had experienced. If rayon was back, the enterprise of occupational health protection needed to be ready for business, too.

Luckily, Italy was not only a major center of postwar rayon manufacturing but was also ideally positioned to take a leading role in championing the cause of worker health and safety. This work came to be personified by a leading figure in the field, Dr. Enrico Vigliani. In an editorial under his name that appeared in the very first postwar issue of the profession's main journal there, *La Medicina del Lavoro*, Vigliani noted that "today we can say freely" that the Fascist regime had been an obstacle to controlling occupational disease.[52] Vigliani was announcing officially that it was time to reset the agenda. And one of the diseases that Vigliani most wanted to put on the docket was carbon disulfide poisoning in Italy's viscose industry.

Vigliani hit the ground running. His first priority was to document his accumulated experience of 100 carbon disulfide poisoning cases he had treated as an occupational physician based in Turin in the years 1940–41. This work appeared in *La Medicina del Lavoro* in May 1946.[53] An abbreviated English-language version was published in 1950 in the U.S. journal *Industrial Medicine and Surgery*.[54] The latter effectively summarizes the original's findings of severe intoxication, documenting the astoundingly high carbon disulfide exposures that accounted for the disease (up to 750 ppm in one manufacturing department). But it lacks the vivid details of the Italian original, in particular its virtual roll call of stricken workers summarized in a litany of case vignettes (nerve damage with polyneuritis: Giovanni Spin., age twenty-seven, and Lorenzo Cast., age twenty-four; psychosis: Giordano Feltr. age twenty-eight, and Marco Poll,. age fifty-three; optic nerve damage: Battista Pagl. age thirty-four).[55]

In 1954, Vigliani began to publish another cycle of papers, beginning with one presented in Italian at an international congress,[56] followed by publications based on the same material that appeared in a major English-language publication (the *British Journal of Industrial Medicine*)[57] and then a leading German occupational medicine journal.[58] Although he begins these reports with a brief recapitulation of his earlier case experience, Vigliani's new work went significantly beyond the previous analysis, taking it in a very important new direction. In 1942, Vigliani moved from Turin and had no opportunity to learn what became of the 100 cases he had treated there. But starting in 1943, he began to gather new cases in Milan, continuing on through the remainder of the 1940s and into the early 1950s. The patients he treated by that time had experienced years of high exposure. As importantly, enough time had elapsed for later effects of carbon disulfide toxicity to become manifested.

At first, the cases in Milan were similar to those in Turin, with peripheral nerve damage being the most common problem. But then, as the follow-up period lengthened, the pattern of disease began to change. More and more frequently, Vigliani encountered cases where the predominant problem was not loss of sensation in the distal nerves, but rather appeared to be brain disease of vascular origin. At first these were taken to be one-off events, simply coincidental cerebral atherosclerosis—hardening of the arteries unrelated to work. But then commonalities began to appear: "The cases gradually became more and more frequent and showed certain peculiarities, such as occurrence at a relatively youthful age and long-term exposure to dangerous CS_2 [carbon disulfide] concentrations, suggesting a relationship between chronic exposure to CS_2 and the occurrence of the disease."[59]

Between 1942 and 1949, Vigliani saw sixteen such cases; five more presented in 1950; in 1951, seven more; in 1952, another thirteen; and even two more just as he began preparing his analysis in 1953. In his British paper, Vigliani summarized data for forty-three cases of central nervous system (CNS) disease consistent with vascular insult, including confirmatory autopsy data in several cases. Only four of the cases were over age sixty. There were an equal number under forty years of age, including the case of Luciano Mel., who was only thirty-seven but had worked nineteen years as a rayon spinner. By 1946, paralysis of the left arm had put him out of work; in 1947 he experienced stroke-like weakness of his right side as well.[60]

By 1955, when Vigliani's German-language publication appeared, he could add six additional cases to his cohort. In this, as in his previous Italian- and English-language papers, Vigliani acknowledged the work of other investigators who were beginning to recognize that CNS vascular pathology could represent a major long-term hazard for viscose industry workers. This predominantly reflected clinical case reports or small series (mostly from colleagues in Italy, although the first clinical paper on the topic had come out of Switzerland in 1948). There was one key American experimental study to which Vigliani called attention, and it was the earliest of the lot: the groundbreaking work of Friedrich Lewy from the 1940s had first raised suspicion that carbon disulfide exposure might lead to atherosclerosis.[61]

Vigliani had at his disposal a growing scientific literature on the potential role of lipids in atherosclerosis, a topic of inquiry being newly invigorated after languishing for decades.[62] Taking this emerging field into account, Vigliani's 1955 publication included additional experimental laboratory studies examining potential mechanisms by which carbon disulfide might induce atherosclerosis, zeroing in on lipid metabolism. Perhaps as important, what had been, in its English incarnation, blandly presented as "Carbon Disulfide Poisoning in Viscose Rayon Factories" was, in its German debut, decked out boldly as "Clinical and Experimental Research on Atherosclerosis Caused by Carbon Disulfide" [Klinische und experimentelle Untersuchungen über die durch Schwefelkohlenstoff bedingte Atherosklerose].[63]

Thanks to this publication, the problem of atherosclerosis among rayon workers had been placed clearly in view in 1955, just as Parkinsonism had been spotlighted in the industry twenty-five years earlier. Still, when it came to these health hazards, the workers exposed to carbon disulfide, day in, day out, remained largely in the dark. As their nerves and vessels weakened, the industry they worked in became stronger.

This was certainly the case in the United States. Morgan Stanley had made a killing on the initial public offering of the American Viscose Corporation in 1941 (when it had been pried loose from Courtaulds). Eight years later, Morgan Stanley was still bullish on this offspring. In *Memorandum on the Rayon Industry, Particularly American Viscose Corporation, Celanese Corporation of America, and Industrial Rayon Corporation,* the bankers couldn't have been more upbeat: "There are many indications that rayon will continue to expand its market through the development of new products and applications, as well as through advances in manufacturing methods and processes."[64]

The Morgan Stanley analysis takes note of two major DuPont products: nylon, which was well established by 1949, and a far newer synthetic polymer textile going by the brand name Orlon. DuPont's nylon entered into commercial production in late 1939 and benefited greatly from an expanded wartime market. DuPont's original nylon product is technically "nylon 6,6," whose technical chemical name is "poly[imino(1,6-dioxohexamethylene)iminohexamethylene]." This jawbreaker underscores nylon's sophisticated chemical synthesis, a patented process in which two different starting materials are chemically joined to form a single monomer, a molecular starting block that then links up in a repeated chain to form a synthetic polymer fiber. Morgan Stanley, understandably, did not delve into these technicalities or into the background of I. G. Farben's competing and far simpler molecule, nylon 6, which had many applications but did not threaten DuPont's market position.

Orlon, although invented by DuPont scientists in the early 1940s, was only coming online commercially just at the time of Morgan Stanley's assessment.[65] The basic monomer building block of that synthetic polymer is a toxic chemical called acrylonitrile. Although both nylon and Orlon (generically, an acrylic fiber) are wholly synthetic and without any basis in cellulose (unlike rayon), the bottom-line Wall Street assessment seemed to be that all was okay for viscose, even with these competitors on the market. After all, the profit for the American Viscose Corporation for the previous ten years had been 14.6 percent, yielding $5.19 per share.[66]

The analysts at Morgan Stanley hardly could be faulted for failing to follow the "Letters to the Editor" section of the scientific journal *Nature*. It was there that late in 1946 two scientists from the University of Leeds published a brief notice titled "Structure of Terylene."[67] Nor was it likely that a group of Wall Street analysts would have seen a follow-up article in the same journal two weeks later, detailing this material's special chemistry (technically, a polymer of terephthalic acid and ethylene glycol). Written by J. R. Whinfield, a scientist working at the laboratories of the British Calico Printers' Association, the article credits the discovery of terylene as coming out of basic research that earlier had led to nylon's invention. But the scientist ends with a more personal acknowledgment, "To my early association with the late C. F. Cross, the discoverer of the viscose reaction, from whom I first acquired an interest in the chemistry and structure of fibres that has endured for many years."[68]

Even if not devotees of *Nature*, the boys at Morgan at least might have caught the wider news coverage on the new fiber, such as in a "Notes on Science" item from the *New York Times* in 1946 that praised the product in prose equivalent to "it slices, it dices" infomercial hyperbole: "A remarkable new textile which is somewhat like nylon, and which can be stretched five times its length without losing firmness. It can be made in various thicknesses and widths; it can withstand bright light; it will iron, launder, and press without any special precaution, and it is not affected by moisture, chemical mixtures or microorganisms. . . . Its depths will not be plumbed until a new plant has been built and more terephthalic acid has been produced."[69]

At DuPont, the announcement of terylene's invention did not go unnoticed. The company had its own version of a similar polymer, which it called "Amilar" but had never developed. Forced to buy the British patent rights to terylene, DuPont searched for a new name for the wonder fabric, and trademarked it as "Dacron."[70] By the spring of 1953, a spanking new $40 million DuPont plant in South Carolina was making the cutting-edge synthetic.[71] Meanwhile, the British persisted with the product as terylene. From early on, though, the popular name for this product was drawn from its generic chemical polymer structure: it was simply known as "polyester."[72]

For a while in the early 1950s, all these competing products, together with the old standards of cotton, wool, silk, and linen, seemed to coexist in a kind of peaceable kingdom of the natural and synthetic. Or at least participate in a friendly "rivalry in fabrics," a moment captured forever in a bit of fluff reporting out of London as it revved up for the coronation of Queen Elizabeth II:

> In the preparation of fabrics for the coronation year, a race has developed between synthetic and natural fibers to produce interesting materials. Heading the synthetics is terylene, the test tube yarn that promises to rival nylon in popularity....
>
> In the first big coronation fashion display to be held in London, on Oct. 6, terylene will be used for the first time as a diaphanous soft dress material. It is almost uncrushable. Norman Hartnell, the Queen's dressmaker, is designing the dress for Ascot.
>
> Crease resistant qualities to rayon have been enormously improved. The new "dope-dyed dupion" has this quality and in a new type of weave and dye has acquired the flecked qualities usually reserved for tweeds....
>
> Coronation brocades and damasks of rayon intended for peeresses' dresses include a beautiful silk damask by Courtaulds with a motif of English rose, crown and thistle.
>
> New nylon velvets are in the best lustrous velvet tradition as well as being crease resistant and washable. Nylon lace with these same qualities will be popular for evening wear.[73]

The actual coronation took place in early June 1953, a week after DuPont's polyester plant came online. Rayon was not blown out of the water by polyester, but its textile business was threatened. Nylon not only was encroaching on that sector but also threatening rayon's important niche as a tire fiber. At least in that market in the 1950s, lower cost was somewhat in rayon's favor (especially in the automobile sector).[74] Rayon manufacturing depended on wood pulp, whereas the growth of the wholly synthetic textiles, of which polyester was a paradigm, was fueled by the postwar boom in the petrochemicals industry, which in turn was driven by relatively cheap raw petroleum feedstock.

Making viscose depends not only on pulp, but also on carbon disulfide, its other key process ingredient. Carbon disulfide is critical to producing viscose rayon not as a component of the final product, but as a "facilitator," turning the cellulose into a syrup that can be extruded and spun. The carbon disulfide goes in and then comes out. For workers in the viscose industry, the "coming out" has always been the problem. From the manufacturer's perspective, this lost carbon disulfide was simply a cost of doing business. Although technical improvements recaptured some of it, even in the Morgan Stanley memorandum, the estimated net consumption of carbon disulfide was roughly a third of a pound for every pound of rayon yarn manufactured.[75] The assessment included not a single reference to workers' health.

Since the initial discovery of carbon disulfide in the eighteenth century, the chemical had been synthesized using coal and sulfur as its starting materials, with added heat to make the reaction occur. This meant, of course, that making viscose was not simply a matter of cellulose; it required coal as well, which brought to bear the externalities of that commodity market. The 1950s saw the invention and commercialization of a new way of making carbon disulfide, by using methane gas in place of coal.[76] In this way, even though viscose was not a synthetic in the same sense as nylon or Orlon acrylic or Dacron polyester, it could still take advantage of relatively cheap petroleum to lower carbon disulfide costs. That savings could relieve some pressure on manufacturing expenses and spare factories the nuisance of aggressively recapturing off-gassed carbon disulfide, even though doing so would have made the workroom atmosphere safer.

Even with cheaper carbon disulfide, the market picture for rayon was not as rosy as Morgan Stanley seemed to think. The American Viscose Company, still the undisputed leader in U.S. viscose rayon, appeared to understand this. As early as 1930, it had gone into the production of cellulose acetate, otherwise the bailiwick of one of its domestic competitors, the Celanese Corporation of America.[77] That sideline was irrelevant, however, to mitigating the impact of the new synthetics being churned out by its long-term major competitor, DuPont. The options were limited. AVC was in no position to take on DuPont's patented synthetics, and the time was long past for gentlemen's agreements to conveniently set rayon prices.

In any event, DuPont was less and less interested in rayon textiles. Viscose itself, however, was near and dear to its corporate heart. Cellophane had been a cash cow for years, and DuPont remained dominant in the U.S. cellophane market. By 1950, DuPont could celebrate ten years of production at its massive Clinton, Iowa, cellophane plant,[78] which in 1940 had been added to the cellophane trinity of DuPont Cellophane in Buffalo,[79] Old Hickory Cellophane in Tennessee,[80] and Spruance Cellophane in Virginia,[81] each marking twenty-five years of operation in 1949, 1954, and 1955, respectively. The anniversary edition of its in-house Clinton publication, the *Midwest Blend*, trumpets cellophane as part of a long and grand tradition in the "fine art" of packaging, which could be traced back to the Renaissance master Benvenuto

Cellini and had now been updated to reflect DuPont's mission of making "better things for better living through chemistry." Amid its glossy pages, including portraits of the ten-year-service veterans and highlights of its injury-free safety record, the *Midwest Blend* features the new expansion at Clinton: eight additional casting machines, doubling plant capacity.[82]

DuPont's only serious domestic competition in cellophane was the Sylvania Industrial Corporation of Fredericksburg, Virginia. Following its unsuccessful attempt to squash Sylvania for copyright infringement, DuPont had to adjust to the realities of a shared playing field.[83] Wartime profits certainly helped ease any pain. The American Viscose Company had not been a party to the litigation, but with the war's end, it was unwilling to remain on the sidelines any longer. In July 1946, AVC announced a merger with Sylvania. Indeed, Sylvania's president, the chemist Dr. Frank H. Reichel, became the chairmen and president of American Viscose, overseeing all combined operations, including its Sylvania Division.[84]

AVC sidling up to the cellophane feeding trough was not the biggest problem for DuPont. On 17 December 1947, the same day that the *Yale Daily News* announced that the college (along with forty-seven other schools) had received funding for graduate and postgraduate scholarships in science from DuPont,[85] the United States filed an antitrust complaint against DuPont under the Sherman Act. The Department of Justice charged DuPont with monopolizing, attempting to monopolize, and conspiring to monopolize interstate commerce in cellophane. As the case wound its way through the courts, data accumulated detailing DuPont's cellophane business, most saliently a growth in sales from $1.3 million in 1924 to a walloping $55.3 million in 1947, accompanied by an operative return in the last ten years of that period of 31 percent, with a profit of roughly 16 percent net taxes, bonuses, and R&D. The legal arguments involved considerable technical detail, including the role of waterproof cellophane innovations in DuPont's success and whether the presence on the market of other nonviscose wrapping materials diluted the impact of the company's 80 percent share of U.S. cellophane trade.[86]

Ultimately, DuPont fended off the antitrust charges. A U.S. Supreme Court decision upheld a lower-court ruling in the company's favor in 1956; the decision elicited a powerful contrary opinion from Justice Earl Warren, one of his strongest dissents.[87] It was morning in America again, at least for cellophane. After all, 1956 was the year when Arnold Nawrocki invented a method to cellophane-wrap individual cheese slices.[88] But Dow Chemical's Saran Wrap already had been on the consumer market for three years. DuPont had been beaten at its own game: like nylon, Orlon, and polyester, novel Saran

Wrap was a wholly synthetic, petrochemical-based product, in advertising parlance, "the most amazing food wrap ever developed."[89]

The growing presence of synthetic polymers was not the only change in the landscape for rayon and cellophane. Since the first industrial viscose facilities had come into being early in the twentieth century, manufacturers had been given nearly free rein regarding what they did inside their plants. And with rare exceptions, almost no attention was given to what such production might mean to nearby communities or the wider environment. Even the good citizens of Coventry, ground zero for Courtaulds, had gotten nowhere beyond an admission that the smells coming from the factory amounted to a harmless odor nuisance. When Courtaulds constructed a huge chimney at its Foleshill facility in Coventry earlier in the century, the locals, having no other recourse, turned their woes into a rhyme,

> Courtaulds built a chimney, it wasn't built for smoke.
> It took the stink from Foleshill and dropped it over Stoke.[90]

Water pollution, as opposed to airborne releases, was a problem that long had been linked with the wood and pulp industry. In the 1920s, for example, when an outbreak of a mysterious seafood-related disease occurred along a lagoon by the Baltic Sea, cellulose factory waste was initially considered the culprit. This theory ultimately was disproved only when additional clusters of the same syndrome appeared at locations remote from any industrial site. The illness, marked by severe muscle breakdown caused by a natural toxin produced by marine microorganisms then concentrated in the food chain, is nonetheless still called Haff disease, named after a generic term for such lagoons in the Baltic region.[91]

Concerns over pulp factory pollution sporadically extended to the viscose industry. In 1934, for example, opposition (especially among sport fishermen) to locating a rayon factory in Paw Paw, West Virginia, was based on potential pollution of the Potomac.[92] A contested issue of water pollution and viscose manufacturing at another site led to a legal fight, but viscose was the plaintiff, not the defendant. DuPont brought suit to stop waste from being discharged upstream from its Richmond, Virginia, rayon and cellophane manufacturing plant. The polluter produced dyestuffs that DuPont feared might discolor its product by tainting the intake water. A 1936 federal court decision upheld the City of Richmond's right to permit the discharge of industrial wastes by the dyestuff maker, denying the injunction that DuPont had sought. The legal reasoning in 1936 was based on long-standing legal precedents, in particular case law relating to "riparian water rights."[93]

The first sign that a more sustained and rigorous effort to link rayon manufacturing with water pollution appeared in the unlikely venue of the *Bulletin of the Virginia Polytechnic Institute*. Its May 1942 issue was devoted entirely to "A Study of the Stream Pollution Problem in the Roanoke, Virginia, Metropolitan District." The investigators established a series of water-sampling stations over a length of the Roanoke River:

> Station 5 was located at the highway bridge on route 618, 1 1/2 miles below Station 4, for the purposes of ascertaining the effect on the stream of wastes from the American Viscose Corporation's plant. The composite wastes from this industry amounted to 6,000,000 gallons daily, distributed over a period of 16 hours....
>
> The effect of these wastes on the physical properties of the stream was to raise the water temperature and to give the water a milky appearance, a strong hydrogen sulfide odor, and a considerable amount of turbidity. The bottom was carpeted with a thick coat of yellow slime, *Sphaerotilus natans*, at all times during the investigation....
>
> No fish life whatever was found in the waters at Station 5.[94]

The scientists who authored the report also performed laboratory experiments with fish, placing river chubs, darters, and mad toms in tanks filled with American Viscose Corporation effluent. No fish survived for as long as ten minutes. Some lasted five minutes, but then succumbed later, even when put in clean water. The researchers concluded that the river was so polluted that it not only was unfit for fish and human recreation, but also was "injurious to hydraulic machinery."[95]

By 1945, the Tennessee Valley Authority publicly raised questions about industrial waste pollution posing a challenge to its work (juxtaposing this with malaria control as another health priority), citing rayon mills as one of the principal sources of stream pollution in the TVA watershed.[96] In 1948, just six years after the Roanoke River study, the U.S. Senate passed, and President Truman signed, the first major piece of national legislation for water pollution control, the Barkley-Taft Water Pollution Control Act.[97]

In the same year that the Roanoke study on water contamination came out, neighborhood air pollution from viscose factory emissions gained a degree of local attention, at least judging from a *Cleveland Press* photograph in 1948 of a health inspector named Anthony Sidlow investigating damage to the home of Mrs. John C. Anderson, who lived on 100th Street.[98] In the same year, the first major industrial air pollution episode in the United States, which occurred in Donora, Pennsylvania, heightened the public's awareness that the air as well as the water could be threatened by uncontrolled manufacturing discharges: twenty people were killed in the short term by toxic smog, and fifty more died in the weeks following.[99] The Donora episode did not involve viscose or the pulp industry—the polluters were a steel plant and a zinc works. In the near term at least, it was unlikely that ambient air pollution controls for rayon or cellophane manufacturers would be mandated.

The following year, though, another event of an entirely different nature put carbon disulfide in the spotlight. At approximately 8:45 A.M. on 13 May 1949, a truck loaded with more than 48,000 pounds of carbon disulfide packed in 80 fifty-gallon drums entered the Holland Tunnel from the New Jersey side, headed toward New York. At 8:48, a tunnel officer transmitted an amber signal of trouble because of an apparent truck stall. As he ran toward the truck there was a loud blast. The resulting conflagration destroyed the inner walls and ceiling of the south tube of the Holland Tunnel for a distance of 600 feet. Although 650 tons of debris eventually were removed, the tube was partially reopened only two and a half days later, and further intermittent shutdowns for repairs lasted for two months. Miraculously (or as the official report noted, "'Death Takes a Holiday' was brought to mind"), despite twenty-seven persons requiring hospital treatment, there were no fatalities. Three busloads of children had been stopped at the entrance of the tunnel by the amber signal.[100]

There was no precedent for such an event in a vehicular tunnel. In twenty-two years of operation, the most serious event in the Holland Tunnel had been a truck loaded with chewing gum and shoe polish that overturned and burned. Although covered with a "molten mass," the roadway reopened after a few hours without significant damage. The "Holland Tunnel Chemical Fire," as it came to be known, initiated a major review of the transportation of hazardous materials by truck in the United States, in particular the need for greater enforcement of many rules already in existence. The shipment of carbon disulfide, made at the Taylor Chemical Division of the J. T. Baker Company at Cascade Mills, New York, was bound for transport by ocean freighter to South Africa.[101]

The anticipated use of this carbon disulfide shipment to South Africa is unclear, but it was not for the viscose industry. Later, in the early 1950s, Courtaulds and SNIA created a joint venture to establish a wood pulp processing plant in South Africa to make viscose-manufacturing-ready cellulose from eucalyptus feedstock. SNIA's role in the project was to bring its technical know-how to bear on the problem of using eucalyptus wood for this purpose, just as it had for the use of cane at Torviscosa. But once the plant was up and running, Courtaulds bought out SNIA's interest, rejecting its proposal to build a viscose plant to use the cellulose in South Africa.[102] Courtaulds was intent on keeping the facility as a processed-cellulose producer to feed its viscose factories back home. Courtaulds did not shun altogether the prospect of viscose production in the Commonwealth. It had held on to its facility in Cornwall, Ontario, which was untouched by the American Viscose divestiture. And while Courtaulds was developing raw materials in South Africa, it established a large, new viscose plant near Newcastle, Australia, to produce rayon cord specifically for the tire market.[103] Courtaulds was even willing to test the U.S. waters again: in 1952 it opened a new rayon plant in Axis, Alabama, near Mobile.[104]

Beyond its modest Commonwealth and U.S. footholds and its focused Continental interests, Courtaulds did not play a major role in an international expansion of viscose manufacturing in the postwar period.[105] Other manufacturers, however, were certainly interested. The Morgan Stanley assessment of the U.S. rayon industry, for example, took specific note of the Celanese Corporation's viscose operation in Zacapu, Michoacán, Mexico. A producer predominantly but not exclusively of acetate, Celanese had pioneered postwar U.S. industrial expansion into Mexico with a 1945 acetate plant, and followed that in 1948 with a viscose plant producing cellophane.[106]

Before the arrival of Celanese, Zacapu was a rural community. It is located in the heart of what had once been the Tarascan state, which successfully fought off the Aztecs in its day and where the indigenous Purépecha language is still spoken. For the natives of Zacapu, the internationalization of viscose carried a powerful environmental impact, in the broadest sense of the term. Years later, a Swedish graduate student developing a proposal for local ecotourism at Zacapu noted: "The introduction of a factory resulted in that the city became more urban and agriculture got less space. This affected the lifestyle, income and survival of the indigenous people in the communities around Zacapu. The indigenous people did not have enough knowledge to work in the factory."[107]

Over the next decade and a half, as the Zacapu scenario repeatedly played itself out around the globe, a single business entity assumed a pivotal role in the proliferation of an ever more internationalized viscose industry. It was none other than the von Kohorns' operation, which had moved from Central Europe to Japan and then ended up ensconced in America. With the end of the war, the world was their oyster. The family (father and sons) began to develop a remarkable string of investments. They adopted the same methods that had served them so well before: they leveraged their ability to orchestrate production machinery and industrial engineering expertise in order to establish manufacturing plants for a piece of the action, and convincing local investors to put up most if not all of the cash.[108] In little over a decade, the von Kohorns' companies had under their collective corporate belt the Societé Misr pour la Rayonne in a joint venture with Banque Misr and King Farouk, which built a new factory in Kafr el Dawar (a village between Cairo and Alexandria); Rayon Peruana in Lima, the first rayon plant in Peru; Rayonhill, SA, with a plant in Llo Lleo, Chile; and the China Man-Made Fiber Corporation in Towlen, Taiwan. The von Kohorns' operation also supplied machinery and know-how (without an apparent stake in the business) to promote the rayon-manufacturing capabilities for Companhia Nitro Química Brasileira in São Paulo (which had gotten its original boost in the 1930s with a transfer of equipment from a locked-out factory in the United States).[109] In India, a rayon facility was provided for the Birla conglomerate: Century Rayon in Kalyan, West Bengal. Later, the von Kohorns did the same for the Indian Rayon Corporation when it built its plant in Silvassa (Saurashtra, Gujarat), acquiring in the deal a substantial number of shares of company stock. On the lookout for new sources of raw materials, the von Kohorns helped establish a rayon factory in the Philippines that was adapted to bagasse processing, a cellulose by-product of sugarcane, and they went so far as to develop an Ecuadorian sulfur-mining operation in the Andes to guarantee a supply of that carbon disulfide prerequisite.

The global von Kohorn reach did falter at times. The Pakistan Rayon Corporation (with the von Kohorn father and sons on the board) did not build its planned factory: the partition of Bangladesh supervened. Similarly, schemes to develop rayon in Sukarno-led Indonesia during the period of his "Guided Democracy" came to naught. Von Kohorn ascribed the breakdown to Washington's issuance of a new requirement that none of its loans could be used to bribe local officials. A secret cable to Walt Rostow (then special assistant to President Kennedy for national security affairs) indicated that the rayon project would be one of the top subjects on the agenda for a planned visit to Washington by one of Sukarno's close confidants, "who did not consider it appropriate for a private concern like Von Kohorn to arrange financing for the Indonesian government."[110]

But the von Kohorns' most abject failure was an attempt to establish a rayon factory in pre-statehood Israel. Having heard that such a project was under discussion with other investors, the von Kohorns' interest chiefly was driven by concerns they might be scooped on a deal.[111] The limitation was not raw materials: presumably, wood pulp would be imported, and water access was under discussion. It was not security considerations that sunk the deal, either, even though a von Kohorn son had flown to Palestine to negotiate the deal in February 1948, very nearly on the eve of Israel's War of Independence. Rather, it was the von Kohorns' insistence that any factory operation, to be successful, would need to run twenty-four hours a day, every day. He was bluntly informed that there would be no work on the Sabbath. He went so far as to meet with David Ben-Gurion to plead the matter, but it was a nonstarter.

One example of a rayon development that had a more local component, with no apparent connection to the von Kohorn enterprises, was a factory established in Mantanza, Cuba, in 1948. Although its founder was an American investor named Dayton Hedges, he and his sons had taken Cuban citizenship, for tax purposes. Upon his death, Hedges (in cameo image) and an aerial view of his factory were memorialized in a 1958 Cuban postage stamp (8 centavos, air mail). His son Burke inherited the business, and being a crony of Cuban president Fulgencio Batista, he swung a deal in which the rayon plant was bought by the government and then leased back tax-free for what was supposed to be thirty years.[112]

We know almost nothing about the health of the rayon workers employed in these far-flung locales during this period. One exception is a later study of the rayon operations of Nitro Química in Brazil, based on interviews with those that worked there in the 1950s. It documents the terrible conditions there. One worker, José Cecilio Irmão, speaks of inadequate ventilation, work clothes rapidly eaten away, and, on the spinning line, the "unfortunate gas" (*gás infeliz*) and a constant eye irritation for which the treatment was the application of cool slices of potato. No exposure measurements were made at the time, but in one telltale sign of poor control, there was a fatal carbon disulfide explosion at one point.[113] Decades later, when carbon disulfide was finally measured in the factory, exposure levels were found to be extraordinarily high.[114] It is reasonable to assume that it was much the same story elsewhere, and that all the old familiar poisonous effects of carbon disulfide were at work, especially to the nervous system.

In the period that immediately followed the end of World War II, the viscose industry rebuilt and expanded its capacity across Europe as well as in North America and Japan. Over the course of the 1950s, the industry extended itself globally, but the overwhelming weight of production was still in those places where rayon and cellophane had been produced all along. The geographic balance later would shift, dramatically so. But as the 1960s dawned, such a future was by no means obvious. If this point in time was something of a fulcrum, it was preserved perfectly in a motion picture that appeared in 1960. For worldwide marketing, the movie was renamed *The High Life*, but in its original German release, it went by the title of the novel upon which it was based: *Das kunstseidene Mädchen*. More than a quarter century after the first publication of Irmgard Keun's novel of the 1930s material girl clothed in artificial silk, the story finally made it to the big screen.

The pan-European elements of *The High Life* seem an apt metaphor both for much that had transpired in the interim and for all that was about to change for rayon: directed by Julien Duvivier (a Frenchman who had gone to Hollywood during the war); cinematography by Göran Strindberg (bringing a Scandinavian perspective); and a joint German-French-Italian production (Capitole Films, France; Kurt Ulrich Filmproduktion, Germany; Novella Film, Italy; and Société Nouvelle Pathé Cinéma, also of France, whose first production had been *Children of Paradise* in 1945). Most notably of all, the film starred Giulietta Masina in the lead role (*Juliet of the Spirits* would follow five years later).[115]

In the 1960s, the questions that Vigliani's work in Italy had raised earlier still remained to be answered. If carbon disulfide could cause vascular disease, not just nerve damage, how far did this problem extend? Was it only a matter of damaged vessels in the brain leading to strokes, or could there be other manifestations of similar insults to other body organs? Answering such questions was not an easy task, demanding new scientific methods. A major limitation of Vigliani's observational approach was precisely that: he was able to describe a series of cases, but could only infer that the frequency of events seen was higher than it should have been.

To firmly establish a link, more than inference was needed. Vigliani was not alone in his reliance on older methods of clinical narrative. The tradition of such case series stretched back to the very first observations of disease among viscose rayon workers and, before that, in rubber vulcanizers using carbon disulfide. All of Alice Hamilton's work in the United States, for example, was predicated on exactly the same descriptive principles. To convincingly establish that a pattern of disease truly differs from what might have been expected to happen by chance, one must do more than simply tally up the injured. This research consideration is true for infectious diseases, whose outbreaks occur rapidly and are relatively easy to characterize; the need for systematic comparison is all the more important when it comes to slow processes and sporadic events.

Epidemiology is the science of such investigations. Although by the 1950s the discipline could trace its roots back more than a century, much of the work in the field until that time had been devoted to short-term disease outbreaks. Studying other questions required specialized resources, especially access to long-term health records—in particular, accurate death certificates—and the sophisticated statistical expertise necessary to analyze the data appropriately. The methods needed to study the epidemiology of chronic disease at the population level were just coming into their own in 1954 when Vigliani's study came across the desk of Dr. Richard Schilling, the paper's editorial reviewer at the *British Journal of Industrial Medicine*. Reading the manuscript must have caused Dr. Schilling to ask himself this: if carbon disulfide caused vascular stroke, could it also cause heart attack? In hindsight, this may seem like a straightforward line of inquiry, but at the time the leap was far from intuitive.[116]

By midcentury, heart disease and, specifically, fatal blockage of the coronary arteries were widely recognized as major killers in Britain, the United States, and elsewhere in the West. Nonetheless, its underlying causes remained a matter of intense debate. As late as 1965, Dr. Howard Sprague's "Convocation Lecture" to the American College of Cardiology, "Environmental Influences in Coronary Disease in the United States," presented his personal and somewhat idiosyncratic views, but also reflected professional doubts that were more widely shared at the time, for example, his reluctance to believe that dietary factors were in part to blame for the epidemic of coronary disease. In addition, Sprague voiced skepticism that the rise in cigarette smoking, which had paralleled the increase in coronary disease over the previous fifty years, represented a cause-and-effect relationship, since "like many statistical studies, the tobacco-heart disease relation can suffer from the seduction of correlation."[117]

Sprague also included occupation among the environmental factors he considered, but only insofar as the question of emotionally stressful jobs was concerned (he also doubted whether this factor could have changed enough to account for an increase in disease). The possibility that chemical exposure on the job might affect the coronary arteries was completely beyond Sprague's ken. This apparent ignorance was present despite the fact that by 1965 heart attacks had been documented among workers in the explosives industry after exposure to nitroglycerine and similar compounds. Although much of this work had appeared in the foreign-language medical literature, a dramatic report of deaths from sudden heart attacks among employees in a Pennsylvania explosives factory had come out in 1963. The Pennsylvania State Health Department, called in to look into the sudden cardiac death of a young worker, found that two similar fatalities had occurred in the previous fourteen months; during the investigation, there was a fourth such death.[118] For these workers and others, chronic heavy exposure to the nitrate class of chemicals, followed by sudden withdrawal (for example, a long weekend off work), was linked to coronary events. Many additional workers in the same industry experienced cardiac chest pain but were lucky enough not to fatally succumb to heart stoppage.

Richard Schilling was unlikely to have been aware of the early reports of the link between work in the explosives industry and heart attack, but he nonetheless was uniquely positioned to carry out a study of heart disease in viscose workers. As an academically based research physician specializing in occupational disease, Schilling had a number of professional contacts in industry. One of them was an energetic factory physician named John Tiller. Tiller was employed by Courtaulds, and his turf was its three large rayon facilities in North Wales: the Greenfield (Holywell, near Flint), Castle (Flint), and Aber (Flint) works. As Schilling pondered the question of viscose and heart disease, he would have immediately thought of Tiller; they were already collaborating on research on the health of rayon workers. Their initial study concerned a cotton-dust-related occupational lung disease called byssinosis, which causes a decline in lung function among those exposed on the job. This condition was of special interest in Britain because cotton textiles were a major industrial sector there. Tiller and Schilling studied the lung function of the Courtaulds rayon workers precisely because they may have been exposed to textile dust, but it was cotton-free. Published in 1957, their results, as expected, showed that byssinosis was not present in the rayon factories.[119]

Tiller and Schilling set out to investigate possible heart disease deaths among the workers of the three Courtaulds factories in North Wales. From the outset, Schilling recognized that studying a chronic and relatively common condition such as heart disease, by no means specific to any single cause, would require special epidemiological expertise. Once again, Schilling had the ideal professional contact for the project: Dr. J. N. "Jerry" Morris. Morris was a leading twentieth-century figure in the epidemiological study of heart disease. Nearly single-handedly, Morris had scientifically established the association between lack of exercise and increased heart attack risk. His breakthrough study in 1953 compared London Transport drivers, whose jobs were sedentary, with London Transport conductors, who were on their feet all day, climbing up and down the double-decker bus stairs. The less active drivers experienced twice the rate of heart attack.[120] At the time of his collaboration with Schilling in the study of rayon workers, Morris was the director of the Social Medicine Unit of the Medical Research Council (MRC), Britain's nationally supported top scientific research body in the field.

Carrying out the viscose study should have been smooth sailing. Tiller, Schilling, and Morris began by carrying out a preliminary analysis. Because the three factories were in the same area (a municipal borough and the surrounding county), it was possible to check local records for cause of death and for occupation, both of which were routinely registered. Limiting their study to males, they managed to identify nearly 400 deaths among Courtaulds employees, approximately half of whom were process workers (including those employed in the spinning room) and were thus likely to have experienced relatively high carbon disulfide exposures. As a simple first step, the researchers compared the proportion of deaths due to cardiac causes among the exposed workers to the proportion of cardiac deaths among the general population. These data had to be categorized by age group, since cardiac deaths increase with age. By the spring of 1960, Tiller, Schilling, and Morris had completed their initial analysis, and what they found was staggering. There had been 8 deaths from coronary heart disease among all the Courtaulds' rayon process workers 30–49 years old. The proportion of deaths among relatively young men should have been much lower—in fact, one-twentieth as high. In the next age bracket, 50–59, there was a fivefold disparity, and even from age 60 to 80, the proportion was double that of the national data. Since North Wales was believed to have somewhat higher rates of cardiac disease than the rest of Great Britain, the researchers were curious to determine the proportion of deaths among all other local males, including the non-process Courtaulds workers. The younger men in the region manifested a somewhat higher proportion of cardiac deaths, but nothing like that of the group exposed to carbon disulfide.[121]

This kind of analysis is called a "proportionate mortality study" because it studies the mix of deaths, but not the mortality rates for the groups in question. To calculate mortality rates, it is necessary to reconstruct the working population at risk and determine how many employees were on the job over the years of exposure. That cannot be done from death registries, because those records have no information on the living—in essence, they supply the numerator of the fraction but not its denominator. Morris, who had literally written the book on applying such methods (his pivotal text, *Uses of Epidemiology*, was published in 1957), understood that calculating the actual mortality rates and their deviation from expected values would provide an additional and reliable confirmation of the initial proportional findings.[122] Dr. Tiller, back in Wales, found that the needed data indeed did exist. One of the three factories (the Greenfield factory, which had been established in 1930s, the very one for which Samuel Courtauld had lobbied for governmental support to expand at the war's end) had maintained detailed employment records since its opening. The outstanding question was how to gain access to these records.[123]

As soon as they had the results of the initial proportionate study in hand, the team sought the needed cooperation from Courtaulds so that the employment data could be obtained and the more definitive analysis completed. Morris wrote to Tiller's boss, the medical director at Courtaulds, Dr. H. Howard-Swaffield. Howard-Swaffield's reply was a none-too-subtle brush-off, enclosing with it a check to Morris in the amount of twenty pounds sterling, intended to be for services rendered. Morris returned the check, making clear that his work was part of his Medical Research Council duties, not a matter for outside payment, and that the council supported his continued work on the project.[124]

Morris got nowhere fast, but did not give up. Writing to the head of the MRC in August 1960, Morris supplied his boss, Sir Harold Hinsworth, with the same tabular evidence he had provided Howard-Swaffield. In his letter, Morris takes the classic scientific approach that an investigator must set out to disprove a finding and accept it only if it still stands. Having done exactly this, Morris sticks by his results: "However hard we try, we are unable to 'break' the suggestion that the rayon process workers are under special risk of developing coronary heart disease."[125]

Morris especially emphasizes that the highest risk was evident among the youngest workers, the same pattern he had seen in his London Transport workers. This age relationship had been found by other investigators studying different risk groups, for example, findings about smoking British physicians that were at that time just becoming evident. The next day, a cautious memorandum was issued under Hinsworth's signature, with the underlined title *Coronary Thrombosis and Carbon Disulfide:* "This is a peculiar story and it seems to me that it obviously requires further investigation. I doubt whether the firm will welcome this suggestion, and I think we ought to have as strong a case as possible before making approaches to them."[126]

Hinsworth went on to question whether the observations might be explicable by higher rates of coronary disease in North Wales, even though the data strongly supported the interpretation that any such increase, albeit also present in the nonexposed locals, was dwarfed by the higher mortality evident only in the process workers. Hinsworth also wanted to see higher rates in the older workers too, either not grasping Morris's cogent arguments in this regard or choosing to discount the views of his own expert in the field.

A month later, Joan Faulkner, secretary of the MRC and a power to be reckoned with, sent Morris Sir Harold's trepidations.[127] Morris mulled the matter over and answered Faulkner, briefly reiterating his reasoning on the specifics of the higher local rates. In response to doubts about the overall validity of the interim findings and the need to pursue this further, he concluded, "On the general question, my feelings remain the same—I sniff that something real is happening. Any clue, however slight, to a connection between a modern industrial process and coronary disease should, I think, be examined seriously. This will not be possible until Courtaulds' [sic] changes their attitude."[128]

The MRC higher-ups urged Morris to continue his attempts to get cooperation from the Courtaulds medical department (Hinsworth felt it counterproductive to go out of channels and reach out to a personal Courtaulds acquaintance of his "in the City," that is, an unnamed corporate executive contact).[129] Two years later, Dr. Howard-Swinfield, the corporate medical man from Courtaulds, was still stalling, writing in the fall of 1962: "Many thanks for your letter and indeed you have not been forgotten. . . . I do assure that my Directors are aware of the serious nature of this problem and please believe me that the matter is by no means shelved, although we still regard it as highly confidential."[130]

Needless to say, keeping a matter "highly confidential" is not generally compatible with publishing scientific findings in the open literature. In the interim, as Morris and his team continued to try to pry the needed data loose from the clenched jaws of Courtaulds, a one-of-a-kind international conference solely devoted to the health hazards of the viscose industry convened in Prague for three days in September 1966. The International Symposium on Toxicology of Carbon Disulfide represented a unique coming together of a wide range of biomedical researchers and clinicians, all of whom had an interest and experience in viscose.[131] The meeting was sponsored by the Permanent Commission and International Association on Occupational Health. This was the same nongovernmental scientific organization at whose periodic international meetings important work on carbon disulfide had been presented in the past, including early key studies on Parkinsonism in Hungary in 1928, Hamilton's report on the American field studies in Germany in 1938, and after the war, Vigliani's initial presentation of his case series of rayon workers with cerebral vascular disease.

The meeting, organized by a subcommittee of the Permanent Commission focused on the artificial fibers industry, was jointly chaired by a Jaroslav Teisinger, a native Czech, and Heinrich Brieger, an American of Central European origin. They were prestigious leaders for such a conference. Teisinger, a pivotal figure in postwar occupational medicine in Czechoslovakia, had firsthand experience with the health problems of carbon disulfide in the viscose industry there.[132] Brieger had been a politically active public health physician in Silesia before the war, fleeing Germany in 1939. Brieger restarted his career in America, focusing on industrial toxicology, with a particular interest in carbon disulfide.[133] In 1961, for example, he had published an in-depth review of the adverse effects of long-term exposure to carbon disulfide. In it, he refers extensively to the work of Vigliani on vascular disease, acknowledging that although his findings had been contested by some, many Italian and

"Central European" authors supported Vigliani's contention that hardening of the arteries was linked to viscose employment. Brieger, who would not have been aware of the preliminary work being done by Schilling, Tiller, and Morris, regretted that "'carbon disulfide atherosclerosis' has not been given the same attention in England or France, and pertinent reports have not been published in the American literature."[134]

Almost fifty papers were presented at the conference. Remarkably for the Cold War timing, the speakers equally represented Eastern and Western Bloc nations. As might be expected, there were a number of presenters from the host country, but there were also speakers from the USSR, East Germany, Poland, Yugoslavia, and Romania, all countries with industrial rayon production. The Western countries represented in platform presentations were Italy, West Germany, France, Finland, Norway, Netherlands, Spain, and the United States (Brieger and one other presenter), but not Britain. Asia was represented by Japan.

There were, of course, attendees at the conference who did not make a formal presentation. Although no complete list of participants is included in the published proceedings, discussion comments made by members of the audience are documented, providing an additional record of some of those present. When P. G. Vertin of AKU (Algemene Kunstzijde Unie, the 1966 corporate identity of what formerly had been Enka) made his presentation on clinical problems among the employees of two Dutch factories and one in Germany (a Glanzstoff plant), one member of the audience seemed particularly interested in his report. Vertin's conference presentation compared the frequencies of health problems among 302 rayon spinners exposed to carbon disulfide and a similar number of unexposed workers matched for age and other demographic measures, a simple yet reasonably elegant study design.[135] He presented a table showing that among the exposed, there were 32 workers with diseases of the heart and blood vessels, but only 10 such cases in the comparison group, a threefold excess risk. John Tiller spoke up: "I would like to ask if Dr. Vertin would be so kind as to explain how the two groups of people, seen in the chart which shows the differences in their proportional morbidity, were composed."[136]

Vertin reiterated that the comparison group was equivalent in all ways except in its lack of substantial carbon disulfide exposure, reinforcing the likely validity of the exposure association. This was key support for Tiller's work, but Vertin's report was as yet unpublished.

Late in 1968, the long-awaited employment data having finally been made available by Courtaulds, the paper by Tiller, Schilling, and Morris at last appeared.[137] It was published not in an industrial-medical specialty venue, but in the more widely read *British Medical Journal*. The added data supported what the authors had first observed and even amplified those earlier results. In their personnel-records-based analysis, among rayon-spinning operatives age 45–64 the rate of coronary heart disease deaths was just shy of twice that expected, but just as strikingly, among rayon-spinning staff members (rather than workers), who were similarly exposed but of high employment standing, the risk was also doubled. An editorial accompanied the paper, underscoring its significance, even though it was oddly titled "Sulphur and Heart Disease" and veered off into speculation that trace elements such as sulfur might be a generalizable environmental culprit in disease of the coronary arteries.[138]

A year and a half later, Schilling wrote a summary of his study's findings for an opinion piece published early in 1970 in the *American Heart Journal*. Schilling took advantage of the editorial format to pointedly argue that the currently permissible worker exposure limit for carbon disulfide in factory air, then still 20 ppm on both sides of the Atlantic, was unlikely to provide adequate protection against vascular disease.[139] Six months after Schilling's editorial, a major new study carried out in a Finnish rayon factory appeared. It confirmed the British finding of increased risk of heart attack from carbon disulfide.[140]

In 1977, when the U.S. National Institute for Occupational Safety and Health (NIOSH) undertook a systematic review to determine what might really constitute a safe level of exposure to carbon disulfide, its researchers agreed with Schilling that 20 ppm did not offer protection from the insidious effects of longer-term lower-level inhalation of the chemical. NIOSH reviewed Schilling's work and the related research from Finland on heart disease, as well as a host of other studies (many in foreign languages) on other potential toxic effects of carbon disulfide. In the end, NIOSH concluded that a 95 percent reduction was in order, all the way down from 20 ppm to 1 ppm.[141]

Although this extensive assessment drew on far-flung sources of information, NIOSH (based in Cincinnati) did not consider a study that had recently been carried out close to home. Dr. Thomas Mancuso had a long association with the State of Ohio's Division of Industrial Hygiene before going on to the University of Pittsburgh's School of Public Health. As he notes in the introduction to his paper, published in 1972, Mancuso was inspired by the work of Tiller, Schilling, and Morris and set out to complete "the first longitudinal epidemiological study of rayon workers in the United States."[142] Mancuso accomplished this through the use of the employment records, dating back to 1938, of an Ohio viscose manufacturer. Although the employer is not identified in the publication, it was the Industrial Rayon Corporation.

Mancuso was particularly interested in the neuropsychiatric toxicity of carbon disulfide—his analysis concentrated on death by suicide. Using Social Security records, he followed workers' vital status through 1968 and then, laboriously, backtracked to obtain death certificates. His findings justified the effort. The suicide rates were elevated across the board, but in particular among those 35–44 years old. By comparison, rates for motor vehicle deaths, other accidents, and homicides were close to the expected numbers. Moreover, Mancuso noted that his estimate of suicide risk was likely to be an underestimate, since some of the death certificates classified causes such as "carbon monoxide in garage" and even "gunshot wound in mouth" as accidental. Mancuso came across one record in which a suicide victim first murdered his wife.

Shortly after he performed this study, Dr. Mancuso started a question-and-answer column on occupational health for the *Machinist,* a newsletter of the International Association of Machinists and Aerospace Workers. He collected the columns from 1974 to 1975 into a small book titled *Help for the Working Wounded.* A union member wrote in from West Virginia: "A friend of mine works in the rayon plant. He is so terribly depressed, and then he goes into a rage. My wife and his wife are friends, and they are really worried. His doctor sent him to a psychiatrist. Can the job make him sick in the head?"[143] Mancuso begins his answer, "Yes, definitely so," adding, "There have been many medical reports of changes in the emotional and mental balance of workers, so severe, that the men threaten to burn down the house, kill members of their own family as well as themselves."[144]

To these possible scenarios of labile (erratic) behavior might be added the murder of a friend or co-worker. Mancuso did not bring this up to the man whose wife was a friend of his depressed pal. Had he known about the case of Raymond (Buster) McDaniel, he surely would have been even less likely to do so. Thirty-one-year-old McDaniel had been executed in Virginia in 1945. The year before, he had murdered a fellow rayon mill worker, Roy Alfred. The purported motive was an affair with the deceased's wife. The defense attorney and McDaniel's mother and sister had attempted, without success, to convince the court that he "suffered from mental troubles."[145]

NIOSH was well aware of Mancuso's work. In fact, it contracted with him to extend his analysis of the Ohio data to include heart disease deaths. In 1981, four years after NIOSH recommended a more health-protective carbon disulfide standard, Mancuso submitted a fifty-page report, *Epidemiological Study of Workers Employed in the Viscose Rayon Industry.*[146] He found a 40 percent increased risk of death from coronary artery disease among those with ten or more years' employment. Moreover, when he looked at employees from the rayon spinning and twisting departments, the excess in deaths was more than double the expected level. Mancuso's study of Ohio rayon factory workers provided important domestic support for the findings from Great Britain and Finland. Remarkably, beyond the report filed with NIOSH, there is little evidence that the detailed findings of Mancuso's contract work ever were widely disseminated.

Meanwhile, others were also continuing to investigate the link between carbon disulfide and heart disease. By 1979, the same Finnish research group already had accumulated ten years of additional follow-up, finding that the coronary death rate was still more than two and a half times that expected.[147] At virtually the same time, a group from the University of Ghent began investigating heart disease among workers in a Belgian rayon factory. The study was initiated when the Belgian Ministry of Labor measured carbon disulfide overexposure in the factory's spinning room, despite differing (lower) levels reported by the factory itself and in the face of its denials of any health problems.[148] Cursory investigation found that compromised blood supply to the heart was 50 percent more common among those exposed to carbon disulfide relative to a comparison group of other industrial workers. But this pilot study, in a rayon factory in which "working conditions had remained the same since the inception of the plant in 1932," was too small in size to establish a firm statistical link.[149] The full research required could be completed only years later, and only then because the Belgian labor inspectorate forced the issue, after years of stalling by the factory's management.[150]

From the first publication of the Tiller, Schilling, and Morris report, the rayon industry saw the link between carbon disulfide and heart disease as a serious matter. Its response was not to support calls for stricter standards, however, but rather to try to show that any potential problem had already gone away. This was not necessarily a new tactic by industry to deflect the thrust of troubling findings, but the viscose people were particularly adept at employing it.

In 1975, Dr. Vertin, the industrial physician who presented the report at the 1966 Prague meeting that clearly showed a heart disease risk among AKU and Glanzstoff workers, finally published a full paper on the subject. It was not his original study, however. In its place, Vertin came up with a new and complex analysis. It included some of the same younger spinning-room workers with evident heart disease, observations that could not be completely obscured, but undercut those findings by diluting the study group with workers at lower risk and including a series of irrelevant observations.[151] Two years later, in a further analysis that was even more difficult to follow, Vertin mischaracterized his interim study as completely negative; the original, key, highly positive Prague analysis was never mentioned in either follow-up publication.[152]

In the United States, industry took a different tack. The problem had not gone away upon further analysis—it was never there in the first place. A 1975 paper on heart disease in rayon workers appeared, authored by the corporate physician of what had been AVC but by then had assumed the corporate identity of FMC (the Food Machinery and Chemical Corporation). The outside collaborators of the FMC company doctor were two faculty members at the Thomas Jefferson School of Medicine in Philadelphia. Their paper argues that there may indeed be problems due to carbon disulfide elsewhere, but that things were different in the United States, where controls were much better: "No chronic cases of CS_2 poisoning have been reported from the U.S. rayon industry since the late forties. This has not been so in Europe."[153]

Apparently, an extra suicide here or there didn't count. In any event, "voluntary" participation of research subjects in the industry study yielded 100 percent participation in one factory, 97 percent in a second, and a low of 87 percent in a third. These were all current workers; they underwent heart assessments by EKG tracings that, not surprisingly, did not show much in the way of disease. The disabled and the dead, obviously not among the currently employed, were not included.

Even the industry understood that this kind of study would never be terribly convincing. And the stakes were getting higher. NIOSH's recommendation in 1977 for stricter controls was only that, a recommendation. NIOSH had no authority to make any legally binding rules—that was OSHA's jurisdiction. When and if the time came for OSHA to consider new rules, however, a better counterweight than a company doctor and two Thomas Jefferson Medical College clinicians might be needed. The Inter-Industry Committee on Carbon Disulfide of the Man-Made Fiber Producers Association sprang into action, recruiting researchers from the Harvard School of Public Health and giving them support and access to data from four U.S. factories.

The Harvard study, however, once again confirmed that in America, too, heart disease risk was linked to rayon work, especially for those most heavily exposed. Unfortunately, a key analysis specific to younger workers, the group where the risk likely would be clearest, was conveniently absent from the final study findings. Moreover, even what is reported tends to be obscured in a flurry of caveats. The analysis relied on "standardized mortality ratios" (SMRs), that is, the number of observed fatalities divided by the number of expected deaths. This was the same well-accepted approach that Jerry Morris had used once he got the needed personnel employment rolls from the Welsh factory he was studying. Despite the centrality of the SMR measure to standard epidemiologic analysis, the authors of the U.S. industry investigation seem to have balked at this. They undermined the results of their own paper even before presenting them: "Before discussing findings for individual causes of death, it should be noted that the SMRs in this study must be interpreted with caution."[154]

The Harvard study arrived just in time for a series of public hearings held in the summer of 1988 (the paper had been accepted for publication, but had not yet appeared in print). The hearings took place as OSHA was considering new safety limits for carbon disulfide. A decade earlier, NIOSH had recommended a standard one-twentieth that of the then current limits, and OSHA finally had come to agree. OSHA proposed that 1 ppm become the new U.S. legally binding rule, replacing the old 20-ppm default value. This was just one of a set of newly proposed OSHA limits intended to cover hundreds of substances, many of them revisions and others new standards for previously unregulated exposures.

This was meant to be the most sweeping set of regulatory interventions that OSHA had ever considered. Unquestionably, reform was critically called for. Carbon disulfide was one of many chemicals whose old exposure limits had been grandfathered when OSHA was first established in 1970. OSHA's carbon disulfide limit of 20 ppm conformed to the original American Standards Association recommendation dating from the late 1930s; it also agreed with the recommendations of the American Conference of Governmental Industrial Hygienists, which had been in effect from 1946. When OSHA was considering its update, however, the ACGIH had ratcheted down its limit value (in 1980) to 10 ppm.[155]

For more than two weeks in August 1988, the Department of Labor's hearing room at Third and Constitution in Washington, D.C., saw a parade of witnesses. Hardly anyone, whether from industry or labor, was satisfied with the new rules. But of all the presentations, testimony from the Inter-Industry Committee on Carbon Disulfide of the Man-Made Fiber Producers Association was among the most aggressive and highly orchestrated. On 2 August 1988, Joseph Price of Gibson, Dunn & Crutcher, counsel to the Inter-Industry Committee, appeared before administrative law judge Michael H. Schoenfeld. After an introductory section by Price, the presentation by the Inter-Industry Committee's "medical issues panel" takes up more than fifty pages of the printed testimony, not counting follow-up questions. A second panel, on economic and technical issues, came afterward, with testimony amounting to another forty pages of transcript (Judge Schoenfeld found it appropriate to take a ten-minute recess before launching into part two).[156]

Although the technical-economic component of the argument dealt extensively with the engineering feasibility of controls for carbon disulfide, it was the medical presentation that was central to the industry's arguments. The experts contended that exposure should not be lowered to 1 ppm or even 10 ppm. The legal limit for carbon disulfide, they argued, should stay at 20 ppm, right at the sweet spot where it had been for so long. Price was flanked by Dr. Doyle Graham, the dean of Medical Education at the Duke University School of Medicine; Dr. William Hugh Lyle, the corporate medical director for Courtaulds, brought over from the United Kingdom for the occasion; and the Inter-Industry Committee's chief toxicologist, Dr. Ernest Dixon.

Dixon took the lead. He cautioned that the United States should not rely on data drawn from European factory hands, since they have always been more heavily exposed than any worker here at home. This was a well-worn justification to discount reports of disease in the overseas viscose industry. But Dixon gave this tried-though-not-true factoid a new twist: the high levels of carbon disulfide exposure documented in medical studies coming out of Europe were actually underestimates. (He provided no evidence for this claim.) Therefore, because these levels were so extraordinarily elevated, any purported toxic effect of working with carbon disulfide found in Finland or England was simply irrelevant to the domestic American industry, where the levels were much, much lower.

The Inter-Industry Committee's medical panel intended the Harvard study to serve as the coup de grâce that would put the new standard out of its misery. Its findings were dressed up for the occasion as an entirely negative study that would put to rest any concerns that carbon disulfide could cause heart disease, at least at anything short of the terrible conditions uniquely peculiar to other countries. At the end of the day, the Inter-Industry Committee's convoluted medical arguments were rejected by OSHA, although it did back off in part: in a follow-up to the hearings, the new proposed limit for carbon disulfide was revised upward to 4 ppm.[157]

Even as the OSHA deliberations were ongoing, a team from NIOSH was called in to investigate an unusual pattern of deaths at Teepak, Inc., located in Danville, Illinois. Teepak (also a member of the Inter-Industry Committee) employed a modest workforce, slightly fewer than seven hundred souls.[158] Operating since 1957, it made cellophane "Wienie Paks" for skinless franks. Because it was a relatively small facility whose operations spanned a limited number of years, the deaths to be studied were also few in number. But even with that limitation, among those who succumbed to heart disease by age fifty or younger, the odds of dying from having worked with carbon disulfide were elevated more than tenfold.

OSHA did not consider these findings, but the point was moot. The AFL-CIO brought suit against OSHA, not because of the carbon disulfide rule, but in opposition to its package of new standards as a whole, believing that many of them did not go far enough. At the same time, multiple industry groups sued from the opposite point of view. In 1992, the U.S. Court of Appeals for the Eleventh Circuit threw out the new regulations altogether, finding that OSHA was required to provide more detailed analysis of its rationale for every chemical it chose to regulate. The crocodile-tear-stained opinion by Judge Peter T. Fay allows that the court's decision, although not intended to be a block, was likely to grind to a halt OSHA's work at setting new standards: "We have no doubt that the agency acted with the best of intentions. It may well be, as OSHA claims, that this was the only practical way of accomplishing a much needed revision of the existing standards and of making major strides towards improving worker health and safety.... Unfortunately, OSHA's approach to this rulemaking is not consistent with the requirements of the OSH Act. Before OSHA uses such an approach, it must get authorization from Congress by way of amendment to the OSH Act."[159]

The old carbon disulfide standard of 20 ppm was back to stay as the legal limit in the United States.

Meanwhile, a regulatory struggle of a different kind had gone on in Great Britain. In 1987, the Medical Research Council published an additional analysis of the deaths from the Courtaulds factory in Greenfield, North Wales, extending its follow-up for almost twenty years after its original study's cut-off point of 1964.[160] The core findings of the original paper were reaffirmed. Dr. Schilling alludes to this additional study in a section of his memoir from 1998 subtitled "Two Decades of British Negligence of a Serious Health Risk."[161] He starts off by emphasizing that in Britain, coronary disease in a rayon worker has never been officially listed as a work-related condition—a designation referred to as a "prescribed disease." In the absence of such a listing status, only successful legal suits, on a case-by-case basis, can lead to any monetary settlement for a work-related illness.

Schilling tells the story of the widow of a rayon worker who died at age fifty from coronary heart disease (CHD, as Schilling abbreviates it) after nearly thirty years of employment in a spinning room. The rayon worker had suffered attacks of chest pain in 1961 and then again in 1971 (the latter date came after Schilling's study was finally published). The widow's only choice was to bring a legal action: "Counsel acting for the widow stressed the need to get a heart specialist who had studied risk factors in CHD as an expert witness. I approached Professor Michael Oliver, an eminent cardiologist, who declined the invitation. 'I have already agreed,' he wrote, 'to act as an expert witness for Courtaulds and it is inappropriate, therefore, for me to advise [the widow's solicitors] or to recommend an alternate cardiologist.' The widow's solicitors advised her to withdraw the case before it reached the courts on account of escalating costs."[162]

Schilling's anger and frustration over the failure to officially recognize work-related heart disease in rayon spinners would only have been magnified had he been privy to the actual deliberations of the decision makers. On 17 March 1978, Britain's Industrial Injuries Advisory Council, Industrial Diseases Sub-Committee (IDSC, to its staff), met to be briefed on the subject of coronary heart disease and viscose work in order to determine whether it should be listed as a prescribed disease.[163] They had the benefit of a lengthy presentation by Dr. John Tiller, Morris and Schilling's former research collaborator. Tiller's presentation, illustrated with slides detailing his research, as well as other relevant studies, was objectively convincing but doomed to fail.

A key memorandum on the question had already gone out from Dr. W. R. Henwood, of the British Department of Health and Social Security, to Mr. A. J. Collins, serving with the Industrial Injuries Advisory Council–IDSC "secretariat." Well in advance of the meeting and ahead of any presentation of evidence, Henwood summarized what should be decided, come what may:

> Whilst one cannot foresee with any degree of exactitude what Dr. Tiller is likely to say when we meet him . . . it is probable I think that he will say—
> a. that carbon bisulphide is a toxic agent
> b. that subsequent studies by him following his paper of 1968 have not caused him to revise his original opinion . . .
> c. that studies in other countries support his views . . .
> Assuming that this is what he will say, then I feel that the Department's attitude in presenting the picture to the IDSC should be that the evidence does not support prescription as this is a common disease in the general population and a two to three times increase in incidence, especially when based on mortality statistics, comes nowhere near satisfying the requirements for prescription.[164]

Not surprisingly, the well-considered departmental view prevailed with the IDSC. Heart attack among rayon workers was not listed as a recognized occupational disease in Britain. Dr. Tiller's testimony, whatever its scientific rigor, had no appreciable impact. Nor did it matter that Tiller, whom Schilling described as a "very jolly decent sort," was not speaking to the panel as an outsider but rather as a public-sector employee of the same agency to which he was providing evidence. Tiller had taken a job as a staff member of Department of Health and Social Security some years before. In 1967, just as the heart disease study was being prepared for publication, and after nineteen years of employment with Courtaulds, John Tiller had been sacked.[165]

7

Rayon Will Be with Us

Viscose was a pacesetter at the start of the twentieth century, the first major synthetic-fiber success story. In the 1920s and 1930s, rayon led the way as the prototype of a multinational business enterprise, an early model of what would become the dominant modus operandi for large business entities after World War II. Then in the 1970s, just as the updated term "transnational" began to come into vogue, viscose once again was at the forefront of a new business trend. Spin-offs, shutdowns, and offshoring became standard operating practice in the rayon industry for the remainder of the twentieth century.

The prelude to this dismantling was a series of mergers and acquisitions that left the old viscose principals unrecognizable by name. More than that, many of them no longer manufactured rayon as their main product. The first to be transformed through corporate mergers were some of the major U.S. viscose manufacturers. In 1963, FMC (originally the Food Machinery Corporation) acquired the American Viscose Corporation, which formed the basis of a new fibers group operating within its larger conglomerated enterprises. An entity called the Midland-Ross Corporation had taken over the Industrial Rayon Corporation in 1961; the old Elizabethton rayon enterprise had morphed into Beaunit, and in 1967 it was taken over by the El Paso Natural Gas Company, becoming part of another "diversified" portfolio.[1]

European rayon followed suit. In 1968, SNIA merged with another Italian manufacturing group, BPD, and got into the chemical business and the defense industry. In that same year, Hoechst acquired what had originally been Süddeutsche Zellwolle; in 1969, the Dutch-German integrated ENKA-Vereinigte Glanzstoff (AKU since 1929) merged with a paint and chemical manufacturer to become Akzo.[2]

A cycle of plant shutdowns led off the 1970s. It set a pattern for reshufflings and closures that continued unabated for the remainder of the century. Enka as Akzo led the way, announcing in 1972 its intention to close multiple production sites. The viscose workers in the Netherlands did not go along with this plan, occupying Enka's rayon plant at Breda in a Dutch trade union first. This protest caused Akzo to back off its scheme, at least for a time.[3] The workers' anxieties during the uneasy truce that followed are captured in a photo-documentary series of the time, "Unemployment in the Netherlands." It includes an image of a lone worker sitting on a stool facing a row of autonomously operating machines in an Enka plant, juxtaposed with newspaper text discussing workers' fears of impending layoffs.[4] After a decent interval, Akzo did shut down the Breda plants, and this time even a hunger strike by workers did not stop it. Then, in a further step to get out of the cellulose fiber business altogether, Akzo sold off their remaining American Enka subsidiaries to the chemical giant BASF. In 1989, Akzo celebrated its twentieth anniversary by giving its remaining employees logo-branded sports jackets. Five years later Akzo merged with Nobel, the Swedish chemical conglomerate, to assume yet another new corporate identity as AkzoNobel.[5]

Dutch workers had been able to temporarily delay the inevitable, but there was not even a brief respite for the employees of Courtaulds. The blow fell first and hardest at Flint, in Wales, where there was the greatest concentration of Courtaulds viscose workers. In late October 1976, Courtaulds gave notice that its Castle works would close a few weeks after the New Year. It was abundantly clear that Courtaulds was likely over the coming years to eliminate its remaining operations in the area as well. Near midnight on 2 May 1985, with closure of the Greenfield plant announced and the axe about to fall on the Deeside mill too, MP Keith Raffan stood up in Parliament to deliver an impassioned plea for his "blitzed local economy," warning that Courtaulds was not only making its workers redundant but also was likely to walk away and leave behind a "burning toxic tip" (dump), as it had done at the Castle works.[6] As with MP Kelly's parliamentary speeches of more than half a century before, Raffan's words came to naught.

Today, nearly nothing physical remains of what had once been the bustling rayon factories of Flint. All the main buildings have been pulled down: there is virtually no way to know where anything stood unless you are guided by locals who remember what used to be there. There is a retail shopping strip where the Deeside factory had been, and "business parks" for small-scale enterprises at the other sites.

In 2011, I was lucky enough to be taken around Flint by Ken and Brian Davies, who had grown up in the area. Their father, Robert, had started work at the Greenfield Courtaulds facility in 1937. Except for five years of RAF service during World War II, until late 1956 he was a rayon process worker, including time in the churn room. He went on to another job for a couple years and then had his first heart attack one morning in 1958. He was fifty years old. Robert Davies did not live long enough to see the plant closings that were to come—he survived the heart attack, but later experienced a series of disabling strokes, succumbing to his illness in 1968, at age fifty-nine.[7]

Later in the same year as my visit to Flint, I arrived on a warm fall day in Wittenberge Elbe. I wanted to see what had become of the VEB Zellstoff- und Zellwollewerke Wittenberge, the postwar rebirth of Phrix in the former East Germany. In 2001, fifteen years after the shutdowns began in Flint, but even more abruptly, the factory was shuttered, along with almost all the rest of the manufacturing of the once-thriving industrial city. Wittenberge Elbe became something of a poster child for postunification economic blight.[8] Unlike what I saw in Flint, though, many of the original Phrix factory buildings from 1938 were still standing, and most were being repurposed, including one that housed the Alten Zellwolle restaurant and caterer.

I was shown around by a former Zellstoff- und Zellwollewerke worker named Helmut Worbs, along with a local historian and retired schoolteacher, Günter Rodegast. On a green lawn near the old administration building, a group of former workers had established an outdoor museum of selected equipment and parts from the defunct factory. Its centerpiece was a large metal viscose churn, looking like a cross between a cement mixer and a deformed, bulbous cannon. As we walked in the open space between the remaining buildings and the banks of the Elbe, the historian speculated on where the barracks of wartime slave laborers may have stood, no marker of them remaining. There is a de facto memorial to those times that, in its way, is nonetheless moving. Near the center of the factory grounds, a small manmade pond surrounded by trees is graced by a statue that clearly dates from the original Phrix establishment: a nude athlete cast in metal, standing on a pedestal. One can only surmise that at the time of the plant's founding, it was intended to embody the nobility of modern industrial manufacturing, epitomized by rayon.

I visited one other worker-run museum memorializing a defunct rayon operation, the former Svenska Rayon located in Varmland, north of Gothenburg on the west coast of Sweden. The story of Svenska Rayon reflects the hybrid nature of the mixed economic model under which the company operated for much of its existence. The business depended on public-sector underpinning, and direct governmental aid escalated during the 1970s, when rayon was faltering in Britain but was entirely protected in East Germany. Support for Svenska Rayon was driven by a perceived national-security need under the tenuous hypothetical that Sweden might somehow be cut off from imported fibers were its sole rayon producer to go out of business.[9] When governmental support eventually was pulled out, the company continued for a while after a private takeover, but then finally succumbed in 2004.

On the day of my visit, a colleague and I first stopped by to meet with Ragnar Magnusson, a retired Svenska Rayon worker and something of a writer in residence among the retirees, having produced not only a history of the factory but several works of fiction as well. He began working in the factory in 1950, eight years after the plant first opened. Even before starting, he had heard rumors that the working conditions were bad and that some workers had collapsed on the way home from the factory. He remembered that a female physician who began researching this problem was soon transferred. Ragnar reported that eye problems were so common that there was a rule about them: you showed your eye problems at six in the morning to the guard at the door, and if you were judged unfit for work, you could go back home. If someone passed out on the job and had to be pulled out to fresh air, which happened with some frequency, the victim was given milk to drink as an antidote. Ragnar was aware that the vapors had a depressive effect and that some workers "became brooding, committed suicide or attacked the family." Although over the years conditions became better, and air levels were later measured regularly, excess exposure continued (inserting zeroes into the data points brought down the averages to within targets, Ragnar recalled).[10]

We went with Ragnar to meet some of his former workmates and to see the small museum they had created in a room made available by the local community council.[11] There were samples of some of the products that had been made (both filament and staple) as well as photos and even signage from the plant, although nothing on the scale of the Wittenberge artifacts. One of my hosts was still convinced that things had been turning around economically and that the new owner had pulled the plug prematurely. Before we departed we all sat down to have cake and sing happy birthday to Ragnar, who was celebrating his eighty-eighth birthday that day. The rest of the group was not much younger.

In the years that spanned the closures in Flint and Wittenberge and Varmland, the rayon business across Europe and the United States changed radically. In a game of musical chairs, corporate owners and business plans kept changing, and each time the tune switched, another plant ceased operations and more workers were out of a job. Across its British operations, Courtaulds' textile business over those years sputtered and then failed utterly. What hadn't been shut down by 1998 was finally bought out by Enka-cum-Akzo-cum-AkzoNobel. This remnant of a once mighty synthetic textile empire was folded into AkzoNobel's newly constituted Acordis fibers group.

Ironically, the one piece of the cellulose fiber business acquired in the Acordis deal was the vestige of Courtaulds' attempt to reestablish a foothold in America, its rayon staple factory in Axis, Alabama. The plant, operating since 1952, had been augmented with a new production line in 1992. The new product coming out of Alabama was meant to be Courtaulds' great cellulose hope, a novel synthetic fiber called Tencel.

Tencel began to be developed in the mid-1980s as the focus of an effort that Courtaulds remarkably designated its "Genesis project."[12] This may or may not represent a subliminal homage to "Project Genesis," central to the plot of the then-recent film *Star Trek II: The Wrath of Khan* (1982). In the movie, the Genesis device is meant to rearrange matter so that a planet with a hostile environment could be made habitable. The folks in Courtaulds' research and development department meant to rearrange cellulose with a chemical alternative to carbon disulfide in hopes of making a product niche that was economically habitable. The substance at the heart of Courtaulds' Genesis project was capable of putting cellulose into solution, just as carbon disulfide did, but with an entirely different chemical structure: N-methyl morpholine-n-oxide, or NMMO for short.

Courtaulds had been innovating with NMMO for a while, but not at Axis, where for decades it was business as usual: making viscose with carbon disulfide. By the time of the plant's purchase by AkzoNobel in 1998, it had undergone close to twenty years of governmental occupational and safety health investigations. The National Institute for Occupational Safety and Health carried out it first preliminary walk-through at Axis in 1975 and then did a larger study in 1979. Air sampling documented peak values of carbon disulfide that were ten times as high as the OSHA standard (a standard that, even if met, NIOSH had determined was not sufficiently protective of human health). NIOSH did not recommend any specific intervention, temporizing instead with a call for further study of the question.[13]

In 1996, two years before the Axis factory passed into AkzoNobel's hands, Alabama's Department of Public Health reached out to NIOSH for a follow-up, asking for its help in addressing illness among the Courtaulds employees. At nearly the same time, an employee from the factory activated NIOSH's "Health Hazard Evaluation" system by formally requesting governmental action. The next month, NIOSH dispatched a physician and an industrial hygienist to Alabama. The team called in at the plant, made a brief assessment, and suggested, once again, further study. More than a year passed. Finally, NIOSH submitted to officials at Courtaulds an outline of its proposed study.[14]

Even though the process was moving along slowly, it was not necessarily out of line with Health Hazard Evaluation protocols. What followed next, however, was quite remarkable. Courtaulds fired back a negative response to NIOSH's proposal, appending to its own lengthy cover letter another twenty-seven pages of supplemental comments solicited from five contracted consultants. At first, NIOSH backed off. As part of its response to the critique from Courtaulds, NIOSH stated that its survey would be delayed because the physician originally assigned to the project was no longer with the agency.[15] By the end of 1997, NIOSH capitulated completely: there would be no study. NIOSH claimed that the workers knew too much about the hazards of carbon disulfide, thus tainting any possible results from a questionnaire survey. NIOSH did not state how such foreknowledge would change carbon disulfide air measurements or physical health defects objectively documented.[16]

Although all this transpired before the AkzoNobel takeover, the future owners were aware of NIOSH's interest in the Axis facility. Even before buying the plant, AkzoNobel had cosponsored, along with Courtaulds and other members of the Carbon Disulfide Panel of Chemical Manufacturers Association, a critique of NIOSH's exposure data from its original study at the plant in 1979. The panel's hired consultants reached the convenient conclusion it was not carbon disulfide, but rather a carbon disulfide–racial interaction, that explained any effects that had been seen.[17]

As part of its due diligence, AkzoNobel should have been cognizant of any legal liabilities that it was assuming by taking over the Axis operation. For one thing, there was a major personal injury suit winding its way through the courts. The medical details of the case had been published in the high-profile medical journal of the National Institute of Environmental Health Sciences, *Environmental Health Perspectives*.[18] This scientific paper documented the downhill course of a rayon worker who suffered from devastating neurological disease, from his first medical assessment at age sixty-two until his death seven years later. This man had spent more than two decades as a viscose rayon worker, much of it in the spinning room. Although the author of the paper discreetly does not name the employer, a follow-up letter to the editor of the journal cited the legal decision, which in the end had gone against the worker, a man named West Berry Becton. The carbon disulfide he was exposed to was purchased from a chemical company, which was the target of the suit. The factory that employed Becton, alluded to in the proceedings but not at direct legal risk, was none other than Courtaulds.[19]

Courtaulds was in this relatively protected position because workers' compensation statutes across the United States provide insulation from liability exposure: it is an insurance scheme that effectively precludes employees from suing their employers. A fixed schedule of benefits reigns in any potential award in the case of damages, and pain and suffering are not accounted for. Of particular importance, however, an employee is free to sue a "third party" outside the constraints of workers' compensation, for example, the supplier of a toxic substance (the seller of the carbon disulfide) that purportedly led to an illness. This was exactly what proved to be Johns Manville's asbestos Achilles' heel, since it was the product supplier, but not the employer, of most of those injured by the fiber.

It is a different matter entirely when it comes to the general public. Since carbon disulfide is not retained in rayon or cellophane by the time these products are sold as finished goods, consumer product liability has never been of particular concern to the viscose industry. Not so with environmental contamination. Two Alabama residents who lived adjacent to the Axis factory brought this message home to the corporation.

When Rosie's health began to decline in 1991, those treating her could not determine the cause. She was referred to a university center for further evaluation. While there, she improved, only to deteriorate when she went back home. In 1993, a similar condition began taking its toll on another local resident, Whisper. Blood tests taken from Rosie and Whisper as well as from other horses on Horace and Margaret Long's farm confirmed the presence of either carbon disulfide or its chemical breakdown product.[20]

In the court case that ensued, an Environmental Protection Agency's Toxic Release Inventory Fact Sheet for Alabama was put into evidence. It revealed that in 1991, Courtaulds released 42,454,520 pounds of carbon disulfide into the air around Axis, discharged 43,105 pounds of carbon disulfide into the "surface water," and disposed of 430,000 pounds of carbon disulfide on land. Remarkably, this pollution was not trending down over time, but actually had increased by nearly 50 percent from previous levels in the mid-1980s.

The Longs were initially awarded a judgment of one million dollars against Courtaulds, recompense for the lost value of their land and the huge veterinary bills that they had amassed. Even with the workers' compensation firewall, the environmental vulnerability was obvious. The surest solution would have been to recapture more of the carbon disulfide, thus controlling emissions. But that was an expensive prospect. The rayon baths continued to spin, as did corporate management. Nearly as soon as AkzoNobel took over and rebranded the operation as Acordis, even before the end of 1998, it was already touting a newfound commitment to clean living, the documentation for which was a brochure for community distribution. In a presentation to the EPA, Akzo proclaimed: "As part of its corporate culture, Acordis believes in taking leadership positions, including its positions on the environment. The Acordis facility in Mobile was the first local company to provide environmental brochures to the community, in which the company made commitments to future performance."[21]

There was insufficient time to establish whether substantive environmental actions would have followed. The following year, AkzoNobel decided to get out of the business, divesting Acordis to an entity called CVC Capital Partners. (This entity was the venture capital arm of Citicorp.) In the interim, Courtaulds had appealed its unfavorable horse-poisoning verdict. In September 2000, the Alabama State Supreme Court voided the award and ordered a new trial on technical grounds: the jury decided the case on the basis of a nuisance claim that should have been disallowed, because the factory's carbon disulfide pollution was already ongoing at the time the Longs moved onto their farm.[22]

Meanwhile, CVC was seeking to expand its fibers business, attempting in 2001 to purchase the still-independent and functioning rayon business at Lenzing in Austria. When the European Union forbade the sale, CVC reversed course, breaking up Acordis for good in 2004 and selling its Alabama facility to Lenzing. The sale gave the Austrians access to the U.S. Tencel unit as well as to Courtaulds' pilot operation for the fiber (now controlled by Acordis) in Grimsby, England. (Acordis had shut down standard viscose production there in 2001.)

Lenzing Fibers, as it turns out, was something of a standout success story during the overall decline in the European and American viscose business.[23] In fact, it was the last survivor when it came to rayon in the United States. In 1992 Lenzing had acquired the rayon staple operation in Lowland, Tennessee, that BASF had taken off Enka's hands but no longer wanted.[24] American Enka had begun construction of the factory on the banks of the Nolichucky River back in 1944. This was at the height of the war boom in rayon, at a time when Enka's parent company was under Nazi occupation in Holland, BASF was a linchpin in the I. G. Farben combine, and Lenzing was the jewel in the crown of the Thüringischen Zellwolle group.

For a while, Lenzing made a good run of it at Lowland, but in the end it had no more of a long-term interest in a stake in Tennessee rayon staple than had BASF or Enka. In 2002, Lenzing transferred ownership; the operation briefly emerged as Liberty Fibers before going under for good in 2005. Remnants of the classic 1940s design of the long-since-abandoned facility are still evident, in particular its sawtooth-clerestory roofline, standing alone as if dropped into Hamblen County in upper East Tennessee and then forgotten.[25]

Even though Lenzing eliminated the last vestiges of U.S. rayon staple, in Austria it continued with traditional viscose production and expanded further into Tencel manufacturing (under the product label Lyocell). I visited Lenzing in September 2012. On the way in from Vienna, just before coming to Lenzing the road passes through the nearby village of Pettighofen, where the factory's subcamp of Mauthausen had been sited. I arrived a bit early for my appointment, so I stopped in to get a bite at the restaurant-café Amigos situated just outside the factory gates, contemplating over lunch what I might see at the factory. The year before, in a North London assisted-living community, I had met with a survivor of the women's camp at Pettighofen. She too had been to see the modern Lenzing operation as an invited guest of a company sensitive to its wartime record. She remembered little that she could compare it to from her months as a slave laborer, emphasizing the impressive scale of the modern Lenzing operation.

The factory is indeed quite impressive. Up-to-date equipment facilitates all the traditional steps in viscose rayon production, from handling pulp to mixing and aging the viscose solution to spinning out the filament and then, for staple, gathering it into a tow and slicing the rayon into short fibers that are baled like cotton. Key at Lenzing are a system of modern enclosures and a ventilation system that combine to form an effective barrier between the workers and carbon disulfide. A sleek medical department staffed by occupational health specialists closely monitors urine samples for any indication of excess absorption of carbon disulfide by the employees.

Yet some things have not changed that much over time. The tallest building in the complex is a multistory tower for material handling that dates from Lenzing's original construction just before World War II. I climbed the tower with my tour guide, a former plant operative whose current duties include public relations. He proudly noted that their last suicide from that spot had occurred in the 1950s, which I took to mean suicide due to mental disturbance from carbon disulfide toxicity, given its causal association with derangement leading to suicide, well known since the nineteenth century. One other architectural feature struck me. Just as at Wittenberge, to the side of one of the old buildings were a small pond and, beside it, what was left of a small figurative statue in the classical style.

The American Viscose Company added to its glory a cellophane crown when it merged with Sylvania. FMC took over both parts of the business in 1963. One of the major rayon facilities it acquired was in Parkersburg, West Virginia, but when that plant stopped being lucrative, there was no reason to keep it going. One of the former workers there, a man named Roger Mackey, who now runs an antique and clock-repair shop in Parkersburg, described his work and the abrupt plant closure in these words: "I worked in the churn room for a while when I first got hired, they had a Fluid they mixed with crumbs they called CS2—It was some bad stuff. Someone said that if a tank they kept it in ever exploded it would blow up the biggest part of Parkersburg. I do not know if that is true but that is what they were saying. The old FMC plant I worked at closed down November the 14th 1974, the 15th they had the gates locked and was handing out checks through the fence."[26]

Not long after, in 1977, FMC sold off the rest of its rayon to a newly established entity, Avtex Fibers, Inc. It didn't take long for Avtex to shut down the former AVC rayon staple plant in Nitro, West Virginia. That closure came five years after NIOSH had visited the facility. Its report at the time was damning: seven out of eight workers who cut the staple were exposed to carbon disulfide in excess of OSHA's weak standard; remarkably, brief exposures to levels even one hundred times as high as that were documented.[27] This extreme overexposure took its toll. In *Muscle and Blood,* her book from 1974, the investigative reporter Rachel Scott told the stories of some of the rayon staple cutters at Nitro who were made ill. One was a man named Ronald Sayre. Soon after he started work at Nitro in 1969, he began to experience psychiatric symptoms: "I couldn't sleep. It was making me think weird things. I thought everybody was watching me. I remember one night, I punched out at ten o'clock, and I thought my father was outside, in the swamp, waiting for me. I went out and looked for him. He wasn't there."[28]

He was hospitalized not long after; three other workers were institutionalized at the same time with similar complaints. The company did not acknowledge that Sayre's psychotic break was due to carbon disulfide: his workers' compensation claim was denied.

After dispensing with Nitro, Avtex tried to keep going what had once been the flagship factory of American Viscose at Front Royal, Virginia. After a protracted, five-year fight with federal and state regulators over its record of pollution at the site, Avtex found a solution to that nagging problem as well: it declared bankruptcy.[29] The Avtex Fibers name has continued on, however, as the Avtex Fibers Superfund site at Front Royal, which includes carbon disulfide among a long list of contaminants. As the EPA describes it:

The contamination discovered at the Avtex Fibers site was of such magnitude and complexity that the area has been the subject of a number of removal, enforcement, and long-term cleanup actions. Tons of rayon manufacturing wastes and by-products, zinc hydroxide sludge, and fly ash and boiler room solids were disposed of on site in 23 impoundments and fill areas encompassing 220 acres. Waste disposal practices at the plant contaminated the groundwater under the site and in residential wells across the river from the site.... When the plant closed in 1989, the community was left to contend with severely contaminated land and water, the devastation of its manufacturing heritage, and the loss of approximately 1,000 jobs.[30]

In its transactions with Avtex, FMC retained its cellophane business; the Sylvania crown passed down from AVC, or what was soon to be left of it. Six months after the Avtex deal, at the beginning of 1977, FMC announced that it was shutting down cellophane production at Marcus Hook, Pennsylvania, where AVC first had started in rayon.[31] At the same time, FMC committed to expanding production at its remaining factory, in Fredericksburg, Virginia. A year and a month passed before FMC announced it was calling it quits at Fredericksburg too.[32]

FMC was not alone in the U.S. cellophane business. At the start of the 1950s, the Olin Corporation had made its foray into the sector, building off experience in the cellulose trade it had gained through munitions manufacturing. Olin's foothold in cellophane was its facility on the edge of the Pisgah National Forest near Brevard in Transylvania County, North Carolina. It was on the second floor of the Film Division, where the industrial equipment (called a barette in the trade) that churned the mix of cellulose and carbon disulfide was located, that the highest exposures to carbon disulfide occurred. Well into the 1970s, this was an ongoing problem. An Olin employee named Marvin Gaddy described the working conditions there: "Sometimes when we'd open those barettes, you get enough fumes to just about knock you out. We'd then take our scrapers and scrape out all that was stuck and there'd still be a lot of CS_2 in it. The company had given us testing machines to measure the fumes, but they would only go up to 50 parts per million."[33]

Another worker from the second floor, George Sanders, likely had even higher exposures. "He used to empty all these trash cans full of CS_2. Boy did he get a lot of fumes!" recalled a workmate. "I worked around him the week before he died and you could definitely tell that he was in a strain. He was awful bad depressed. He wouldn't say nothing to no one. His wife was pregnant at the time. He died of a shot gun wound one Saturday night. Everybody said it was just an accident."[34]

A team from NIOSH went to North Carolina to interview Olin's workers in 1973, but its conclusions were equivocal at best: "There is no doubt that occasional acute exposures to CS_2 have occurred episodically and that these exposures have provoked the expected medical symptoms. These exposures have not been frequent. There does not appear to be sufficient medical evidence at this time to warrant a conclusion that chronic exposure is occurring in a sufficient degree to provoke illness."[35]

The ten-page NIOSH report makes no mention of any suicides or "accidental" deaths such as that of George Sanders. Olin shut down its operation in 2002. Not long after, the EPA became involved because of concerns about environmental contamination. The concern was not misplaced. In 2008, a redevelopment company purchased the Olin property, and in May of that year a half-mile-long fish kill in the adjoining river resulted from the release of old waste from the abandoned factory. All the plant buildings have since been torn down, but according to its current EPA "Superfund Profile," the site remains unused.[36]

American Viscose/FMC was always in second place in cellophane, and Olin got into the game late. It was for DuPont that cellophane traditionally had been the big moneymaker. But over time, even it abandoned the business. DuPont's cellophane exodus started as far back as 1964, when it shut down its plant in Old Hickory, Tennessee.[37] Its cellophane factory in Richmond, Virginia, came next, in 1976.[38] Then DuPont converted (and downsized) its cellophane operation in Davenport, Iowa, to one making a noncellulose shrink film. Finally, in 1986 it sold the remnant of its cellophane business, DuPont's old factory in Covington, Kentucky, and a state-of-the-art factory in Tecumseh, Kansas.[39]

The assets went to an Atlanta corporation called Flexel. A decade later, Flexel went belly up. Covington was finished for good, but the Tecumseh plant was bought out of bankruptcy by a Brussels-based corporation, the UCB group, that had already gobbled up all the cellophane makers in Britain.[40] In fact, it had capped off its transparent-film buying spree with the purchase of British Cellophane, along with the rights to that company's coveted trade name. Eventually, all this morphed into something called Innovia Films, Ltd., which took on a refound British corporate identity for its international chain of factories. This included Tecumseh, by that point the only remaining facility manufacturing classic cellophane in the United States.[41]

Cellophane may have retreated to Tecumseh, but viscose film was on a forward march in the United States. Cellulose film for processed-meat casings, cellophane by another name, has occupied a strong market niche because it is an ideal product for industrially prepared sausages. The casing serves as a cooking mold that is peeled off by the manufacturer before sale to the ultimate consumer. The manufacture of cellulose casings may be a source of worker exposure to carbon disulfide, but it makes the skinless weenie possible. The U.S. standard-bearer in this industry has been Viskase, founded as Visking back in 1925, adjacent to the Chicago stockyards. Today it has a multinational presence as a manufacturer of synthetic polymers as well as viscose-based casing products. The latter is still a mainstay, especially Viskase's trademarked Nojax brand.[42]

The casings trade was lucrative enough to engender competition. A new corporate identity, Viscofan, emerged in 1975. Also based in Illinois, it too evolved into a major multinational sausage-casing enterprise. In 2006 it gobbled up Teepak, the old Danville casings maker that NIOSH once had studied. This was only one of multiple acquisitions by Viscofan. In addition to cellulose, Viscofan makes casings out of collagen and plastic and has manufacturing plants on four continents.[43] ▬

The trajectories of rayon, cellophane, and cellulose casings tell much of the U.S. viscose story, but not all of it. In the late 1940s, three former DuPont scientists hit upon, and patented, a manufacturing process through which myriad small air pockets could be introduced into the carbon disulfide–cellulose regeneration process. The building blocks of this novel method were oxygen, cellulose, and, for good measure, oxygen again. O-Cel-O was born. It was an immediate hit, a success "tied up with the attractive, pastel-colored cellulose sponges which have been so popular with housewives (and their car-washing husbands) ever since World War II."[44] O-Cel-O also benefited from a timely natural sponge blight in the Caribbean and the aftermath of wartime shortages for Mediterranean supplies. In 1952, the founders sold out to General Mills for a hefty $3.4 million (more than $30 million today, inflation adjusted). The company changed hands again when 3M purchased the Tonawanda plant outside Buffalo, New York, in 1990. In 2006, 3M went on to acquire its only major U.S.-based sponge competitor, the Nylonge Company of Elyria, Ohio (formerly the Sponge Company of Cleveland).[45]

O-Cel-O's real competitor, however, has been in Europe, where Spontex, founded in France in 1932, predates its American rival.[46] Spontex has a leg up in another way too: it has an adorable mascot. The Spontex hedgehog has gained fame worldwide, including in the United States. One of the best-known television spots with the mascot, released in 2000, featured the creature humping a Spontex sponge in a manner evoking the scandalous choreographic climax to Nijinsky's ballet "L'après-midi d'un Faune."[47]

However cute Ernie the Hedgehog may be, viscose sponge making may be every bit as efficient a source of carbon disulfide exposure to its workers as rayon or cellophane. A 1984 study in a sponge factory in France, the only one of its kind in the open scientific literature, found short-duration exposures to carbon disulfide to be as high as 100 ppm.[48] Inhalation of fumes at this high level certainly would be dangerous for any prolonged work, even if not a heavy enough dose to induce the hypersexual behavior that Delpech had chronicled among the poisoned rubber workers of Paris more than a century before.

As rayon went offline in much of Europe and North America in the last decades of the twentieth century, the jobs lost in that sector of the viscose industry might have been matched by a parallel decrease in the number of manufacturing workers exposed to carbon disulfide. This was not the case. There still was money to be made in viscose rayon, simply not in Flint or Coventry or Nitro or Front Royal.

Japan really set the business benchmark for offshoring rayon back in 1964, when it began exporting its viscose industry to South Korea—lock, stock, and leaky barrel. Underwritten with Korean governmental loans, old equipment from the Toray Company, formerly Toyo Rayon, was transferred to the Heunghan Synthetic Fiber Company, which went on to become the Wonjin Rayon Company. The lessons that easily could have been learned from Japan's experience with carbon disulfide poisoning in previous decades were conveniently ignored, most importantly the need to adequately enclose the machinery and use appropriate ventilation to draw away the toxic fumes.[49]

The first case at Wonjin was noted in 1981, although its cause was misattributed to sulfur dioxide, a chemical not present in the work environment. Up until that time, it seems, Korean doctors had never heard about the hazards of carbon disulfide. This changed in no small part because of the activist role of a key clinician involved in the workers' care, Dr. Rok Ho Kim.[50]

Wonjin closed in 1993, but the chronic, long-lasting effects of carbon disulfide poisoning did not cease. By 1997, there were 600 known cases of rayon-related disease among the workers of the former factory; by 2004, over 900 in total had received financial compensation. The Wonjin Foundation for Victims of Occupational Diseases was founded; it established the Wonjin Green Hospital (in Kuri City, South Korea) both to treat the many patients who needed care and to disseminate new medical findings.[51]

The Wonjin tragedy yielded one document that is unique in the long history of carbon disulfide. Kyung-Yong Kwon was hired at Wonjin in 1977, initially became ill in 1985 (including features of psychosis), and in 1988 was pensioned out on full disability. In 1991, Kyung-Yong Kwon committed suicide. What is remarkable is that his suicide note was made public. Writing to his son, Uoon Chun, Kyung-Yong urged the family to continue to collect his disability payment (the equivalent of just over $200 a month) and, in order to do so, not to report his death until he would have been ninety years old: "It is okay. I am not shameful about it because I acquired illness from working at Wonjin Rayon and die from the illness after suffering for many years. Fight with Wonjin and fight with the Department of Labor. Keep up the fight."[52]

The Wonjin factory left another legacy. Its old equipment changed hands again, destination this time: the People's Republic of China. The PRC was the next to join the Asian rayon-manufacturing boom, and in a big way. One perspective on the changing geography of the rayon business comes from the vantage point of Ing. A. Maurer, SA. This industrial engineering firm, based in Berne, Switzerland, provides industrial design expertise to rayon and cellophane manufacturers. Maurer maintains a detailed list of its clients, dating back to 1931.[53] Maurer has done work around the world, servicing, at one time or another, Enka, Courtaulds, Glanzstoff, Lenzing, Mitsubishi, and Svenska Rayon. In the mid-1960s, it became involved in setting up the Formosa Chemicals & Fibre Corporation, which has remained an active rayon staple producer in Taiwan.[54] Since 2000, though, much of Maurer's business has been in China, working with a series of entities whose names come less trippingly off the tongue: Sateri (Fujan) Fiber Company; Longda (JianXi) Differential Fibre Company; Shandong Yamei Sci-tech; and the Weifang Henglian group (a cellophane manufacturer).

International production figures bear this out. Viscose manufacturing capacity in China quadrupled in the first decade of the twenty-first century and now accounts for approximately 60 percent of worldwide production.[55] Much of the equipment used in China may be modern, but that does not guarantee safety. That depends on how it is used and maintained. Thus, the health status of the Chinese viscose-making labor force may be compromised, but the true extent of the problem can only be guessed at. The smattering of medical reports that have appeared indicate that carbon disulfide exposures routinely exceed officially set limits. For example, a 2012 study from the Occupational Disease Prevention Hospital in Nanjing, China, found that more than 10 percent of the rayon workers examined complained of numbness, and a similar proportion showed abnormalities during electrical nerve testing. So much for disease prevention.[56]

Of the conditions inside viscose factories elsewhere in Asia, even less is known. India, Indonesia, and Thailand have become major rayon-manufacturing hubs; in the aggregate, not as large as China, but nonetheless accounting for one quarter of international production.[57] Aditya Birla Group, based in India, is emblematic of today's viscose industry. Birla has powerful rayon staple (Grasim) and rayon fiber (Aditya Birla Nuvo Limited) manufacturing arms; its reach further extends to subsidiaries in China, Indonesia, and Thailand; and to guarantee a supply of raw materials, its holdings encompass joint ventures in Canadian and Swedish pulp, and development is in the works for a new pulp plantation in Laos.[58]

There are virtually no published biomedical data on the health of these South Asian and East Asian workers, including those employed by the Birla group. But there was one fascinating report from the world of postcolonial studies: a paper by the anthropologist Christopher Pinney, ponderously titled "On Living in the Kal(i)yug: Notes from Nagda, Madhya Pradesh."[59] Nagda is an industrial town where Grasim has a major rayon staple plant. *Kaliyug*, the author explains, is our current epoch, or *yug*, the final one of a repeated cycle, marked by the moral decay of the modern, machine age. Thus, the rayon plant is important not for its very real hazards, but rather for its symbolism:

> Pollution and health and safety issues in and around the factory have been key concerns for several decades and this, combined with the continuous shift system and division of labour, embodies industry as the apparently (negative) antithesis of the rural. This is certainly the perspective of local high-caste village employers who articulate a very negative view of the factory, seeking to project it as part of the degenerate kaliyug which is associated with machinery, the goddess Kali, and a dangerous and unstable modernity. Local village-resident factory workers, however, value the comparatively high industrial wages, shorter working hours, and their liberation from the oppressive expectations of rural "patronage." The complex everyday predicaments of living in the kaliyug are explored through a variety of different voices which suggest the inadequacy of trans-local narratives of industrialisation.[60]

Pinney reports that air levels of carbon disulfide in the Nagda churn room had been documented at one point to be twelve times as high as the Indian legal limit. If that was the case, the exposure in the spinning room, which typically far exceeds that around the enclosed churns, must have been off the charts. The overexposure is not surprising: at Nagda, Gasim was running old, hand-me-down British equipment, just as the Koreans (and later the Chinese) had done with Japanese factory machinery. Pinney also comments that among the complaints linked with the facility by the local populace were paralysis, heart attack, mental illness, and, prominently, impotence. This last problem is a classic manifestation of carbon disulfide toxicity. But according to Pinney, impotence is just symbolic. Taking a page from Professor Charcot's work on hysteria and then deconstructing it with a Derridean flourish, Pinney interprets this illness as not so much a real condition as a representation of a projected concern of "higher-castes' anxieties which they choose to map on the bodies of factory workers."[61]

Whatever attention has been given to rayon in the region, there has never been an exposé of Asian viscose manufacturing on a par with the scrutiny given to microelectronics in *Nightline*'s "Trip to the iFactory" segment in 2012, which reported on the suicide nets placed by Foxconn to catch the assembly-line workers driven by despair to jump from the factory roof.[62] Microelectronics manufacturing has its share of toxic chemicals, but carbon disulfide is not among them. All the same, there is a resonance between Foxconn in our own time and the defenestration of crazed vulcanization workers in the nineteenth century. Indeed, the legacy of vulcanization lives on in other ways, too. Long since phased out as a direct agent of vulcanization, carbon disulfide nonetheless has persisted in the rubber industry. This is because the chemical is used to make the specialty accelerator chemicals that replaced carbon disulfide in its pure form.

Rubber industry workers synthesizing accelerators from carbon disulfide continue to be threatened. In 2014, a scientific study of an otherwise unnamed "rubber and plastic manufacturing plant in New York State" was released.[63] Compared to other employees of the factory, shift workers exposed to carbon disulfide while making rubber-vulcanizing accelerators were afflicted by more than double the expected rate of death from heart disease. The factory in question has been the subject of many medical investigations over the years because of the multiple toxic chemicals used there. In 2013, for example, an investigative report from the Center for Public Integrity traced the history of a bladder cancer outbreak there due to a chemical called ortho-toluidine. That report did not refrain from naming the facility, the Goodyear Tire and Rubber in Niagara Falls, even noting that its workers have a nickname for the place: they call it the "ginch."[64] This term of endearment may or may not be related to slang originating from Canada meaning men's briefs.

When it was discovered that certain substances used to make rubber accelerators not only promoted vulcanization but also interfered with the human body's capacity to fully break down alcohol, an entirely new carbon disulfide problem emerged.[65] Although a perturbation in the metabolism of alcohol might seem like a clever way to extend a binge, the intermediate chemical that builds up not only is not very nice (good-case scenario: severe vomiting), but can even lead to a life-threatening fall in blood pressure. This very adverse reaction is the rationale for deliberately administering this chemical to alcoholics, generically known as disulfiram and marketed by prescription as Antabuse.

Over and above its acute intended reaction, there are some peculiar "side effects" of long-term Antabuse use. A prominent one is damage to the distal nerves of sensation; questions also have been raised about brain effects and even about an increased risk of atherosclerosis.[66] It has been argued that these problems merely reflect residual damage from long-standing heavy alcohol intake, the wages of past abuse. There is a catch, though. Antabuse itself is broken down like any other medication, metabolized on the way to its excretion. The salient issue for disulfiram is that in the human body, it breaks down to carbon disulfide. The metabolic fate of disulfiram is so predictable, in fact, that it has formed the basis of a simple breath test designed to detect carbon disulfide exhaled by those taking the medication, in order to better gauge adherence to the drug regimen.[67] The commercially marketed testing device that does this is named the Zenalyser, an unintended invocation of the "Moment of Zen" segment that closed the satirical television program *The Daily Show*, often highlighting the intersection of the absurd and the tragic. Disulfiram, as it turns out, may be an effective carbon disulfide delivery device, but recent evidence shows that it may not work that well to stop alcohol abusers from drinking.[68] Nonetheless, disulfiram continues to be approved for human use, although carbon disulfide was itself removed from the medical pharmacopeia long ago, and even its limited use in veterinary practice (primarily to treat parasites in horses) fell out of favor precisely because of its toxicity.[69]

Carbon disulfide has made another comeback as well, not through pharmaceuticals but via agribusiness. True, the phylloxera epidemic is long gone, and even though carbon disulfide held sway as the gopher killer of choice for many years, that use too eventually fell away. Last to go was the application of carbon disulfide as a grain storage fumigant. One of the preferred carbon disulfide grain treatments mixed it in a 1:4 ratio with the liver toxicant carbon tetrachloride, a product commonly known as "80/20."

This fumigant cocktail posed a real danger. In the 1980s, a research group at the University of Wisconsin put out a call for 80/20-exposed grain workers with neurological symptoms to come in for evaluations. In the end, twenty-one affected persons were recruited. Those that had worked in grain elevators had experienced some of the worst exposures. Case 5 in the series, for example, was a sixty-three-year-old grain elevator worker who had retired after twenty-five years of labor, the first nine years spent doing active fumigation with 80/20 on up to ten railroad cars at a time: "While he would stand on the grain surface inside the car, he would reach over a temporary wall and grasp a 19-L (5-gal) pail of undiluted, liquid fumigant that was passed up to him by another worker on the ground. He would then wade through the grain to one end of the car and dump the pail across the width of the car. After retrieving the second pail, he would repeat the procedure at the other end and then 'scramble over the wall and out of the boxcar.' "[70]

On medical examination, case 5 showed classic findings of Parkinsonism as well as other evidence of peripheral nerve damage. In comparison, case 11 was a little better off in the extent of his neurological disease, but then again, he was younger. A twenty-nine-year-old, he had replaced the previous worker when he retired; the younger man benefited from a technical innovation in fumigant application: he used a spray hose rather than carrying a pail. It was not until 1985 that 80/20 finally was pulled off the U.S. market. This was after the EPA required new company tests—not over concerns about workers' health, but out of fear that chemical residues in consumer products might be too high. Margie Williams, a spokeswoman at the time for the National Association of Wheat Growers, voiced her clients' displeasure at losing a popular product: "[It is] easy to use by a farmer. . . . He can get a good bug kill." [71]

Exit the direct use of carbon disulfide for grapes, gophers, and grains—enter metam sodium (aka metham or carbam, depending on the manufacturer or seller). Metam sodium is a chemical pesticide fumigant used to treat soil to eliminate a variety of pests of a number of different crops. Just as carbon disulfide was finally being eliminated in the 1980s, metam sodium came quietly into commerce. Then, one night in 1991, the derailment of a train that included the fumigant in its cargo dispersed a toxic fog that settled over the small town of Dunsmuir, California, and liquid metam sodium percolated into the adjoining Sacramento River. Virtually all plant, fish, and insect life along a forty-mile stretch of the river was killed. Humans along the banks suffered from intense irritation of the respiratory tract, and some went on to suffer from persistent asthma.[72]

As it turns out, once released into the environment, metam sodium quickly breaks down to form several new toxic materials. One of these breakdown products, a strong chemical irritant, appears to account for the asthma outbreak at Dunsmuir. It is called methyl isothiocyanate, a variant of the far more lethal poison methyl isocyanate (the release of MIC from a Union Carbide chemical plant in Bhopal, India, in 1984 killed thousands). But there is another important breakdown by-product of metam sodium, in fact the chemical precursor from which the pesticide is synthesized in the first place—carbon disulfide.

Metam sodium has many potential uses, which means that there are many scenarios for associated carbon disulfide contamination. The long list of EPA-approved target crops for soil pretreatment with the agent runs the gamut from alfalfa and amaranth to yams and zucchini.[73] And whenever metam sodium comes in contact with the open environment, its breakdown products are inevitably released—the mix seems to depend on specific conditions. When a farmer in France, for example, disposed of some unneeded metam sodium by putting it down a residential sewer, multiple residents on the same street suddenly experienced a foul odor coming from their toilets. The air measured above the sewer manhole detected more than 100 ppm carbon disulfide.[74]

As if the metam sodium story wasn't enough, only a few years ago the EPA decided to pull the plug on another soil fumigant that didn't break down just *some of the time* to yield carbon disulfide; it was designed to turn into the toxic chemical 100 percent of the time. Technically called sodium tetrathiocarbonate, this carbon disulfide delivery agent went by the catchy commercial name Enzone. Originally a DuPont product, in 2004 the manufacturing rights were taken over by a pesticide company based in Japan that was looking to expand its portfolio.[75] That company, Arysta LifeSciences, was attempting to ramp up its soil fumigant business in a big way. Along with Enzone, it was also behind a product named Midas. The fumigant with the golden touch was a cancer-causing, neurotoxic chemical called methyl iodide. In 2011, the EPA responded to the "voluntary" withdrawal of sodium tetrathiocarbonate by Arysta with the formal cessation of its approved use.[76] This had included application on grapes to treat none other than (blast from the past) phylloxera. In 2012, Arysta took methyl iodide off the U.S. market, just before it was about to lose an embarrassing legal battle in California over the manner in which it had gained approval by state regulators.[77]

These have been only minor setbacks for the company. Its market is still wide open internationally. In fact, Arysta LifeSciences' global presence was strengthened with its 2015 purchase by Platform Specialty Products Corporation, linking it with two other major players in the agricultural chemical field (Agriphar Group and Chemtura) to create an "integrated agricultural chemicals segment" with targeted annual sales of $2.1 billion.[78]

Carbon disulfide and all the products and applications connected with it have had their ups and downs over the years. Despite that inconstancy, viscose rayon has demonstrated considerable staying power as a cultural icon, if nothing else. Indeed, that somewhat tawdry *kunstseidene Mädchen* flashiness once connected with rayon seems to have persisted, albeit overlaid with an element of retro playfulness. Or as the Miami Standard Hotel hostess and lifelong local resident Brittany Marissa was quoted in a *Hemispheres Magazine* photo spread in 2012: "Miami used to be all about rayon and tight dresses, very clubby. Now it's a bit hipper. I feel it's laissez-faire—anything goes."[79]

Brittany seems to have a fictional soul mate in Elle Woods, the heroine of *Legally Blonde*. When a saleswoman at a boutique underestimates her ("There's nothing I love more than a dumb blonde with daddy's plastic") and tries to foist on Elle last year's model as current, she asks, "Is that low-viscosity rayon? With a half-loop top stitching on the hem?" The saleswoman attempts to bluff her, "Of course. It's one of a kind." To which Elle replies, "It's impossible to use a half-loop stitching on low-viscosity rayon. It would snag the fabric. And you didn't just get it in—I saw it in the June *Vogue* a year ago. So if you're trying to sell it to me for full price, you've picked the wrong girl."[80]

Even if rayon as a cultural reference now can be glossed as hip in a feminine context, viscose rayon still retains an old connotation of cheesiness for men, epitomized by Kent in *Animal House*, the would-be legacy pledge who wears a 90 percent rayon clip-on tie his mother bought for him. This same theme was carried through in a 2004 Stephen Colbert bit on Pervez Musharraf's pardon of a physicist who had traded in nuclear secrets; the comedian equated "atomic bombs from Pakistan" with "rayon leisure suits from Guatemala," both being part and parcel of the same global economy.[81] And for men at least, the sinister element that infused the Mussolini-period *Agent in Italy* was retained by Dexter Morgan, the fictional forensics-lab-worker-cum-serial-killer with a tendency to wear rayon: "The crime-lab gang didn't wear suits. Rayon bowling shirts with two pockets was more their speed. I was wearing one myself. It repeated a pattern of voodoo drummers and palm trees against a lime green background. Stylish, but practical."[82]

The popular-culture gender gap for rayon is most explicit in a *Law and Order* episode in which assistant district attorney Jamie Ross is subjected to the sexist musings of an aging Judge Marks. She exploits his Neanderthal tendencies by wearing an alluring silk blouse to obtain a favorable ruling. When her boss, Jack McCoy, asks whether she would model silk again for Marks if the need arose, Jamie replies, "Trust me, he would have been just as big a fan of rayon."[83]

Cellophane also has retained a symbolic presence. In John Updike's pivotal *Rabbit, Run*, cellophane makes multiple appearances, including as a covering for laundered shirts, cigarettes, and hotdogs. Harry "Rabbit" Angstrom balances at the fulcrum of the sixties, when cellophane still retained the power to make the everyday object seem just a little bit special. But cellophane was about to enter a new age, going from moderne to postmodern and even taking on a bit of a psychedelic tint. In the same year that Updike's novel appeared (1960), Frank O'Hara's poem "Second Avenue" was published (although written earlier), spinning the image of "tabletops of Vienna carrying their bundles of cellophane to the laundry."[84]

By the time the 1960s were in full swing, cellophane had gone psychedelic, implicitly in the polychromatic flowers of "Lucy in the Sky with Diamonds" and explicitly in "Cellophane Symphony" by Tommy James and the Shondells. Meanwhile, the old political imagery of cellophane, first popularized by Postmaster Farley in the 1930s, still had legs, transubstantiated into "Mr. Cellophane" from the musical *Chicago*, but then applied metaphorically to any politician without substance or to too much governmental transparency.[85] Cellophane has even been political fodder for Sarah Palin, who, in eat-what-you-kill mode, noted on her self-hosted television reality show, "Our meals happen to be wrapped in fur, not cellophane."[86]

Even though cellophane and viscose rayon are produced by the same chemical process, both entirely dependent on carbon disulfide, this enabling poison remains virtually unknown by the general public. Carbon disulfide does not stand as an icon nor serve as any kind of popular-culture metaphor. The chemical makes a cameo appearance in the technical details provided to underpin an invented Soviet viscose plant called Solkemfib in a postmodern novel from 2010 called *Red Plenty*.[87] But the closest carbon disulfide ever came to a leading role was in the soap opera *General Hospital*. Lord Larry Ashton, it seems, was scheming to corner the market on carbon disulfide, conveniently produced at a Port Charles cannery. The nefarious plot was eventually thwarted by the Port Charles Police Department, and the dangerous carbon disulfide stockpile was destroyed.[88]

Carbon disulfide may have been removed entirely from the mythical Port Charles, but the chemical is still very much a part of an ongoing and indeed expanding viscose-manufacturing industry. Worldwide, rayon production doubled in the twenty years from 1990 to 2010, led by Asia but with an important European presence.[89] Novel, high-tech applications of specialized forms of rayon have filled new market niches, for example, Outlast fiber, which has special thermal properties, was first developed for NASA, or as the promotional materials read, "Tested in Space, and Right Here on Earth."[90] Indeed, NASA has been no stranger to high-tech rayon. Viscose fiber was the basis of a carbonized derivative that was essential to insulating the nozzles of reusable solid rocket motors. The rayon for this formerly came from the North American Rayon Corporation (NARC). When NARC went out of business, NASA stockpiled its product and then embarked on a "NARC Rayon Replacement Program." NASA settled on Avtex; the *Challenger* disaster in 1986 and the consequent interruption in this lucrative product line was cited as one of the drivers in Avtex's failure in 1988.[91] Proving that even the modest sponge can be cutting-edge too, biotech has been exploring a viscose application, this one based on a specialized viscose cellulose material used to cover wounds.[92]

Another relatively new application for viscose is to turn rayon fibers into very finely chopped-up pieces, smaller than those in staple. As it turns out, this process brings with it its own novel hazard. In 2007, respiratory problems were linked with work in a factory using this form of fiber, news that was particularly disturbing because rayon had never been known to cause lung disease. In the trade, such finely cut fibers are known as flock. A major use of flock is the coating of cloth or paper to produce a faux-velvet surface: the greeting card industry is a big market for flock-treated paper. Flock can be made from any number of synthetics, including rayon, nylon, and polyethylene. The workers at greatest risk from rayon flock were the ones who used air hoses to blow away the flock waste that collected in the machinery every day. It is small comfort that this potentially debilitating condition is not specific to rayon—nylon and polyethylene can cause this problem too. When the human respiratory system is the machinery that gets clogged, the disease is simply known as "flock worker's lung." The medical report on rayon-caused flock worker's lung shied away from identifying the factory where this outbreak occurred, but a governmental report pinpoints it as a Hallmark facility in Lawrence, Kansas.[93]

Technological innovation is only one newfound selling point for viscose. In a real tour de force of corporate chutzpah, today rayon is marketed as an eco-friendly, nearly green product. Austria's Lenzing, for example, trumpets the ecological aspects of its trademarked Modal line of viscose. The wood pulp for Modal comes from renewable beech trees, and Lenzing proudly displays a Programme for the Endorsement of Forest Certification logo in its promotional materials. Lenzing touts its manufacturing system as being so environmentally responsible that the company even has named the process "Edelweiss technology," a carbon dioxide–neutral production method "as pure as the edelweiss flower," using "renewable and natural raw material" (that would be the wood pulp, not the carbon disulfide).[94] This may indeed be a relatively well-controlled production process that minimizes worker and environmental exposure, but at the same time, Edelweiss invokes the eco-shopping-driven cha-ching of cash registers, which must be the sound of music to Lenzing's ears.

Unlike Model, the other major Lenzing product, Lyocell (Tencel), is made by a process free of carbon disulfide, but it is not an entirely natural product either. Just as the word "synthetic" is avoided in its marketing, Lyocell's key synthetic solvent component, NMMO, may have a clean bill of health so far, but that is only a default status. It has been so little studied to date that its potential toxicity may become evident only with later investigation.[95] Lenzing also has made a cooperative agreement with a German-based company, Smartfiber AG, that allows it to modify Lyocell. Modify how, you might ask? According to the Smartfiber website: "You will be surprised to learn that textiles can include the best of seaweed, care for your skin, and protect your body; all while also feeling incredibly soft."[96] Along with this promotional copy, the web page includes the image of a female figure sitting in a yoga position on a beach at dusk (she may be naked, but her back is to the viewer and the photograph was taken a bit too far past sunset to be sure). The message: by the addition of pulverized seaweed, SeaCell rayon not only is a boon for the ecosystem, but is even healthy for you! It is enough to make an edelweiss blush.

Cuprammonium rayon—the old Bemberg silk—also does not use carbon disulfide. The industry leader in its manufacture is the fiber division of the Japanese conglomerate Asahi-Kasei. The manufacturing of Bemberg, also marketed as "cupro," may be free of carbon disulfide, but it nonetheless is a chemically intensive process. That does not stop Asahi-Kasei from promoting it, too, as eco-friendly, emphasizing that Bemberg uses cotton feedstock as its source of cellulose, and omitting any mention of other chemicals.[97] And Bemberg is perceived to be anything but the cheesy viscose of a menswear bowling shirt or cheap necktie. Indeed, it has made something of a presence for itself as the ne plus ultra of suit linings, preferably bespoke. Thus, the *New York Times* described one 2004 creation, available for only a mere $2,750, as a "flawlessly constructed cascade of inside pockets on a Bemberg-lined Oxxford suit."[98]

Bamboo-derived pulp has been used as another pitch to appeal to the eco-friendly shopper. But when marketers leave out the fact that the "green" bamboo-based fiber they are pushing still is rayon, they cross a legal line that the U.S. Federal Trade Commission (FTC), in a crackdown on such false product claims, labeled "bamboo-zling."[99] At least one eco-marketing website hit back at the FTC, implying that that agency's motivation was to protect government-subsidized cotton, which is a far less environmentally friendly crop. Carbon disulfide is never named, although Greener Ideal does acknowledge that "unfortunately, in order to produce fabric that is at all soft from either bamboo or cotton, chemicals (and some harsh chemicals at that) are required."[100]

Nor does cellophane wish to be left out in the eco-unfriendly cold. The industry leader, Innovia Films, continues to make cellophane in plants at Tecumseh, Kansas, and Wigton, United Kingdom (still insisting on putting a ™

on the long-since-generic product name). But it also has introduced a new product line, NatureFlex. Innovia Films promotes this packaging material for its biodegradability (organic food sellers have signed on to its use). In July 2015, the company took this one step further, announcing the same endorsement received by Lenzing: "Innovia Films now uses only FSC (Forest Stewardship Council) and PEFC (Programme for the Endorsement of Forest Certification) certified wood pulp in the manufacture of its NatureFlex product."[101]

Elsewhere in its promotional materials, Innovia acknowledges that NatureFlex, just like traditional cellophane, is manufactured through the viscose process. Any mention of carbon disulfide, however, is discreetly omitted. Unsurprisingly, Innovia did not announce with similar fanfare that OSHA had slapped it with fines totaling $112,500 in January 2014 for a series of serious violations at Tecumseh. Many of these involved the plant's handling of carbon disulfide, although it is unclear whether inspectors took any air measurements. OSHA seemed more worried that workers might fall off a carbon disulfide tank ("Employee(s) connecting a metallic transfer hose on top of the Carbon Disulfide tanker were exposed to a fall hazard of approximately 12 feet due to no fall protection system being provided") or get caught in a solvent-caused conflagration ("Employee(s) were exposed to release of extremely flammable vapors to atmosphere that historically result in fire/explosion hazards leading to injury and death to persons in the workplace").[102]

Even the old viscose-based bottle-capping product known in the trade as a viskring in Britain and a cellon in the United States has reinvented itself as a protector both of the consumer and of the environment, asking its potential customers, "Do you have problems with tamper evidence or counterfeit issues? Are you an environmentally conscious producer looking for recyclable packaging?"[103]

More than anything else, the key to success for any "rayon is green" or "cellophane is biodegradable" public relations campaign is the self-focus that is axiomatic to the consumerist equation. As long as the final product is safe, why does it matter that there may be a hazard in manufacturing it? The very first question I am asked, invariably, whenever I talk about carbon disulfide and viscose is "Is rayon safe to wear?" Interestingly, an atoms-for-peace promoter in the later 1940s suggested adding radioactive sulfur to carbon disulfide, and thus any residual chemical in the manufactured rayon could be traced with a Geiger counter.[104] Not to worry. Any carbon disulfide that might have been in the fiber would have long since vaporized into the workroom back at the factory well before it ever got to the savvy shopper.

The consumer blind spot that keeps the wounded worker out of sight is not new. In 1823, a children's book meant to inform juveniles about how everyday products were made saw the need to educate on this subject. After describing the Staffordshire pottery trades, *Little Jack of All Trades, or, the Mechanical Arts Considered in Prose and Verse Suited to the Capacities of Children* cautions its young readers to be cognizant of the health toll that this "pernicious" work takes on those employed in making beautiful china. Any admiration these objects inspire should be mixed with regret: "In viewing any article which contributes to our pleasure or domestic advantage, we should enquire of ourselves whether it has caused great inconvenience and painful confinement to the maker, or been the cause of shortening his life one moment."[105]

Remarkably, even after two centuries of human experience with carbon disulfide and extensive research spanning the course of modern biomedical investigation, new knowledge about its hazards continues to emerge, and resistance to the dissemination of such knowledge continues just as strongly as ever. In 2000, for example, a case of progressive kidney disease, a newly reported toxic effect, was linked to carbon disulfide exposure in a rayon factory spinning room. The treating physicians had to resort to the Freedom of Information Act in order to obtain the measurements that documented the excess exposure the worker had endured.[106]

That report appeared just in time to be included in the latest extensive re-review of carbon disulfide safety by the American Conference of Governmental Industrial Hygienists (ACGIH). The ACGIH finally had come to agree with NIOSH, in 2006 accepting as its exposure limit a reduction by a factor of ten, down to 1 ppm.[107] This non-legally-binding recommendation has had no impact on the OSHA standard of 20 ppm, which today is still the law of the land in the United States. Perhaps not entirely without coincidence, this degree of legal laxity is matched only by the workplace standards of India, Indonesia, and Thailand, three of the big four Asian rayon producers.[108] Even China, at just below 2 ppm, is far closer to the NIOSH/ACGIH-recommended level, at least officially.[109]

Concern has also been raised regarding carbon disulfide's potency to harm the nervous system as it matures from the period of fetal growth through childhood. The systematic study of this question, drawing on a discipline known as developmental neurotoxicology, has yet to be applied to carbon disulfide.[110] Even for the hazards of carbon disulfide that are clearly established in epidemiologic studies, in particular for central nervous system effects and vascular disease, the precise mechanisms by which this chemical does its damage remain largely a matter of conjecture. For example, the capacity of carbon disulfide to induce hypersexual behavior, noted among the very earliest observations of Delpech in Paris 150 years ago, is only now starting to come together with deeper understanding of brain neurotransmitters. This is especially relevant to dopamine pathways when they are manipulated in the treatment of Parkinsonism, a disease process that is itself a well-established end point of long-term toxic carbon disulfide exposure.

We also now can show what we long suspected but did not have the tools to see: that brain damage from carbon disulfide can progress with even relatively low exposure. A sophisticated study enrolled workers from eleven Japanese viscose factories and performed two magnetic resonance imaging (MRI) examinations six years later. They found increasing numbers of changes indicative of small areas of "silent" brain infarction.[111] By now, all but one or two of those eleven factories are closed.[112] But the workers made redundant are likely still to be experiencing the effects of their past employment.

One other piece of new knowledge about carbon disulfide is not particularly reassuring. New slow-motion camera footage has captured the fluctuating nature of the flame eruption that underlies the carbon disulfide–based "barking dog experiment." The "barking dog" requires a glass cylinder or tube, nitrous oxide (which may be obtained from a spray can of whipped cream), a lighter, some carbon disulfide, a cavalier chemistry teacher, and a classroom of students. On a YouTube slow-motion video of this phenomenon, a University of Nottingham narrator's voice-over informs us that the only thing that ever went wrong while doing the experiment (for him) happened when the bottom of the tube blew out once. The video, released in June 2013, had over 435,000 viewings the last time I checked.[113]

So it seems that we can count on chemistry students continuing to have the opportunity to have fun with carbon disulfide. Their exposures, barring an occasional blowout, spill, or other misadventure, likely will be low and fleeting. Pesticide breakdown and misuse, or an adverse change of wind direction, will put others at risk. Those taking the alcohol aversive Antabuse are guaranteed to get a pharmacologically active dose of carbon disulfide. Retired workers will experience the aftereffects of their past employment, the lucky survivors falling well into the age where vascular disease and nervous system decline take their greatest toll. The neighbors of viscose facilities can track the factory emissions online via the handy EPA Toxic Release Inventory website: in 2014, for 3M Tonawanda, 424,000 pounds; for Innovia in Tecumseh, 910,000 pounds; for Vikase in Loudon, Tennessee, 2,000,000 pounds; and since sausage is king, for Viscofan in Danville, Illinois, 3,845,000 pounds of smokestack carbon disulfide, amounts that represent substantial releases.[114] Finally, the current and growing worldwide viscose labor force is more directly in the line of fire through close workplace contact with carbon disulfide, whether making cellophane or rayon or cellulosic sponges, by whatever trade name, and no matter what claim of Gaia guardianship is made by their manufacturers. As a further testament to this globalized risk, eight workers in China's northern Shanxi province were killed in May 2015 by a carbon disulfide "leak."[115]

Our dystopian futures still hold out the prospect of rayon, too. This is explicit in *Brave New World,* set hundreds of years ahead, and implicit in *The Hunger Games,* set at an unspecified postapocalyptic future date. It is not clear precisely what textiles are manufactured in District 8, but when the movie adaptation's costume designer had to clothe Katniss for the reaping, rayon rather than cotton seemed to do the trick.[116] Not all future vistas are dystopic. The Slovakian architect Eva Bellakova has proposed that one of the former Svit factory warehouses be converted into a "Museum of Industrial Architecture": "Visitors of the museum can go through all the floors and have an opportunity to experience the entire width and length of the original factory open space. . . . Visiting the museum ends on the roof of the building which provides a lovely view of the town and especially the gorgeous High Tatras."[117]

The 2010 documentary film *Ultrasuede: In Search of Halston* underscores the designer's move into mass self-marketing with a brief video clip of an advertisement that Halston made for ITT Rayonier, Inc. This was during a brief period in the 1970s when International Telephone and Telegraph, through its subsidiary Rayonier Canada, was venturing into the wood pulp business. The conglomerate was hoping to make a profit from viscose-industry feedstock production, but it soon got out of that business. Ultrasuede is a polyester synthetic, but Halston was not averse to working with rayon either. In the ITT Rayonier promo, Halston speaks suavely and, looking straight into the camera, notes with assurance, "Rayon. It's going to be with us a long, long time."[118]

Notes

Chapter 1. In the Beginning

1. Gerd Grabow, "W. A. Lampadius, ein vielseitiger Wissenschaftler und Wegbereiter bei der Einführung der ersten Gasbeleuchtungsanlage auf dem europäischen Kontinent," *Bergknappe* 32 (2008): 40–41.

2. W. A. Lampadius, "Etwas uber Flussigen Schwefel, und Schwefel-Leberluft," *Chemische Annalen (Lorenz von Crell)* 2 (1796): 136–37.

3. R. Chenevix, "On the Action of Platina and Mercury upon Each Other," *Philosophical Transactions of the Royal Society London* 95 (1805): 104–30.

4. Humphry Davy, "New Analytical Researches on the Nature of Certain Bodies, Being an Appendix to the Bakerian Lecture for 1808," *Philosophical Transactions of the Royal Society London* 99 (1808): 450–70.

5. Henry M. Leicester, "Berzelius" in *Dictionary of Scientific Biography,* ed. Charles C. Gillispie (New York: Charles Scribner's Sons, 1970), 2:90–97.

6. Emilie Wöhler, "Aus Berzelius' Tagebuch während seines Aufenthaltes in London im Sommer 1812: Aus dem Schwedischen," *Zeitschrift für angewandte Chemie* 19 (1906): 187–90.

7. Jöns Jacob Berzelius and Alexander Marcet, "Experiments on the Alcohol of Sulfur, or Sulphuret of Carbon," *Philosophical Transactions of the Royal Society London* 103 (1813): 171–99.

8. Johan Erik Jorpes, *Jac. Berzelius: His Life and Work* (Berkeley: University of California Press, 1970), 59–60.

9. Marjan J. Smeulders et al., "Evolution of a New Enzyme for Carbon Disulphide Conversion by an Acidothermophilic Archaeon," *Nature* 19, no. 7369 (19 October 2011): 412–16.

10. Sarah L. Jordan et al., "Novel Eubacteria Able to Grow on Carbon Disulfide," *Archives of Microbiology* 163 (1995): 131–37.

11. Thomas P. Jones, *New Conversations on Chemistry, Adapted to the Present State of Science on the foundations of Mrs. Marcet's "Conversations on Chemistry"* (Philadelphia: John Grigg, 1831), 164–65. Even through the thirteenth edition of Marcet's book (1853) there is no mention of sulphuret of carbon; see Jane Marcet, *Conversations on Chemistry* (London: Longman, Brown, Green, and Longmans, 1853), 251–56.

12. Alexander Marcet, "Experiments on the Production of Cold by the Evaporation of Sulphuret of Carbon," *Philosophical Transactions of the Royal Society London* 103 (1813): 252–55.

13. Daniel H. Robinson and Alexander H. Toledo, "Historical Development of Modern Anesthesia," *Journal of Investigative Surgery* 25 (2012): 141–49.

14. James Young Simpson, "Notes on the Anaesthetic Effects of Chloride of Hydrocarbon, Nitrate of Ethyle, Benzin, Aldehyde, and Bisulphuret of Carbon," *Monthly Journal of Medical Science* 8 (1848): 740–44, quotation on 743.

15. John Snow, "On Narcotism by the Inhalation of Vapours," *London Medical Gazette* 6 (16 June 1848): 1074–78, cited passage on 1077. Both Snow and Simpson alluded to unsubstantiated reports earlier in the year on the use of carbon disulfide for anesthesia in Christiana, Norway, Snow specifying that this had been attributed to a practitioner named Harald Thanlow.

16. William Gregory, *Outlines of Chemistry, for the Use of Students* (London: Taylor and Walton, 1845), 122–23, quotation on 123.

17. John S. Haller, "Sampson of the Terebinthinates: Medical History of Turpentine," *Southern Medical Journal* 77, no. 6 (June 1984): 750–54.

18. Paolo A. Porto, "'Summus Atque Felicissimus Salium': The Medical Relevance of the Liquor Alkahest," *Bulletin of the History of Medicine* 76, no. 1 (Spring 2002): 1–29.

19. Chauncey D. Leake, "Valerius Cordus and the Discovery of Ether," *Isis* 7, no. 1 (1925): 14–24.

20. August Sigmund Frobenius, "An Account of a Spiritus Vini Aethereus, Together with Several Experiments Tried Therewith: By Dr. Frobenius, F.R.S," *Philosophical Transactions* 36 (1729): 283–89, cited passage on 286.

21. R. B. Prosser, "Parkes, Alexander (1813–1890)," in *Dictionary of National Biography*, ed. Leslie Stephen and S. Lee (London: Oxford University Press, 1959–60), 15:292–93.

22. Harriet Martineau, *Health, Husbandry, and Handicraft* (London: Bradbury and Evans, 1861), 406–15, quotation on 409. This work includes "Magic Troughs at Birmingham," which originally appeared in *Household Words* 25 (October 1851), 113–17.

23. Kathryn Jones, "'To Wed High Art with Mechanical Skill': Prince Albert and the Industry of Art," essay presented at a study day held at the National Gallery, Lon-

don, 5 and 6 June 2010, www.royalcollection.org.uk/sites/default/files/V%20and%20 A%20Art%20and%20Love%20(Jones).pdf.

24. Martineau, *Health, Husbandry, and Handicraft.*

25. Alexander Parkes, "Patent No. 11,147, March 25, 1846 [Commissioners of Patents]," in *Abridgments of Specifications Relating to Preparation of India-Rubber and Gutta Percha* (London: Eyre and Spottiswoode, 1875), 27–28.

26. Thomas Hancock, *Personal Narrative of the Origin and Progress of the Caoutchouc or India-rubber Manufacture in England* (London: Longman, Brown, Green, Longmans & Roberts, 1857), quotation on iii.

27. Ibid., unnumbered plate, "Domestic articles," interleaved between 116 and 117.

28. "French letter," s.v., *Oxford English Dictionary,* which dates to 1844 (although questioning the cited source) the first English-language appearance of this term as a colloquialism for condom.

29. Anselme Payen, *Précis de chimie industrielle* (Paris: Hachette, 1849), 74n.

30. Anselme Payen, *Précis de chimie industrielle,* 2nd ed. (Paris: Hachette, 1851), 675–89. This edition includes in its subtitle "Augmentée de Chapitres sur le Sulfure de Carbone." Payen alludes to carbon disulfide's explosive capacity.

31. [Brongniart, Pelouze, and Dumas reporting], "Rapport sur un mémoire de M. Payen, relatif à la composition de la matière ligneuse," *Comptes Rendus Hebdomadaires des Séances de l'Académie des Sciences* 8 (14 January 1839): 51–53.

32. Guillaume-Benjamin Duchenne de Boulogne, "Étude comparée des lésions anatomiques dan l'atrophie musculaire progressive et dans la paralysie générale," *L'Union Medicale* 7, no. 51 (30 April 1853): 202–3 [203 misnumbered as 303], summarizing a presentation to the Société médico-chirurgicale de Paris on 11 March and 8 April 1853. Duchenne briefly describes a case he had observed on the service of Dr. (Gabriel) Andral at the Charité.

33. [Auguste L. Dominique Delpech], "IV. Sociétés Savantes . . . Académie de Médecine," *Gazette Hebdomadaire de Médecine et de Chirurgie* 3 (18 January 1856): 40–41. This notice summarized a meeting of the Académie de Médecine (15 January 1856) communicated by Michel Levy, Grisolle, and Bouchardat. See Apollinaire Bouchardat, *Traité d'Hygiène Publique et Privée Basée sur l'Etiologie,* 2nd ed. (Paris: Ballière, 1883), 775, in which he notes, "dans mon cours de 1852, j'ai exposé, d'après ce que j'ai observé dans la fabrique de Gariel, les effets sur les ouvriers de inhalations contines du sulfure de carbone."

34. Delpech's 1846 doctoral thesis included comments on work-related palsies; see Auguste L. Delpech, *Des Spasmes Musculaires Idiopathiques et de las Paralysie Nerveuse Essentielle* (Paris: Rignoux, Imprimeur de la Faculté de Médecine, 1846), 94–96.

35. Auguste L. Delpech, "Accidents que développe chez les ouvriers en caoutchouc: L'inhalation du sulfure de carbone en vapeur," *L'Union Medicale* 10, no. 60 (31 May 1856): 265–67.

36. Auguste L. Delpech, "Accidents produits par l'inhalation du sulfure de carbone en vapeur: Expériences sur les animaux," *Gazette Hebdomadaire de Médecine et de Chirurgie* 3 (30 May 1856): 384–85. This material was part of his initial Académie de Médecine presentation, 15 January 1856.

37. Auguste L. Delpech, *Accidents que développe chez les ouvriers en caoutchouc: L'inhalation du sulfure de carbone en vapeur* (Paris: Labe, Libraire de la Faculté de Médecine, 1856). In this monograph (with the same title as his *Union Medicale* article), Delpech explicitly acknowledges that Bouchardat, in his lectures on hygiene at the faculty, first called his attention to the illness among rubber workers.

38. Auguste L. Delpech, "Industrie du caoutchouc soufflé: Recherches sur l'intoxication spécial que détermine le sulfure de carbone," *Annales d'Hygiène Publique*, 2nd. ser., 19 (1863): 65–183.

39. Ibid., 122–24, 165–70. Delpech identifies the inventor of the glove box device (his case 19) as Monsieur D, dwelling at rue Pradier, Paris-Belleville. A later report confirmed the inventor by name as M. Deschamps, from Belleville (*Congrès International d'Hygiène* [Paris: Imprimerie Nationale, 1878], 1:623). Belleville was a working-class sector of Paris long known for its militancy. Rue Pradier was the site of one of the last Belleville barricades to fall in the Paris Commune, May 1871; see "Une des dernières barricades de Belleville qui résiste à l'avancée des versaillais," www.weekisto.fr/plan/dept/paris-75/commune-de-paris-1871-19eme-arrondissement.php.

40. H. Masson, "Moyen de prévenir les accidents que développe chez les ouvriers l'inhalation du sulfure de carbone en vapeur" (meeting of 5 April 1858), *Comptes Rendus Hebdomadaires des Séances de l'Académie des Sciences* 46 (January–June 1858): 683–84. Other early reports on carbon disulfide toxicity contemporary with Delpech's include: Emile Beaugrand, "Action du sulfure de carbone," *Lancette Française: Gazette des Hôpitaux Civils et Militaires* 29 (14 July 1856): 331–32 (the case of a sixteen-year-old boy named Bois, which includes a passing remark on his ongoing sexual impotence), and Frederick Duriau, "Intoxication par le sulfate de carbone—Varioloïde intercurrente—Guérison," *Lancette Française: Gazette des Hôpitaux Civils et Militaires* 31 (27 May 1858): 241.

41. Louis Huguin, *Contribution à l'étude de l'intoxication par le sulfure de carbone chez les ouvriers en caoutchouc soufflé* (thesis) (Paris: Imp. A. Parent, 1874). Two earlier medical theses at Paris, by J.-B. Tavera (1865) and Paul Gourdon (1867), also covered this topic.

42. Abel Marche, *De l'intoxication par le sulfure de carbone* (thesis) (Paris: A. Derenne, 1876).

43. R. R. O'Flynn and H. A. Waldon, "Delpech and the Origins of Occupational Psychiatry," *British Journal of Occupational Medicine* 47 (1990): 189–98. For biographical details on Delpech, see the website of the Bibliothèque interuniversitaire de Santé, http://www2.biusante.parisdescartes.fr/bio/?cle=5338.

44. Peter E. Dans, "The Use of Pejorative Terms to Describe Patients: 'Dirtball' Revisited," *Baylor University Medical Center Proceedings* 15 (2002): 27–30.

45. Jean Martin Charcot, *Leçons du Mardi a la Salpêtrière: Policlinique, 1888–1889, notes de cours de MM. Blin, Charcot, Henri Colin* (Paris: Progrès Médical, 1889), 43–53. The case of carbon disulfide intoxication was presented as the first of two cases of the third lesson of the series, Tuesday, 6 November 1888.

46. Mark S. Micale, "Charcot and the Idea of Hysteria in the Male: Gender, Mental Science, and Medical Diagnosis in Late Nineteenth-Century France," *Medical History* 34 (1990): 363–411.

47. The term "Charcot's carbon disulfide-hysteria" continued to appear until the fourth decade of the twentieth century; see Karl B. Lehman and F. Flury, *Toxicology and Hygiene of Industrial Solvents,* trans. Eleanor King and Henry F. Smyth Jr. (Baltimore: Williams & Wilkins, 1943), 303.

48. Pierre Marie, "Hystérie dans l'Intoxication par le sulfure de carbone," *Bulletins et Mémoires de la Société Médicale des Hôpitaux de Paris* (9 November 1888): 1479–80. Marie's report was published three days after Charcot's lesson of 6 November.

49. G. Gilles de la Tourette, *Traité clinique et thérapeutique de l'hystérie* (Paris: Librairie Plon, 1891), 101–9. Some years before this, Gilles de la Tourette reported on the work of an Italian investigator of carbon disulfide toxicity; see G. Gilles de la Tourette, "De l'intoxication aiguë par le sulfure de carbone: Recherches expérimentales, par le Dr. Arrigo Tomassia, Professeur de Médecine Légale a l'Université de Pavie," *Annales d'Hygiène Publique et de Médecine Légale* 7, 3rd ser. (1882): 292–97.

50. George Guillain and V. Courtellemont, "Polynévrite sulfo-carbonée," *Revue Neurologique* 12 (1904): 120–23.

51. Rudolf Laudenheimer, *Die Schwefelkohlenstoff-Vergiftung der Gummi-Arbeiter unter besonderer Berücksichtigung der psychischen und nervösen Störungen und der Gewerbe-Hygiene* (Leipzig: Verlag Von veit, 1899).

52. The clinic was founded in 1901; see Hans Laehr, *Die Anstalten für Psychisch-Kranke in Deutschland, Deutsch-Österreich, der Schweiz und den Baltischen Ländern* (Berlin: Reimer, 1907). Franziska zu Reventlow, a prime figure in the Munich avant-garde, was among the patients staying at the clinic; see Franziska zu Reventlow, *Sämtliche Werke,* vol. 5, *Briefe 2: Briefe 1893 bis 1917* (Hamburg: Igel-Verlag, 2010), 182, 347.

53. Children's Employment Commission, *Commissioners Appointed to Inquire into the Employment of Children and Young Persons in Trades and Manufactures Not Already Regulated by Law: First—Sixth Report of the Commissioners* (London: Her Majesty's Printing Office, 1863–67).

54. "Gassed," s.v., *Oxford English Dictionary,* 2nd ed. (Oxford: Clarendon Press, 1989), 6:385. The online *OED,* under the transitive for "gas," meaning "to poison or asphyxiate," has an 1896 entry as the earliest citation for "gassed."

55. [Ophthalmological Society], "Reports of Societies: Ophthalmological Society of the United Kingdom; Nettleship: Amblyopia and Nervous Depression from the Vapour of Bisulphide of Carbon and Chloride of Sulphur," *British Medical Journal,* 18 October 1884, 760. The Ophthalmological Society appointed a committee made up of Drs. Nettleship, Adams, Frost, and Gunn to investigate this problem further. One year later, the committee summarized findings from thirty-three cases of carbon disulfide poisoning, of which twenty-four had involvement of the optic nerve; see "Ophthalmological Society," *Lancet,* 17 January 1885. Additional publications include W. B. Hadden, "A Case of Chronic Poisoning by Bisulphide of Carbon," *Proceedings of the Medical Society of London* 9 (1886): 115–17, summarized in *Lancet,* 2 January 1886, 18); and A. M. Edge, "Peripheral Neuritis, Caused by the Inhalation of Bisulphide of Carbon," *Lancet,* 7 December 1889, 1167–68.

56. James Ross and Judson S. Bury, *On Peripheral Neuritis: A Treatise* (London: Charles Griffin, 1893), 183–95, quotation on 188. This case is further reported in

David Little, "Toxic Amblyopia—Bisulphide of Carbon," *Transactions of the Ophthalmological Society of the United Kingdom* 7 (1886–87): 73–76.

57. Benjamin Ward Richardson, "Euthanasia for the Lower Creation: An Original Research and Practical Result," *Asclepiad* 1 (1884): 260–75.

58. Benjamin Ward Richardson, *On Health and Occupation* (London: Society for Promoting Christian Knowledge, 1879), 51.

59. Thomas Oliver, "Indiarubber: Dangers Incidental to the Use of Bisulphide of Carbon and Naphtha," in *Dangerous Trades,* ed. Thomas Oliver (London: John Murray, 1902), 470–74, quotation on 473–74.

60. Frederick Peterson, "Three Cases of Acute Mania from Inhaling Carbon Bisulphide," *Boston Medical and Surgical Journal* 128 (1892): 325–26.

61. Ibid., 326.

62. F. C. Heath, "Amblyopia from Carbon Bisulphide Poisoning," *Annals of Ophthalmology* 11 (1902): 4–8, quotation on 7. Dr. Heath originally read his paper before the Marion County Medical Society in Indianapolis in October 1901; it was later abstracted in *The Post-Graduate: A Monthly Journal of Medicine and Surgery* 17 (1902): 506–7.

63. Auguste Millon, "Mémoire sur la nature des parfums et sur quelques fleurs cultivables en Algérie," *Journal de Pharmacie et de Chimie* 30 (1856): 407–21.

64. R. D. Macgregor, "Supposed Poisoning by the Daily use of Bi-sulfide of Carbon," *Australian Medical Journal* 14 (15 December 1892): 622–24.

65. A. H. Douglas, "Case of Poisoning by Bisulphide of Carbon—Recovery (Under the Care of Dr. Davidson)," *Medical Times and Gazette,* pt. 2 (21 September 1878): 350. A later fatal case of carbon disulfide ingestion was reported in a shoemaker; see W. M. Foreman, "Notes on a Fatal Case of Poisoning by Bisulphide of Carbon with Post-Mortem Appearances and Remarks," *Lancet,* 17 July 1886, 118–19.

66. R. Eglesfeld Griffith, *A Universal Formulary* (Philadelphia: Blanchard and Lea, 1859), 451.

67. Valery Mayet, "Phylloxera" (trans. from the French), in *Report of the Board of Viticultural Commissioners for 1893–94* (Sacramento: State of California, 1894), app. C, 78–104.

68. Cephas L. Bard, "Cases in Practice: Malignant Pustule and Insanity Due to Bisulphide of Carbon," *Southern California Practitioner* 7 (1892): 476–85, quotation on 481–82. Dr. Bard was the brother of U.S. senator Thomas R. Bard.

69. Ibid., 483–84. Bard also comments on the use of carbon disulfide by jockeys to lower the value of a horse by causing kicking and rearing.

70. Display advertisement, "Read and Foster's Carbon Bisulphide," *Los Angeles Times,* 26 May 1883.

71. Display advertisement, "Death to Squirrels and Gophers!," *Los Angeles Times,* 14 July 1883.

72. "Foster Acquitted," *Los Angeles Times,* 23 August 1883. The arrest was reported as "Rum and Revolver: The Handy Pistol Again Brought into Play; Charles Foster, in a Drunken Frenzy, Attempts to Shoot A. H. Judson—But Makes a Signal Failure of it—The Particulars," *Los Angeles Times,* 14 July 1883. Dr. Bard alluded to this case.

73. Oliver, *Dangerous Trades*. Oliver discusses new workplace rules on carbon disulfide.

74. Briau, "Y A-T-Il une 'folie du cuir'? Empoisonnement chronique par le sulfure de carbone," *Lyon Médecine* 109 (24 November 1912): 897–901.

75. Gutta-percha was harvested from an Asian tree species (*Palaquium gutta*) and was also commercially prominent in the nineteenth century.

76. R. Bernard, "Different Imitations of Natural Silk," *Journal of the Society of Dyers and Colorists* 21 (1905): 166–68 (June), 190–91 (July), 215–16 (August); this is a translation of "Sur les diverses imitations de la soie naturelle," *Le Moniteur Scientifique-Quesneville*, no. 761 (May 1905): 321–30.

77. "Scatter Acorns That Oaks May Grow," http://libraries.mit.edu/archives/exhibits/adlittle/history.html; see also E. J. Kahn Jr., *The Problem Solvers: A History of Arthur D. Little, Inc.* (Boston: Little, Brown, 1986).

78. "Death of Mr. Roger B. Griffin," *World's Paper Trade Review* 19 (5 May 1893): 30.

79. Arthur D. Little, "Carbon Filament and Method of Manufacturing Same," U.S. Patent No. 532,568 (15 January 1895).

80. Arthur D. Little, "Report to Daniel C. Spruance, Esq., on the Technical Development of Viscose on the Continent of Europe and in Great Britain," 10 October 1899, Papers of Arthur D. Little, box 214, MSS 30312, Library of Congress.

81. Ibid., cited passage on 118.

82. Edwin J. Beer, *The Beginning of Rayon* (Shorton, U.K.: Phoebe Beer, 1962). Beer draws on entries from a contemporary journal. Eye troubles are mentioned in entries for 1898 (66), 1899 (78), and 1900 (91).

83. Little, "Report to Daniel C. Spruance," 394–98. An appendix shows Louis D. Brandeis as an attorney for Cross and Bevan. Brandeis later drafted the corporate charter for Little's consulting company.

84. H. D. Jump and J. M. Cruice, "Chronic Poisoning from Bisulphide of Carbon," *University of Pennsylvania Medical Bulletin* 17 (1904–5): 193–96, quotation on 193; initially presented in *Transactions of the College of Physicians* 26, 3rd ser. (1904): 314.

85. A. P. Francine, "Acute Carbon Disulphide Poisoning," *American Medicine* 9 (1905): 871. The paper does not refer to the cases described by Jump and Cruice.

86. Jump and Cruice, "Chronic Poisoning from Bisulphide of Carbon," 196.

87. E. M. Mogileveskii and A. P. Pakshver, "Fifty Years of Viscose Rayon Production in the USSR," *Fibre Chemistry* 19 (1977): 110–15; see also K. E. Perepelkin, "Ways of Developing Chemical Fibres Based on Cellulose: Viscose Fibres and Their Prospects; Part I: Development of Viscose Fibre Technology; Alternative Hydrated Cellulose Fibre Technology," *Fibre Chemistry* 40 (2008): 10–23.

88. Dimitri Ivanovich Mendeleev, *Uchenie o promyshlennosti* [Studies on industry] (St. Petersburg: Vstuplenie v biblioteka promshlennykh, 1900), vol. 1, pt. 1, n. 5, 54–55. In this lengthy footnote, Mendeleev discusses cellulose as a cheap raw material that could be converted to continuous filaments, according to the work of the pioneering viscose chemists Cross and Bevan, a process that he had become aware of only the year before (1899). Mendeleev also describes the technical basics of viscose process (for example, referring to xanthate).

89. Vasily Konstantinovich Khoroshko, "Group poisoning of the nervous system by carbon bisulphide: Professional disease" (in Russian), *Meditsinkoe Obrozrenie* 79 (1913): 848–59.

90. Donald Cuthbert Coleman, *Courtaulds: An Economic and Social History*, vol. 2, *Rayon* (Oxford: Clarendon Press, 1969), 18–19, 82–83, 105–8.

91. Emile George Perrot, *Discussion on Garden Cities: An Industrial Village on Garden City Lines, Being Built at Marcus Hook, Penna., for the American Viscose Company by Ballinger & Perrot, Architects and Engineers, Philadelphia* (Philadelphia: Mitchell, 1912), quotation on facing page of fourth leaf. Perrot took the Brownville estate in England as his prototype.

Chapter 2. The Crazy Years

1. Hansard, *Parliamentary Debates*, Commons, 5th ser., vol. 214 (15 March 1928), col. 2078.

2. "Parliamentary Intelligence," *Lancet*, 24 March 1928, 630–33. The *Lancet* piece omits a March 15 subheading, implying, in error, that the debate took place on the previous day, March 14.

3. Coleman, *Courtaulds*.

4. H. A. Taylor, *Jix, Viscount Brentford: Being the Authoritative and Official Biography of the Rt. Hon. William Joynson-Hicks, First Viscount Brentford of Newick* (London: S. Paul, 1933). Jix took on the Joynson surname the year following his marriage.

5. David Cesarani, "The Anti-Jewish Career of Sir William Joynson-Hicks, Cabinet Minister," *Journal of Contemporary History* 24 (1989): 461–82.

6. Ibid.

7. Hansard, *Parliamentary Debates*, Commons, 5th ser., vol. 214, col. 2078.

8. Ibid., vol. 213 (23 February 1928), col. 1828.

9. Ibid.

10. Ibid., vol. 214 (29 February 1928), cols. 399–401. Further debate on wages in the artificial silk industry also occurred on 6 March 1928; see ibid., cols. 975–77.

11. William A. Robson, "The Factory Acts, 1833–1933," *Political Quarterly* 5 (1934): 55–73.

12. Hansard, *Parliamentary Debates*, Commons, 5th ser., vol. 189 (10 December 1925), col. 642.

13. "Industrial Poisoning (Medical Notes in Parliament)," *British Medical Journal*, 14 December 1925, 1205.

14. "Sir Thomas Legge, C.B.E., M.D." (obituary), *British Medical Journal*, 14 May 1932, 913–14.

15. Chief Inspector of Factories and Workshops, *Annual Report of the Chief Inspector of Factories and Workshops for the Year 1925* (London: His Majesty's Stationery Office, 1926), 70.

16. Ibid., 121.

17. Thomas M. Legge, *Industrial Maladies* (London: Oxford University Press, 1934), 150.

18. Chief Inspector of Factories and Workshops, *Annual Report of the Chief Inspector of Factories and Workshops for the Year 1926* (London: His Majesty's Stationery Office, 1927), 86. There was only a single case of carbon disulphide poisoning documented in the *Report*, attributed to exposure in a worker manufacturing the chemical, not in a viscose worker.

19. "Factory Inspection Reports for Courtauld's Works, Flint, 1922–1935," D/DM/656/1, Flintshire Record Office, Hawarden, Wales; includes notes of churn room visits on 7 September 1922, 7 November 1922, 31 August 1923, 4 October 1923, 31 October 1924, 31 July 1925, 14 October 1926, and 20 January 1927.

20. Chief Inspector of Factories and Workshops, *Annual Report of the Chief Inspector of Factories and Workshops for the Year 1927* (London: His Majesty's Stationery Office, 1928), 85.

21. The "full Ginsburg" is named for William H. Ginsburg, the attorney for Monica Lewinsky, who, on 1 February 1998, was the first to complete the feat.

22. Thomas Oliver, "Some Achievements of Industrial Legislation and Hygiene," *British Medical Journal,* 19 September 1925, 530–31. There was a simultaneous publication (with slight editorial variants) in the *Lancet,* 19 September 1925, 630–32.

23. Edward William Hope, William Hanna, and Clare Oswald Stallybrass, *Industrial Hygiene and Medicine* (London: Baillière, Tindall and Cox, 1923), 176.

24. "Cause di avvelenamento nelle industrie della seta artificiale," *La Medicina del Lavoro* 16 (1925): 418–20.

25. Ibid., 419.

26. Karl B. Lehmann, "Experimetelle Studien über den Einfluss technisch und hygienisch wichtiger Gase und Dämpfe auf den Organismus. Thiel VII: Schwefelkohlenstoff und Chlorschwefel," *Archiv für Hygiene* 20 (1894): 26–77.

27. Yandell Henderson and Howard W. Haggard, *Noxious Gases and the Principles of Respiration Influencing Their Action* (New York: Chemical Catalog Company, 1927), 169.

28. David R. Shreeve, "Dr. Arnold Renshaw (1885–1980): Manchester Pathologist and Forensic Pathologist with a Clinical Interest in Rheumatoid Arthritis," *Journal of Medical Biography* 17 (2009): 225–30.

29. [Local sections reports—Manchester], *Institute of Chemistry of Great Britain and Ireland: Journal and Proceedings,* pt. 2 (1925): 128.

30. Royal Institute of Chemistry, Manchester and District Section, minute book, 1918–27 (ref: M90/1/1), Manchester Archives and Local Studies, Manchester, U.K.

31. "Obituary: Dr. Arnold Renshaw, Leader in Forensic Medicine," *Times* (London), 7 June 1980, 14.

32. Antonio Ceconi, "Polineuriti," *Minerva Medica* 6 (1925): 267–84, quotation on 282.

33. Ibid.

34. Giovanni Loriga, "Le condizioni iginieniche nell'industria della seta artificale," *Bollettino del Lavoro (Studi, Rapporti e Incheieste)* 43, no. 4 (1925): 85–95.

35. Ibid., 86.

36. Giovanni Loriga, "Le condizioni iginieniche nell'industria della seta artificale," *La Medicina del Lavoro* 16 (1925): 309–14 (same title as the more complete report in note 34 to this chapter).

37. Piero Redaelli, "Sull'anatomia patologica dell'avvelenamento cronico da solfuro di carbonio," *Bollettino della Società Medico-Chirurgica, Pavia* 37 (1925): 133–40; the case was initially presented on 6 February 1925. M. Arezzi, "Osservazioni sperimentali sull' avvelenamento da sulfuro di carbonio," *Bollettino della Società Medico Chirurgica, Pavia* 37 (1925): 141–47.

38. Giovanni Bignami, "Modificazioni del sangue nell' avvelenamento da sulfuro di carbonio," *Bolletinno della Società Medico-Chirurgica, Pavia* 37 (1925): 745–55.

39. Emmanuele D'Abundo, "Psicosso sensoriale melanconica de avvelenamento per sulfuro di carbonio," *Revista Italiana di Neuropatogia, Psichiatra e d'Elettrotearpia* 16 (1923): 155–57 (published in Catania).

40. Thomas Oliver, "The Sulfur Miners of Sicily: Their Work, Diseases, and Accident Insurance," *British Medical Journal*, 1 July 1911, 12–14.

41. Giarrizo Giarrizzo, *Catania: La città moderna, la città comporanea* (Catania: Domenico Sanfilippo, 2012), 36.

42. "Die Tagung der Arbeitsgemeinshaft Deutcher Gewerbärtze, Berlin" *Social Praxis* 33 (1924): 1060–62. The article summarizes a conference held on 7 July 1924, including a report on the outbreak by Dr. Beltke of Weisbadden.

43. Wauer [first name unstated], "Gesundeheitsschädigungen in der Kunstseideindustrie," *Zentralblatt für Gewerbehygiene und Unfallverhütung* 12 (1925): 67–71; paper presented at the Deutchen Gesellschaft für Gewerbehygiene, Würzburg, 30 September 1924.

44. Alberto Trossarelli, "Die geistigen Störungen bei Arbeitern der Kunsteidenindustrie," *Psychiatrisch-Neurologische Wochenschrift* 31 (5 January 1929): 1–6; reprinted in the same year in *Monatschrfit für Textile-Industrie (Leipzig)* 44 (1929): 5–8, and abstracted in *Journal of Industrial Hygiene* 12 (1930): 138. Two years later, Aristide Ranelletti published a summary of the Italian experience as "Die berufliche Schwefelkohlenstoffvergiftung in Italien: Klinik und Experimente," *Archiv für Gewerbepathologie und Gewerbehygiene* 2 (17 December 1931): 664–75.

45. Josef Witt, "Die deutsche Zellwolles-Industrie" (diss., Friedrich-Alexanders Universität, Erlangen 1939), 74–75.

46. Emil Neumann, "Status of German Fibre Substitutes," *Textile World* 61 (20 May 1922): 25–26, quotation on 26.

47. Ibid.

48. [Snia Viscosa], *Mezzo secolo di Snia Viscosa* (Milan: Pan Editrice, 1970), 15–17; see also Martino Orsi, "L'evoluzione della Snia Viscosa tra gli anni Venti e Trenta," *Imprese e Storia* 19 (1999): 7–46.

49. Coleman, *Courtaulds*, 268–69.

50. The first detailed report on the industrial hygiene of carbon disulfide use in viscose appeared in France as Chevalier [last name only, as published], "La manipulation du sulfure de carbone dans les fabriques de viscose et de la soie artificielle viscose," *Bulletin de l'Inspection du Travail et de l'Hygiène Industrielle* 6 (1922): 166–74. It emphasizes explosiveness but is equivocal on toxicity, describing this question as

"highly controversial." Chevalier was an inspector in the French Department of Labor at Charleville. A general review of carbon disulfide toxicity appeared in 1924: C. Mattei and J. Sédan, "Contribution a l'étude de l'intoxication par le sulfure de carbone de l'opportunité de l'inscrire parmi les maladies professionnelles prévues par la loi du 29 Octobre 1919," *Annales d'Hygiène Publique, Industrielle et Sociale*, n.s., 2 (1924): 385–430. Its only mention of artificial silk (390–91) is upbeat, describing the "great future" of viscose. A thesis that appeared in 1923—Emilee Janicot, *Considérations sur un cas d'intoxication par le sulfure de carbone* (Paris: Les Presses Universitaires de Frances, 1923)—centers on a rubber vulcanization worker treated at Necker Hospital for chronic carbon disulfide poisoning. The thesis does not mention viscose in its background review. As late as 1925, a "lesson" (a clinical case presentation) by Jean Hallé reviewed the Necker Hospital experience with carbon disulfide poisoning, using one of Dr. Janicot's cases. He, too, does not mention the artificial silk industry. See Jean Hallé, "Les formes cliniques du sulfcarbonisme (Leçon faite le 5 Novembre 1925)," *La Semaine des Hôpitaux de Paris* 3 (1926): 90–95.

51. Emile Chalencon, *Du sulfocarbonisme professionnel dans l'industrie de la soie artificielle (préparation de la viscose)*, thesis (Firminy, France: F. Chelancon, 1927), 11.

52. Ohara Institute for Social Research, Hosei University, "Industrial Welfare Association Posters," http://oohara.mt.tama.hosei.ac.jp/sangyofukuri/enpos1345.html.

53. Masatane Takuhara, "[Occupational diseases in an artificial silk factory]," *Sangyo Fukuri* [Industrial Welfare] 4 (1929): 73–82. A later paper covered the technical measurement of hydrogen sulfide; see Masatane Takuhara "[Occupational diseases in an artificial silk factory: Hydrogen sulfide in an artificial silk factory]," *Sangyo Fukuri* 5 (1930): 46–57. The international presentation by Takuhara summarizing the outbreak appeared as Masatane Takuhara, "Uber die Berufskrankeiten in der Kunstseide-Industrie," in *Proceedings, Sixth International Congress for Work-Related Disease (ICOH)* (Geneva: ICOH, 1931), 758–59. At the same international meeting, a presentation was made on workers' compensation data for Japan for 1929 that included 3 cases of carbon disulfide poisoning and 542 additional cases of eye disease in the artificial silk industry; see B. Koinuma, "Gewerbliche Berufskrankheiten in Japan," ibid., 757–58.

54. "The Business World: Name Committee Meets Thursday," *New York Times*, 15 April 1924, 28; "'Rayon a Substitute for 'Glos,'" *New York Times*, 18 April 1924, 34; "Retailers to Adopt 'Rayon,'" *New York Times*, 26 April 1924, 24; "To Continue Use of 'Glos,'" *New York Times*, 2 May 1924, 33; "Silk Association Favors Rayon," *New York Times*, 11 June 1924, 34.

55. [E. I. du Pont de Nemours and Company], *Leonard A. Yerkes: Fibersilk to Fiber "A"* (Buffalo, N.Y.: Keller, 1945).

56. Coleman, *Courtaulds*, 18–19, 82–85, 104–19, 140–41, 145–47, 302.

57. Ibid., fig. 10, opposite 172.

58. George Martin Kober, "The Button, Horn, Celluloid and Allied Industries," in *Industrial Health*, ed. George M. Kober and Emery Roe Hayhurst (Philadelphia: Blakiston's and Son, 1924), 273–77.

59. Alice Hamilton, *Industrial Poisons in the United Sates* (New York: Macmillan, 1925), 360–69; reviewed in the *Boston Medical and Surgical Journal*, 23 July 1925, and the *Journal of the Medical Association*, 15 August 1925.

60. Hamilton, *Industrial Poisons*, 369.

61. Alice Hamilton, day pocket diary for 1923, box 1, vol. 1923, Hamilton Family Papers, Schlesinger Library, Radcliffe Institute for Advanced Study, Harvard University. The diary indicates that Hamilton had attended a professional dinner the previous evening in Buffalo with Dr. Wright.

62. Richard L. Cameron, "Two Case Reports," appearing with R. S. McBirney, "Carbon Bisulphide Poisoning," *Industrial Hygiene Bulletin* (New York State Department of Labor) 2 (August 1925): 12. A synopsis of the Cameron report was published the following year in "Abstracts," *Journal of Industrial Hygiene* 8 (1926): 74–75.

63. Wade Wright's employment with Metropolitan Life was announced on 22 January 1924, with a start date deferred until that summer to allow him to "complete his term an Instructor in the Harvard Medical School" (Daniel May, company archivist, MetLife, e-mail to the author, 23 June 2008). Wade died in 1936 at the age of forty-seven, succumbing to pulmonary tuberculosis; see *Journal of the American Medical Association*, 14 November 1936, 1652. In a physician, such a condition would very likely have been occupationally acquired.

64. [Policyholders Service Bureau, Metropolitan Life Insurance Company], *Rayon: A New Influence in the Textile Industry* (New York: Metropolitan Insurance Company, n.d. [c. 1928]).

65. Ibid., 21.

66. Ibid., 11.

67. Coleman, *Courtaulds*, 265.

68. "High Court of Justice, Chancery Division, Nuisance Caused by Gas Fumes, Attorney-General v. Rayon Manufacturing Company (1927), Limited," *Times* (London), 20 July 1928, 5. W. J. Bonner, "The Textile Industry in Surrey," *Annual of the Dorking High School for Boys O.B.A.* 5 (1927): 18–20.

69. Hansard, *Parliamentary Debates*, Commons, 5th ser., vol. 215 (22 March 1928), cols. 545–46; 29 March 1928, col. 1330; 3 April 1928, col. 1795; vol. 216 (17 April 1928), cols. 16–17; 18 April 1928, col. 180; vol. 217 (9 May 1928), cols. 252–53; 16 May 1928, cols. 1038–39; 17 May 1928, cols. 1184–85.

70. Hansard, *Parliamentary Debates*, Commons, 5th ser., vol. 224 (4 February 1929), cols. 1418–19; 5 February 1929, col. 1604. "Ailments of Workers in Artificial Silk (Medical Notes in Parliament)," *British Medical Journal*, 9 February 1929, 278.

71. "Correspondence with Workers Union and National Union of Textile Workers Including Deputation to the Home Secretary, 1 March 1929," MSS.292/144.54/6, Trades Union Council Archives, Modern Research Center, Warwick University.

72. Chief Inspector of Factories and Workshops, *Annual Report of the Chief Inspector of Factories and Workshops for the Year 1928* (London: His Majesty's Stationery Office, 1929), 36–37.

73. Ibid., 79.

74. Chief Inspector of Factories and Workshops, *Annual Report of the Chief Inspector of Factories and Workshops for the Year 1929* (London: His Majesty's Stationery Office, 1930), 46.

75. Ibid., 77–78. There were five notified cases of carbon disulfide poisoning in the artificial silk industry in 1929, all from the same facility, the Rayon Manufacturing Co., Ltd, of Leatherhead (Surrey).

76. Home Office and Ministry of Labour and National Service and Successors, Factory Inspectorate and Factory Department, *Registers of Lead, etc, Poisoning and Anthrax Cases,* The National Archives, United Kingdom, LAB 56/30. One of these listed workers appears to match a case presented at a clinical meeting held in November 1929; see F. R. R. Walshe, "Carbon Disulphide Intoxication," *Proceedings of the Royal Society of Medicine* 23 (1929–30): 89–90.

77. In 1929, a case of amblyopia from carbon disulfide in the artificial silk industry was first noted in France in Pierre Moulinié, *De L'amblyopie par le sulfure de carbone* (thesis) (Lyon: Imprimerie Bosc frères, M. et L. Riou, 1929). Case 1 of the twenty-one cases covered in this thesis worked in artificial silk and was treated by Dr. Moulinié's mentor, Dr. Jacques Rollet, noticed in *Lyon Medical* 144 (15 September 1929): 320.

78. J. Strebel, "Durch SO_2 verursachte Augenschädigungen (spez. zentrale punktörnige Viskoseverätzung der Hornhäute)—Schutz durch Maskenbrille mit Zinkkohlefilter," *Schweizerische Medizinische Wochenshschrift* 53 (1 June 1923): 560–61.

79. [Société de la Viscose Suisse], *50 Jahre Viscose Emmenbrücke—1906–1956* (Emmenbrücke, Switzerland: Société de la Viscose Suisse, 1956). In addition to Emmenbrücke, other Swiss producers in this period were located in Rheinfelden near Basel and at Reidekon-Ulster.

80. C. Bakker, "Oogziekten in kunstzijdefrabrieken," *Nederlandsch Tijdschrift voor Geneeskunde* 67 (1923): 576–84. In 1924, *Il Lavoro* called attention to this report linking conjunctivitis in artificial silk manufacturing to hydrogen sulfide exposure: "Affezioni oculari in una fabrica di seta artificiale," *Il Lavoro* 15 (1924): 244. Also abstracted by *JAMA:* "Eye Affections in Silk Factories," *Journal of the American Medical Association* 82 (8 March 1924): 808.

81. Willem Reinier Hubert Kranenburg, "Oogziekten in kunstzijdefrabrieken" [letter], *Nederlandsch Tijdschrift voor Geneeskunde* 67 (1923): 822.

82. W. R. H. Kranenburg and H. Kessener, "Schwefelwasserstoff- und Schefelkohlenstoff-vergiftungen," *Zentralblatt für Gewerbehygiene und Unfallverhütung* 12 (1925): 348–50. There were three artificial silk factories operating in the Netherlands: two owned by the Eerste Nedelandsche Kunstzjijdefabriek (ENKA, at Arnhem and Ede) and one owned by the Hollandsche Kunstjide Industrie (KHI, at Breda). The two Dutch firms merged into a single entity in 1928.

83. Eva Klein, "Les lésions oculaires dans les fabriques de soie artificielle," *Archives d'Ophtalmologie* 45 (1928): 686–93. The only viscose plant located in Strasbourg was the Soieries de Strasbourg, founded in 1924 by Dr. Emile Bronnert.

84. David Rankine, "Artificial Silk Keratitis," *British Medical Journal,* 4 July 1936, 6–9.

85. Ministry of Health, *Report on an Investigation Regarding the Emission of Fumes from Artificial Silk Works* (London: His Majesty's Stationery Office, 1929). Long before this there was a local police-blotter entry on pollution from Courtaulds for "effluvium which constituted a nuisance." Courtauld received a six-month extension for

correction; see *Coventry Standard,* 24–25 May 1912. In November of that year, Courtaulds received yet another postponement (*Coventry Standard,* 8–9 November 1912).

86. "Amblyopia and the Artificial Silk Industry," *British Medical Journal,* 1 June 1929, 1008–9.

87. Ibid., 1009.

88. Thomas Morison Legge, *Shaw Lectures on Thirty Years' Experience of Industrial Maladies: Delivered Before the Royal Society of Arts, February and March, 1929* (London: Royal Society of Arts, 1929), 16–17.

89. Correspondence, Workers Union and National Union of Textile Workers, including deputation to the Home Secretary, Trades Union Council Archives, Modern Research Center, Warwick University.

90. Gustavo Quarelli, "Intossicazione da solfuro di carbonio nella lavorazione della seta artificiale," *International Congress on Occupational Health: Opera Collecta; Congressus V, Internationalis Medicorum pro Artificibus Calamitate Afflictis Aegrotisque, Budapest, 2–8 Sept. 1928* (Budapest: Victor Hornýanszky, 1929), 805–18.

91. Gustavo Quarelli, "L'azione del sulfuro di carbonio sul sistema nervosa vegitivo," *IVe Réunion de la Commission Internationale Permanente pour l'étude des Maladies Professionnelles, Lyon, 3–6 Avril 1929,* vol. 2, *Comptes Rendus des Séances et Communications Diverses* (Lyon: Trévoux, 1929), 200–204.

92. Gustavo Quarelli, "Spasmo di torsione ed avvelenamento da sulduro di carbonio," *Proceedings, VII Congresso Nazionale di Medicina del Lavoro—Napoli, 10–13 October 1929,* 52–68. See also G. B. Audo Guinotti, "Sul tremor nell'avvelenamento professionale da solfuro di carbonio," *La Riforma Medica* 45 (21 September 1929): 1275–78.

93. Philip Williamson and Edward Baldwin, eds., *Baldwin Papers: A Conservative Statesman, 1908–1947* (Cambridge: Cambridge University Press, 2004), 216.

94. Hansard, *Parliamentary Debates,* Commons, 5th ser., vol. 230 (25 July 1929), cols. 1459–60.

95. Karl Bonhoeffer, "Über die neurologischen und psychischen Folgeerscheinungen der Sehwefelkohlenstoffvergiftung," *Monatsschrift für Psychiatrie und Neurologie* 75 (1930): 195–206.

96. Gustavo Quarelli, "Del tremor parksinosimile dell'intossicazione cronica da solfuro di carbonio," *Medicina del Lavoro* 21 (28 February 1930): 58–64.

97. Fedele Negro, "Les Syndromes parkinsoniens par intoxication sulfo-carbonée," *Revue Neurologique* 2 (November 1930): 518–22.

98. Gustavo Quarelli, *Clinica delle Malattie Professionali* (Turin: Union Tipografico–Editrice Torinese, 1931), 197–200, photographic illustration of the man with hand spasm, opposite 196.

99. Hansard, *Parliamentary Debates,* Lords, 5th ser., vol. 83 (10 November 1931), cols. 33–37.

100. "Major J. S. Courtauld, M.P., Business, Politics and Racing" (obituary), *Times* (London), 21 April 1942.

101. Thomas M. Legge, "An Industrial Danger," *Statesman and Nation,* n.s., 2 (15 August 1931): 190. This was one of Legge's last publications before his death the following year.

102. Ibid.

103. William Henry Willcox and G. Roche Lynch, "The Artificial Silk Industry" [letter], *New Statesman and Nation*, n.s., 2 (26 September 1931): 371.

104. E. Grey-Turner, "The Detective-Physician: The Life and Work of Sir William Willcox" (book review), *Medical History* 17 (1972): 104–6; "G. Royce Lynch, O.B.E., M.B., F.R.I.C., D.P.H." (unsigned obituary), *British Medical Journal*, 13 July 1957, 105.

105. Willcox and Roche, "Artificial Silk Industry."

106. Ibid.

107. "Chancery Division, Artificial Silk Factory: Sequestration Writ Suspended, Attorney-General v. Rayon Manufacturing Company (1927), Limited," *Times* (London), 21 October 1931, 9.

108. This small competitor of Courtaulds shut down in October 1932; see Coleman, *Courtaulds*, 332.

Chapter 3. Wrapped Up in Cellophane

1. Coleman, *Courtaulds*, 174–76; see also Jeffrey Harrop, "Growth of the Rayon Industry in the Inter War Years," *Bulletin of Economic Research* 20 (1968): 71–84.

2. "Major Developments in Teijin's History," www.teijin.com/ir/library/annual_report/pdf/ar_04_08.pdf. Teikoku went on to be absorbed into the Teijin industrial group.

3. Toray, "Corporate Brand," www.toray.com/aboutus/brand.html. Toyo morphed into Toray.

4. Grace Hutchins, *Labor and Silk* (New York: International Publishers, 1929), 63–81.

5. Ibid. The diagrammatic figure "International Connections in the Rayon Industry" is on 68.

6. Mira Wilkins, *The History of Foreign Investment in the United Sates, 1914–1945* (Cambridge, Mass.: Harvard University Press, 2004), 216.

7. Alan M. Wald., *Exiles from a Future Time. Forging of the Mid-Twentieth-Century Literary Left* (Chapel Hill: University of North Carolina Press, 2002), 76–80.

8. Alexander Trachtenberg, *The History of Legislation for the Protection of Coal Miners in Pennsylvania, 1824–1915* (New York: International Publishers, 1942).

9. Albert Maltz, *The Way Things Are, and Other Stories* (New York: International Publishers, 1938). The short story "Man on the Road" originally appeared in the *New Masses* (8 January 1935).

10. "To Buy Oxygen Company; Union Carbide Arranges for Deal with the International," *New York Times,* 12 September 1930. The company's name was the International Oxygen Company.

11. Hutchins, *Labor and Silk,* 120–25.

12. Ibid., 79.

13. Julia M. Allen, *Passionate Commitments. The Lives of Anna Rochester and Grace Hutchins* (Albany: State University of New York Press, 2013), 152.

14. "Esther Shemitz Chambers; Widow of Man Who Played Key Role in the Alger Hiss Case," *Los Angeles Times,* 28 August 1986. Shemitz died at age eighty-six.

15. Hutchins, *Labor and Silk,* 164–66.

16. E. M. Mogileveskii and A. P. Pakshver, "Fifty Years of Viscose Rayon Production in the USSR," Fibre Chemistry 19 (1977): 110–15. K. E. Perepelkin, "Ways of Developing Chemical Fibres Based on Cellulose: Viscose Fibres and Their Prospects; Part I. Development of Viscose Fibre Technology: Alternative Hydrated Cellulose Fibre Technology," *Fibre Chemistry* 40 (2008): 10–23.

17. Mogileveskii and Pakshver, "Rayon Production in the USSR," 110.

18. Ibid.

19. Coleman, *Courtaulds,* 172–73.

20. Oxford English Dictionary Online, s.v. "cellophane," "diaphane," and "diaphanous." In addition to its direct meaning as transparent, diaphane was also a nineteenth-century descriptor of a specific weave of silk characterized by the presence of transparent colored figures.

21. [DuPont], *Leonard A. Yerkes.* This twenty-two-page corporate publication was in homage to Yerkes ("Len" or even "LAY"), a chemical engineer and chief architect of DuPont's rayon and cellophane operations.

22. Heinrich Voelker, *75 Jahre Kalle: Ein Beitrag zur Nassauischen Industrie-Geschichte von Universitätsprofessor Dr. Heinrich Voelcker, 1863–1938* (Wiesbadden-Biebrich: Kalle & Co. Aktiengesellschaft, n.d. [1938]).

23. Marin Tillmanns, *Bridge Hall Mills: Three Centuries of Paper and Cellulose Film Manufacture* (Tisbury, U.K.: Compton, 1978), 104–9.

24. Ibid., 109.

25. Cyril Henry Ward-Jackson, *The "Cellophane" Story: Origins of a British Industrial Group* (Edinburgh: British Cellophane Limited, 1977). Ward-Jackson worked for British Cellophane from 1948. In 1941, he wrote *A History of Courtaulds: An Account of the Origin and Rise of the Industrial Enterprise of Courtaulds Limited and of Its Associate the American Viscose Company* (London: Curwin, 1941).

26. C. P. Atkinson, "Manufacture, Dyeing, and Application of Viscacelle (Regenerated Cellulose Film)," *Journal of the Society of Dyers and Colourists* 5 (1934): 132–38, quotation on 132.

27. Coleman, *Courtaulds;* see also Ward-Jackson, *"Cellophane" Story.*

28. [Courtaulds Ltd.], *National Rayon Week* (London: Courtaulds Ltd., 1936). This was an eleven-page promotional pamphlet.

29. "'Say Rayon' Week; Many Uses and Endless Variety," *Straits Times* (Singapore), 25 June 1936.

30. "Our Overseas Trade," *Advertiser* (Adelaide, Australia), 22 April 1936.

31. William Sunners, *American Slogans* (New York: Paebar Company, 1949), 71, 73, 190.

32. James A. Farley, "Address of Chairman James A. Farley of the Democratic National Committee at the Washington Day Banquet of the Kansas Democratic Club at Topeka, to be Broadcast over the Nationwide Hookup of the National Broadcasting Company, Feb. 22, 1936," Papers of James Aloysius Farley, box 61, file 1, Library of Congress; quotation is on 6; emphasis in the original.

33. "Democracy Saved, Farley Declares; He Says Roosevelt Leadership Has Thwarted Enemies of Best Governing Plan," *New York Times*, 23 February 1936.

34. "Wrapped up in Cellophane," *New York Post*, 30 January 1936.

35. On 6 November 1947, Farley was a guest on the first television broadcast of *Meet the Press*; photographic image at www.nbcnews.com/id/34344875/displaymode/1247?beginSlide=1.

36. Gail Collins, "The Milt Romney Pardon," *New York Times*, 1 December 2011.

37. Kevin N. Kruse, "For God so Loved the 1 Percent...," *New York Times*, 17 January 2012.

38. Daniel Scroop, *Mr. Democrat: Jim Farley, the New Deal, and the Making of Modern American Politics* (Ann Arbor: University of Michigan Press, 2006), 131.

39. Coleman, *Courtaulds*, 389–97. An extended key U.S. Senate debate on the rayon tariff can be found in the *Congressional Record*, 72nd Cong., pt. 3, 27 January 1930, 2423–47.

40. Coleman, *Courtaulds*, 400–406.

41. "Eleanor du Pont Engaged; Daughter of Irenee du Pont to Wed Philip S. Rust [sic]," *New York Times*, 4 January 1931.

42. "Eleanor du Pont's Bridal; Chooses Attendants for Marriage to P. G. Rust on May 8," *New York Times*, 6 April 1931.

43. "Eleanor F. du Pont Wed to Philip G. Rust; Bishop Philip Cook Officiates at Wilmington, Del.," *New York Times*, 10 May 1931.

44. Philip G. Rust, "Means for Treating Yarn," U.S. Patent Office, application files, 2 August 1928, serial no. 297,075; patented 27 October 1931, Patent 1,829,678.

45. "Philip Goodenow Rust Jr." (obituary), *Delaware News Journal*, 31 October 2010. His Georgia ranch was called Winnstead Plantation.

46. Frank W. Taussig and Harry Dexter White, "Rayon and the Tariff: The Nature of an Industrial Prodigy," *Quarterly Journal of Economics* 45 (1931): 588–621.

47. Ibid., 588. Although Taussig and White were early promoters of the term "duopoly," an earlier usage is credited to Arthur Cecil Pigou, *Economics of Welfare* (London: Macmillan, 1920), 232; see *Oxford English Dictionary*, s.v. "duopoly."

48. Frank W. Taussig and Harry Dexter White, *Some Aspects of the Tariff Question: An Examination of the Development of American Industries under Protection*, 3rd ed. (Cambridge, Mass.: Harvard University Press, 1931).

49. Gottfried Haberler, "Taussig, Frank W.," in David L. Stills, ed. *International Encyclopedia of the Social Sciences* (New York: Macmillan/Free Press, 1968), 15:516–18.

50. Benn Steil, *The Battle of Bretton Woods: John Maynard Keynes, Harry Dexter White, and the Making of a New World Order* (Princeton, N.J.: Princeton University Press, 2013), 20–21.

51. Taussig and White, "Rayon and the Tariff," 609.

52. Coleman, *Courtaulds*, 406–8. Coleman consistently misspells "Isidor" as "Isidore."

53. "Salvage: Cable to Johnson Courtaulds Coventry," 20 March 1932, exhibit 1, Federal Trade Commission Docket 2161, Federal Trade Commission (Record Group 122), U.S. National Archives.

54. "Isidor Reinhard: Cables to Henry Johnson," 11 August 1932 (exhibit 394); 17 August 1932 (exhibit 395) and 19 August 1932 (exhibit 393), Federal Trade Commission Docket 2161, Federal Trade Commission (Record Group 122), U.S. National Archives.

55. "Charge Not Fought by 7 Rayon Makers; Federal Trade Commission Concludes Three Years of Conspiracy Hearings," *New York Times,* 13 May 1937.

56. Coleman, *Courtaulds,* 402.

57. "Rayon Group Ordered to Stop 'Price Fixing'; Federal Commission Charges Ten Corporations Entered into a 'Conspiracy,' " *New York Times,* 7 July 1937.

58. Coleman, *Courtaulds,* 408n3. Covington, Burling, Rublee, Acheson & Shorb has been a powerhouse firm in Washington, D.C., since 1919 (now Covington & Burling).

59. "Mystery: The American Viscose Corp.; A U.S. Investment of $930,000 by Some Shrewd and Close-Mouthed Britishers Yields a Stupendous Net in Twenty-Six Years, Estimated by *Fortune* at $300,000,000-odd," *Fortune,* July 1937, 39–43, 106, 108, 110, 112. The striking, modernist photographic images that illustrate the piece were made by William M. Rittase, available at the website The Visual Telling of Stories, www.fulltable.com/VTS/f/fortune/photos/rittase/mna.htm.

60. *Fortune,* "Mystery: The American Viscose Corp.," 110.

61. Ibid., 112. Sir Esme Howard at that time was Britain's ambassador to the United States.

62. Ibid., 112.

63. The important role of the cablegrams in the case was emphasized in the contemporary press; see "Rayon Price Fixing Charged," *Wall Street Journal,* 2 June 1934. The "smoking-gun" cable from Salvage to Johnson served as the FTC's exhibit 1 among 1700 (see note 53 to this chapter).

64. Coleman, *Courtaulds,* 408.

65. Eli Reinhard, personal communication with the author, San Jose, California, 18 December 2013.

66. "Sylphrap," letter from Lorillard Tobacco, 13 June 1935, item no. 04357363, Truth Tobacco Industry Documents, University of California, San Francisco, https://industrydocuments.library.ucsf.edu/tobacco/docs/ggjno166. This was an internal communication on the performance of the Sylvania product.

67. Du Pont Cellophane Co., Inc., v. Waxed Products Co., Inc., No. 6839, District Court, E.D. New York. 6 F. Supp. 859 (11 May 1934).

68. Du Pont Cellophane Co., Inc., v. Waxed Products Co., Inc., No. 266, Circuit Court of Appeals, Second Circuit, 85 F.2d 75 (17 July 1936), heard before Judges Learned and Augustus N. Hand, the opinion by the latter.

69. Du Pont Cellophane Co., Inc., v. Waxed Products Co., Inc., 304 U.S. 575 (1938). Harry D. Nims, cocounsel with Covington, was an expert on trademarks. DuPont followed with legal action targeted directly at Sylvania, which also failed; E.I. Du Pont de Nemours & Co. v. Sylvania Industrial Corporation, 122 F.2d 400 (4th Cir. 1941).

70. Taussig and White, "Rayon and the Tariff," 619. Their observation resonates with Thorstein Veblen's views expressed in "Dress as an Expression of the Pecuniary Culture" (chap. 7 of Veblen's *Theory of the Leisure Class* [1899]).

71. Irmgard Keun, *Das kunstseidene Mädchen* (Berlin: Universitas Deutsche Verlags-Aktiengesellschaft, 1932). The British translation appeared in 1933 under the Chatto and Windus imprint, translated by Basil Creighton.

72. Richard Denia Charques, review of *The Artificial Silk Girl, Times Literary Supplement,* 7 December 1933, 874. A brief notice in *New Stateman and Nation* referred to Keun's Doris character as "a half-hearted and amateurish little hetaira" (21 October 1933, 486).

73. Yvette Florio Lane, "'No Fertile Soil for Pathogens': Rayon, Advertising, and Biopolitics in Later Weimar Germany," *Journal of Social History* 44 (2010): 545–62. See also the work of Maria Makela, a historian at the California College of the Arts, www.cca.edu/academics/faculty/mmakela.

74. Keun, *Das kunstseidene Mädchen,* 78, 83, 154, 158 for *Bembergerseide.*

75. Irmgard Kuhn, *The Artificial Silk Girl,* trans. Karthie con Ankum (New York: Other Press, 2002), 94.

76. Felix Stössinger, "Die verwandelte Tauentzien: Umschichtung im Berliner Westen," in *Glänzender Asphalt: Berlin im Feuilleton der Weimarer Republik,* ed. Christian Jäger and Erhard Schütz (Berlin: Fannei & Walz, 1994), 107–11; anthologized from *Vossische Zeitung,* 17 April 1932.

77. Keun, *Artificial Silk Girl,* 134; the original wording is *echt Vulkanfieber* (Keun, *Das kunstseidene Mädchen,* 154).

78. Thomas Taylor, "Improvement in the Treatment of Paper and Paper Pulp," U.S. Patent Office, Patent No. 114,880, 16 March 1871. The patent was for the treatment of paper with zinc chloride.

79. "The Business Girl and Lux" (advertisement), *Times* (London), 25 August 1926.

80. Carl A. Alsberg, "Economic Aspects of Adulteration and Imitation," *Quarterly Journal of Economics* 46 (1931): 1–33, quotation on 5. Alsberg, a physician, was with the Food Research Institute at Stanford University. He was first chief of the Bureau of Chemistry at the Food and Drug Administration before taking his post at Stanford.

81. Ibid., 6.

82. Joseph Bersch, *Cellulose, Cellulose Products, and Artificial Rubber,* trans. William T. Brannt (Philadelphia: Baird, 1904), 323–28. White factice required disulphur dichloride, which could be diluted with carbon disulfide.

83. Three cases of carbon disulfide toxicity were reported from a factice (in German, *kautschukerzatz*) USSR operation using rapeseed oil; see A. E. Kulkow, "Beiträge zur Klinik der gewerblichen Vergiftungen," *Zeitschrift für die gesamte Neurologie und Psychiatrie* 103 (1926): 435–54. Kulkow was based at the V. A. Obukh Institute for Occupational Diseases, the premier Soviet center for this specialty.

84. *Oxford English Dictionary,* s.v. "factice."

85. The earliest proponent for the use of carbon disulfide in perfume manufacturing seems to have been Auguste Millon in Algeria; see chapter 1, note 63.

86. Thomas Brough, "Artificial Silk," *Journal of the Royal Society of Arts* 75 (10 December 1926): 97–115.

87. *Imitesan cilkku celai centamilc cintu* (Kirusnakiri, India: Sri Cuppiramaniya Vilassam Piras, 1930); this is an eight-page song cycle in Tamil.

88. Aldous Huxley, *Brave New World* (London: Chatto and Windus, 1932), 17.

89. Ibid., quotations on 139; acetate is also mentioned in chapters 3, 9, and 13.

90. Ibid. Viscose textiles are mentioned in chapters 2, 3 (twice), 7, 8, 9, 12, and 18.

91. Ibid., 164. In addition, Huxley's Lenina sports a "morocco surrogate" (i.e., synthetic leather) cartridge belt for her contraceptives (chapter 3).

92. American Medical Association, Department of Investigations, folder 899 Violetta-Viscose Treatment (inclusive); folders 0884-08, Viscose Treatment, Special Data, 1924–67; 0885-01 Viscose Treatment, Correspondence, 1926–30; 0885-02 Viscose Treatment, Correspondence, 1931–68.

93. Federal Trade Commission, "Stipulation of facts with vendor-advertiser and agreement to cease and desist; False and misleading advertising a treatment for varicose veins," stipulation no. 0340, 17 July 1933; AMA, Department of Investigations, Violetta-Viscose Treatment, folders 0884-08, item 58. The cited claim referred to is on the first page of the ten-page order.

94. P.N.L., "So-Called Viscose," 30 June 1927, AMA, Department of Investigations, Violetta-Viscose Treatment, folders 0884-08, item 54. The two-page report, initialed "P.N.L.," was transmitted to Dr. Cramp at the AMA. It was reproduced verbatim within "'Viscose' for varicose veins" in the "Propaganda for Reform" column of the *Journal of the American Medical Association* 29 (16 July 1927): 225.

95. Cole Porter, "You're the Top," 1934; see Cole Porter, *The Complete Lyrics of Cole Porter*, ed. Robert Kimball (New York: Da Capo, 1992), 170. Of note, the P. G. Wodehouse adaptation of the musical *Anything Goes* for the London stage, which did make alterations to some of the lyrics of "You're the Top" (for example, notoriously adding Mussolini to the list of top things), retained the cellophane lyric; see Hal Cazalet and Sylvia McNair, *The Land Where Good Songs Go: The Lyrics of P. G. Wodehouse* (audio recording), Harbinger Records.

96. Edward Heyman (words) and Richard Myers (music), "If Love Came Wrapped in Cellophane," © 8 February 1935, Library of Congress, Copyright Office, Catalogue of Copyright Entries, Part 3 Musical Compositions, 1935, n.s., vol. 30, no. 2, 167 (entry 3440).

97. Billie Madsen (words and music), "My Cellophane Baby," © 2 March 1937, Library of Congress, Copyright Office, Catalogue of Copyright Entries, Part 3 Musical Compositions, 1937, n.s., vol. 32, no. 3, 321 (entry 6794).

98. Steven Watson, *Prepare for Saints: Gertrude Stein, Virgil Thomson, and the Mainstreaming of American Modernism* (New York: Random House, 1998), 169, 276.

99. [Gertrude Stein], *Four Saints in Three Acts* (New York: Aaronson and Cooper, 1934); souvenir booklet for the stage production.

100. Ring Lardner, "Quadroon," *New Yorker,* 19 December 1931, 17–18, quotation on 18.

101. "Le Cellophane," in "Talk of the Town," *New Yorker,* 30 January 1932, 11.

102. E. B. White, "Alice Through the Cellophane," *New Yorker,* 6 May 1933, 22–24, quotation on 22.

103. [Brooks Paper Company], *How to Make Things with Cellophane: A Book of Suggestions* (St. Louis: Brooks Paper Company, 1932), 17.

104. Ibid.

105. Dashiell Hammett, *The Thin Man* (1934; repr., New York: Vintage, 1992), 192.

Chapter 4. Body Count

1. Petricha E. Manchester to Dr. Alice Hamilton, telegram, 11 March 1933, Special Collections, Linda Lear Center for Special Collections and Archives, Shain Library, Connecticut College (hereafter cited as Special Collections, Lear Center).

2. Alice Hamilton, *Industrial Poisons Used in the Rubber Industry*, Bureau of Labor Statistics Bulletin 179 (Washington, D.C.: Government Printing Office, 1915). The "cold cure" using carbon disulfide was sometimes also referred to as the "acid cure."

3. Ibid., 26–32.
4. Ibid., 30.
5. Ibid., 31.
6. Ibid., 19.
7. Ibid., 34, 38.

8. "The Dangers of Rubber Manufacture," *Journal of the American Medical Association* 66 (29 January 1916): 356–58. The unsigned editorial refers to Ohio and thus may have been authored by Emery Hayhurst, head of the Division of Occupational Diseases of the Ohio State Board of Health; see, for example, Emery Roe Hayhurst, *A Survey of Industrial Health-Hazards and Occupational Diseases in Ohio, Transmitted February 1, 1915* (Columbus, Ohio: Heer, 1915).

9. Alice Hamilton, "Marcus Hook, Nov. 28th 1919," typed three-page report of a factory site visit, Special Collections, Lear Center. Although Hamilton does not identify the name of the facility, she lists "Dr. C. E. Ford, 25 Broad St. New York" first among a group joining her on the visit, a link to Benzol Products Co. (later National Aniline and Chemical Company Inc.), which had a manufacturing site in Marcus Hook.

10. George Edmund de Schweinitz, "Concerning Some Varieties of Toxic Amblyopia, with Illustrative Cases, Being a Clinical Communication," *Transactions of the College of Physicians of Philadelphia*, 3rd ser., 47 (1921): 216–19, also reported in the *American Journal of Ophthalmology* 5 (1922): 381–88 (de Schweinitz's cases on 386–88). De Schweinitz was an expert on the condition; see G. E. de Schweinitz, *The Toxic Amblyopias* (Philadelphia: Lee Brothers, 1896), 105–16.

11. Alice Hamilton, "Inorganic Poisons, Other than Lead, in American Industries," *Journal of Industrial Hygiene* 1 (1919): 89–102. This paper makes clear that the insane-asylum-committed case that Hamilton reported in 1915 was from a Detroit, Michigan, factory.

12. Hamilton, *Industrial Poisons in the United States*, 360–69.

13. Ibid., 368–70.

14. Ibid., 369. Hamilton refers to a fatal case of "carbanilid" poisoning. There are variant terms and spellings for thiocarbanilid(e), carbanilid(e), and sulpho-carbanilide (preferred technical name, N,N'-diphenylthiourea).

15. Alice Hamilton, "Nineteen Years in the Poisonous Trades," *Harper's Magazine*, 1 October 1929, 580–91.

16. Ibid., 587.

17. Ibid., 591.

18. Alice Hamilton to Mrs. Petricha E. Manchester, 11 March 1933, Special Collections, Lear Center.

19. Ibid.

20. Alice Hamilton, *Exploring the Dangerous Trades* (Boston: Little, Brown, 1943), 391–92.

21. Petricha E. Manchester to Dr. Alice Hamilton, telegram, 11 March 1933.

22. "Industrial Poisons a Hazard to Workers," *Bulletin of the Consumers' League of New York* 2, no. 1 (1923): 3.

23. Barbara Sicherman, *Alice Hamilton: A Life in Letters* (Cambridge, Mass.: Harvard University Press, 1984), 281.

24. Hamilton, *Exploring the Dangerous Trades,* 390. Hamilton notes: "Other histories came to me from Estelle Lauder, of the Philadelphia Consumers' League, but I had no way of checking up on these stories." A. Estelle Lauder was a leader with the Consumers' League of Eastern Pennsylvania and active in occupational safety and health in this period.

25. Pierre S. du Pont to Petricha Manchester, 22 November 1927, Hagley Library and Museum, Pierre S. du Pont Papers, LMSS 10/A/File 614. The records of the National Consumers' League held at the U.S. National Archives contain records of later correspondence from Manchester: to Mary Dublin (then general secretary of the National Consumers' League and a personal friend of Hamilton's) on 18 December 1939, and to Elizabeth Magee on 25 March 1944, when she was general secretary (Library of Congress, Manuscript Division, National Consumer League Records, reel 15). Thus, Manchester was still active at the time that Hamilton published *Exploring the Dangerous Trades.*

26. U.S. Department of Labor, Division of Labor Standards, *Reports of Committees and Resolutions Adopted by Third National Conference on Labor Legislation, November 9, 10, 11, 1936* (Washington D.C.: Government Printing Office, 1936), 5–6, 9–11.

27. In August 1944, Hamilton assumed the honorific post of national president of the National Consumers' League; see Sicherman, *Alice Hamilton,* 375.

28. Alice Hamilton, day pocket diary, entry for 3 March 1938, Hamilton Family Papers 86–133, Schlesinger Library, Radcliffe Institute for Advanced Study, Harvard University. The diary entry notes the other diners as "Edison of Navy" and "Quaker in Spa.(nish) Relief." Edison may be Charles Edison (son of the inventor Thomas Edison), who at that time was undersecretary of the navy.

29. The Delaware Rayon Company was one of only three of the ten charged companies that continued to fight the case until the end; see "Charge Not Fought by 7 Rayon Makers; Federal Trade Commission Concludes Three Years of Hearings; Only Three of Ten Appear; Indication Is That Orders Will Be Issued to Stop Fixing of Companies' Prices," *New York Times,* 13 May 1937.

30. "John Pilling Wright; President of Textile Companies Dies in Newark, Del., at 66," *New York Times,* 19 April 1947. At the time of his death, Wright was the president of Delaware Rayon, New Bedford Rayon, and the Continental-Diamond Fibre Com-

Notes to Pages 84–87   243

pany. Wright's home later was given to the University of Delaware to become its presidential residence. See "Out of the Attic," NewarkPostOnline.com, www.newarkpostonline.com/news/local/article_253e7b44-2aa8-56a6-9612-b1dc2ede0155.html?mode=jqm.

31. Theresa Hessey, *Newark* (Mount Pleasant, S.C.: Arcadia, 2007), 109.

32. "Historical Note: Inventory of the New Bedford Rayon, Inc., Records in the New Bedford Whaling Museum Research Library," New Bedford Whaling Museum, www.whalingmuseum.org/explore/library/finding-aids/mss25.

33. "Old Steel Plant Sold; Shell-Loading Works of Bethlehem Corp., near Wilmington, Bought by Delaware Rayon Co.," *Wall Street Journal*, 16 February 1926.

34. Delaware Federal Writers' Project, *Delaware: A Guide to the First State* (New York: Viking, 1938), 467. A 1955 revision of the guidebook still calls attention to the "Delaware Rayon Plant" in its tour number 10 but, consistent with Alice Hamilton's experience, adds a parenthetical note "*(no admission)*" not in the original (italics in text).

35. Hamilton, *Exploring the Dangerous Trades*, 184.

36. Bowing v. Delaware Rayon Co., Superior Court of Delaware, New Castle County, 38 Del. 206; 190 A. 567; 1937 Del. LEXIS 24; 8 W.W. Harr. 206 (15 February 1937).

37. Ibid.

38. Bowing v. Delaware Rayon Co., Superior Court of Delaware, New Castle County, 38 Del. 339; 192 A. 598; 1937 Del. LEXIS 34; 8 W.W. Harr. 339 (2 April 1937)

39. Ibid.

40. Susan M. Hartmann, "Herrick, Elinore Morehouse," in Barbara Sicherman and Carol Hurd Green, eds., *Notable American Women: The Modern Period* (Cambridge, Mass: Harvard University Press, 1980), 335–57.

41. Elinore Herrick, "Rough Draft of Chapters for Memoir," Elinore Morehouse Herrick, Papers, 1931–1964, file 154, Schlesinger Library, Radcliffe Institute, Harvard University, "Chapter II. "I BECOME A BUREAUCRAT" (capitalization and quotation marks in original), quoted passage on 8.

42. Ibid., "Chapter One. I BREAK A STRIKE" (capitalization in original), quoted passage on 4.

43. Victoria Enos, "Two Weeks at the Industrial Rayon Corporation in February 1930," National Women's Trade Union League of America Records, 1910–1934, B-16, folder 30, pp. [1]–[8] (seq. 139–46), Schlesinger Library, Radcliffe Institute, Harvard University, Cambridge, Mass. (hereafter cited as National Women's Trade Union League of America Records). This most likely but not definitively describes the Industrial Rayon Corporation in Covington, Virginia.

44. Ibid., 4.

45. Four-page typed manuscript, National Women's Trade Union League of America Records.

46. Frieda Schwenkmeyer to Elizabeth Christman, 18 September 1930, National Women's Trade Union League of America Records. This letter transmits the typed report on Industrial Rayon cited in note 43 to this chapter.

47. Brownie Lee Jones to the General Council, Trade Union Congress [U.K.], 17 October 1929, Trade Union Congress Archives, Modern Records Center, University

of Warwick. Brownie Lee Jones set up the Industrial Department for the Denver YWCA and then served in that role in Flint, Michigan, before working in Richmond, Virginia, from 1928 to 1932. When Jones came to the Richmond YWCA, Lucy Randolph Mason was its director; she then went to the National Consumers' League; see "Oral History Interview with Brownie Lee Jones," conducted by Mary Fredrickson, 20 April 1976, Program on Women and Work, Institute of Labor and Industrial Relations, University of Michigan—Wayne State University, vitae (unpaginated) and p. 22–29, which focus on her time with the YWCA in Richmond.

48. Joseph A. Fry, "Rayon, Riot, and Repression: The Covington Sit-Down Strike of 1937," *Virginia Magazine of History and Biography* 84 (1976): 3–18.

49. Marie Tedesco, "North American Rayon Corporation and American Bemberg Corporation" and "Elizabethton Rayon Plants Strikes, 1929," *Tennessee Encyclopedia of History and Culture,* http://tennesseeencyclopedia.net/entry.php?rec=1005 and http://tennesseeencyclopedia.net/entry.php?rec=1277.

50. The school was founded in 1923 and was originally called First High School; see the website for Carter County Schools, https://sites.google.com/a/hvhs.carterk12.net/happy-valley-high/home/archives.

51. "Dr. Mothwurf Arrested; Chief Chemist of Bayer Plant Will Be Examined Here," *New York Times,* 24 August 1918; see also Mira Wilkens, *The History of Foreign Investment in the United Sates, 1914–1945* (Cambridge, Mass.: Harvard University Press, 2004), 124, 702n357.

52. "38 Germans are Sent to Internment Camp," *New York Times,* 12 November 1918.

53. ZoomInfo, "Dr. Arthur Franz Mothwurf," www.zoominfo.com/p/Arthur-Mothwurf/1207715184.

54. James A. Hodges, "Challenge to the New South: The Great Textile Strike in Elizabethton, Tennessee, 1929," *Tennessee Historical Quarterly* 23 (1964): 343–57.

55. Jacquelyn Dowd Hall, "Disorderly Women: Gender and Labor Militancy in the Appalachian South," *Journal of American History* 73 (1986): 354–82.

56. "Miss Margaret Bowen" [address], *Report of the Proceedings of the Forty-Ninth Annual Convention of the American Federation of Labor, Held at Toronto, Ontario, Canada* 49 (October 1929): 276–67.

57. Bessie Edens, "My Work in an Artificial Silk Mill," in *Scraps of Work and Play,* Southern Summer School for Women Workers in Industry, Burnsville, North Carolina, July 11–August 23, 1929, 21–23, American Labor Education Service Records, Kheel Center for Labor Management Documentation and Archives, Cornell University, quotation on 21.

58. Alfred Hoffmann, "The Mountaineer in Industry," *Mountain Life and Work* 5 (1930): 2–7, quotation on 3.

59. Christine Galleher [sic], "Where I Work," in *Scraps of Work and Play,* 23. Her named is spelled elsewhere as Galliher.

60. [Ida Heaton], *Scraps of Work and Play,* 17. Among a group of unsigned brief personal reports, this appears to have been written by Ida Heaton, who also worked at Glanzstoff.

61. Matilda Lindsay, "Rayon Mills and Old Line Americans," *Life and Labor Bulletin* 7, no. 71 (1929): 3–4.

62. Ibid., 4.

63. "The National Guard and Elizabethton," *Life and Labor Bulletin* 7, no. 74 (1929): 1–2.

64. "North American Rayon (Bemberg & Glanzstoff) Strike of 1929 and Preparations for a Visit by Herbert Hoover," unnumbered series (series A), American Bemberg Corporation (American Glanzstoff Corporation) (North American Rayon Corporation) Film and Video Collection, State of Tennessee, Department of State, Tennessee State Library and Archives.

65. Rozella Hardin, "Elizabethton Once Had Its Own Chewing Gum Factory," *Elizabethton (Tennessee) Star*, 19 January 2009.

66. "Lehman Quit as Director of Rayon Concerns in Protest on Troops in Southern Mill Strike," *New York Times*, 15 May 1929.

67. Sherwood Anderson, "Elizabethton, Tennessee," *Nation*, 1 May 1929, 526–27.

68. Press photograph, uncredited, stamped "Referred-E. Dept. October 5 29 N.E.A.," collection of the author.

69. "Rayon Mills' Head Dies by his Own Hand; Konsul W. C. Kummer Breaks Down after Taking Charge at Elizabethton, Tenn.," *New York Times*, 2 October 1929; see also "Charles Wolff Succeeds Mothwurf," *New York Times*, 1 February 1930.

70. *Scraps of Work and Play.*

71. Hall, "Disorderly Women."

72. *Scraps of Work and Play*, n.p. This appears in a "Jokes" section that follows the main body of the text numbered 1–49.

73. Anderson, "Elizabethton, Tennessee," 527.

74. Alice Hamilton, *Industrial Toxicology*, Harper's Medical Monographs (New York: Harper & Brothers, 1934), 241–44.

75. Ibid., 244n. Hamilton uses footnotes sparingly in the book—there are fewer than 36 in 271 pages of text. The bibliography, by contrast, has 655 citations. In the book's preface, Hamilton notes that the citations reflect the medical literature up to January 1933 (that is, a year before the 1934 publication). Thus, it is likely that Hamilton had completed her work on this handbook before the March 1933 telegram from Manchester (see note 1 to this chapter).

76. Alice Hamilton to Petricha Manchester, 11 March 1933.

77. Sicherman, *Alice Hamilton*, 357.

78. Alice Hamilton, "Some New and Unfamiliar Industrial Poisons," *New England Journal of Medicine* 215 (3 September 1936): 425–32.

79. Ibid., 427.

80. Alice Hamilton, "The Making of Artificial Silk in the United States and Some of the Dangers Attending It," in *Discussion of Industrial Accidents and Diseases: 1936 Convention of the International Association of Industrial Accident Boards and Commission, Topeka, Kansas,* Division of Labor Standards, U.S. Department of Labor, Bulletin 10 (Washington D.C.: Governmental Printing Office, 1937), 151–59.

81. Ibid., 151.

82. Ibid., 154.

83. Ibid., 159.

84. Hamilton, "Nineteen Years in the Poisonous Trades."

85. Sicherman, *Alice Hamilton,* 249.

86. Mary Ann Dzuback, "Women and Social Research at Bryn Mawr College, 1915–40," *History of Education Quarterly* 33 (1993): 579–608. Mildred Fairchild Woodbury was Kingsbury's longtime collaborator and successor as department chair at Bryn Mawr.

87. Susan M. Kingsbury to Dean Helen Taft Manning, Bryn Mawr College, 26 February 1935, Archives, Bryn Mawr College. Although it is not clear how Cohn became connected with the Byrn Mawr research project, Kingsbury's letter leaves little doubt that Alice Hamilton was not acquainted with her directly before this. Other than Cohn's medical school dean, the other recommender that Kingsbury quotes is Dr. Henry Kessler, medical director of the New Jersey Rehabilitation Clinic. Previously, Kessler had been heavily involved in occupational medicine, studying illness as well as injury. A Dr. Kessler (likely the same) was noted by Hamilton as being present on her visit to the Marcus Hook aniline factory (see note 9 to this chapter).

88. Department of Labor and Industry, Commonwealth of Pennsylvania, *Report to Governor Georg H. Earle by Ralph M. Bashore, Chairman Governor's Commission on Occupational Disease Compensation and Supplements* (Harrisburg, Penn.: Department of Labor and Industry, 1937). This typed, mimeographed report includes a memorandum of transmission to Pennsylvania's governor Earle from Bashore (5 pp.), followed by four supplements, each paginated separately: "Draft of Commission's Bill for Occupational Compensation" (8 pp.); "Report of the Medical Committee" (5 pp.); "Preliminary Report of the W.P.A. Survey of Industrial Disease Hazards" (3 pp.); and "Bryn Mawr Study of Occupational Disease in Pennsylvania" (83 pp. with a 2-page appendix). The results of the lead poisoning investigation appear on pp. 6–61 of the Bryn Mawr Study report.

89. Ibid., "Bryn Mawr Study of Occupational Disease in Pennsylvania," 5–5a.

90. Ibid., 64.

91. Ibid., 68.

92. Ibid., 70. This case is case 1 in Cohn's series in the Bryn Mawr Report (the coworker hospitalized at the same time appears as case 32).

93. Ibid., 78. This was case 28.

94. Ibid. The suicide attempt is case 33, the worker who shot off his arm is case 36. Spinning room (as opposed to churn room) workers with mental illness include cases 16 and 30 and the brother of case 12.

95. Ibid.; memorandum of transmission from Bashore, 12 March 1937.

96. Gerald Markowitz and David Rosner, eds., *"Slaves of the Depression": Workers' Letters About Life on the Job* (Ithaca, N.Y.: Cornell University Press, 1987), 43–45. In a brief introduction to the letter referred to in the text, viscose manufacturing is mischaracterized as a part of "the developing plastics industry."

97. Ibid. The letter from Bashore's department was signed by Raymond J. Nicaise.

98. Dan Cupper, *Working in Pennsylvania: A History of the Department of Labor and Industry,* ch. 2, "1913—1940: From Small Beginnings: The Department of Labor

Notes to Pages 97–99   247

and Industry Is Created," Pennsylvania Department of Labor and Industry, www.portal.state.pa.us/portal/server.pt?open=514&objID=621479&mode=2. The occupational disease statute was one part of a wider legislative agenda for the improvement of workers' compensation, which was central to the labor reforms of the Pennsylvania "Little New Deal"; see Richard C. Keller, *Pennsylvania's Little New Deal* (New York: Garland, 1982), 269–71.

99. Occupational Disease Prevention Division, *Survey of Carbon Disulphide and Hydrogen Sulfide Hazards in the Viscose Rayon Industry,* Bulletin 46 (Harrisburg, Penn.: Department of Labor and Industry, August 1938); see, in particular, Lillian Erskine, "Scope of the Viscose Rayon Survey," 7–9.

100. Ibid., 1, transmittal letter from Bashore, 31 August 1938.

101. Ibid, 7. Erskine describes her federal position as "Special Agent."

102. Hamilton, *Exploring the Dangerous Trades,* 392.

103. Bernd Holdroff, "Friedrich Heinrich Lewy (1885–1950) and His Work," *Journal of the History of Neurosciences* 11 (2002): 19–28.

104. See F. H. Lewy, "The Application of Chronaximetric Measurement to Industrial Hygiene, Particularly to the Examination of Lead Workers," and Ronald E. Lane and F. H. Lewy, "Blood and Chronaximetric Examination of Lead Workers Subjected to Different Degrees of Exposure: A Comparative Study," *American Journal of Industrial Hygiene* 17 (1935): 73–78, 79–92. Lewy's work with chronaximetric measurement began in Germany; see F. H. Lewy and Stefan Weisz, "Chronaxieuntersuchungen an schwach- und starkgefährdeten Bleiarbeitern," *Archiv für Gewerbepathologie und Gewerbehygiene* 1 (1930): 561–68. Although Hamilton notes in her autobiography that Lewy had done previous work on carbon disulfide psychoisis *and* palsy, this appears to have been related work of Weisz's limited to palsy (i.e., neuropathy). See Stefan Weisz, "Chronaximtriche Untersuchungen uber die Wirkung verschiedene Gewerbegifte," *Deutsche Medizinische Woschenschrift* 55 (1929): 782–83.

105. Patrick J. Sweeney, Mark Frazier Lloyd, and Robert B. Daroff, "What's in a Name? Dr. Lewey and the Lewy Body," *Neurology* 49 (1997): 629–30.

106. Hamilton, *Exploring the Dangerous Trades,* 393.

107. Occupational Disease Prevention Division, "Neurologic Aspects of CS2 Intoxication," in *Survey of Carbon Disulphide and Hydrogen Sulfide Hazards,* 31–37. Lewy went on to publish his neurological findings in a major medical journal: F. H. Lewey [sic], "Neurological Medical and Biochemical Signs and Symptoms Indicating Chronic Industrial Carbon Disulphide Absorption," *Annals of Internal Medicine* 15 (1941): 869–83. The mental health findings from the study were published the following year by the lead psychiatrist on the team: Francis J. Braceland, "Mental Symptoms Following Carbon Disulphide Absorption and Intoxication," *Annals of Internal Medicine* 16 (1942): 246–61. Braceland became a major figure in U.S. psychiatry. In addition, an article on the study's ophthalmologic findings was published: Robb McDonald, "Carbon Disulfide Poisoning," *Archives of Ophthalmology* 20 (1938): 839–45.

108. "Head of Medical Department Retires at American Viscose," *Delaware County (Pennsylvania) Daily Times,* 9 August 1969; see also "Dr. J. A. Calhoun; Medicine Pioneer" (obituary), *Delaware County Daily Times,* 11 February 1976.

109. On Drinker's close relationship with the American Viscose Corporation, see Philip Drinker, "Industrial Hygiene with Illustrations Drawn from the Rayon Industry, Thursday, February 8, 1940, Northeastern Section and Rhode Island Section of the American Chemical Society," *Nucleus* 17 (1940): 103; see also George M. Reece, Ben White, and Philip Drinker, "Determination and Recording of Carbon Disulfide and Hydrogen Sulfide in the Viscose-Rayon Industry," *Journal of Industrial Hygiene and Toxicology* 22 (1940): 416–24. Financial support by the AVC is not formally acknowledged in that publication, although White was an AVC employee.

110. Alice Hamilton to Verne Zimmer, 4 May 1939, U.S. National Archives, Division of Labor Standards, Department of Labor, Classified General Files, Record Group 100, box 28, 1937–1941, file 7-0-6-5 Viscose Study (Rayon), hereafter cited as U.S. National Archives, Viscose Study (Rayon). Zimmer's response, approving Hamilton's plan to visit the American Viscose Corporation "under the invitation of Dr. Calhoun and Mr. Drinker," was sent on 8 May 1939.

111. Alice Hamilton, *Occupational Poisoning in the Viscose Rayon Industry*, U.S. Department of Labor, Division of Labor Standards, Bulletin 34 (Washington D.C.: Government Printing Office, 1940). The document was formally transmitted by Verne Zimmer to Frances Perkins, the secretary of labor, on 17 November 1939.

112. Lillian Erskine, "Appendix: A Study of Cases of Psychosis among Viscose Rayon Workers," in ibid., 63–76.

113. Ibid., 72–73. This is Erskine's case 15.

114. Lillian Erskine to Verne Zimmer (Division of Labor Standards), 5 May 1938, U.S. National Archives, Viscose Study (Rayon). The cover memorandum states, "Attached hereto are Cases I, II, and V of the Supplementary Lewistown Survey, which together with those in your possession (Cases II, IV and VII) complete the files of the Lewistown Supplementary Survey." The case numbering does not correspond to the numbering Erskine used in her published appendix; the case in question is case 2 among the transcribed records.

115. Erskine, "Appendix: A Study of Cases of Psychosis," 73. These are Erskine's cases 5 (p. 69), 6 (70), 2 (68), and 16 (73).

116. Jean Alonzo Curran, "Transcript of Taped Interview, Nov. 29, 1963, with Alice Hamilton, M.D., A.M. (hon.), S.D. (hon.), Assistant Professor of Industrial Medicine, Emerita, Harvard University of Public Health," Jean Alonzo Curran Papers, Special Collections, Francis A. Countway Library of Medicine, Center for the History of Medicine, Harvard University. A *JAMA* editorial also emphasized the importance of pathological data from the study: "Industrial Carbon Disulfide Poisoning," *Journal of the American Medical Association* 112 (7 January 1939): 51–52.

117. Pennsylvania Historical and Museum Commission, Pennsylvania Governors, "Governor George Howard Earle III," www.portal.state.pa.us/portal/server.pt/community/1879-1951/4284/george_howard_earle/469117.

118. George H. Earle to Franklin D. Roosevelt, 27 November 1933, in Edgar B. Nixon, ed., *Franklin D. Roosevelt and Foreign Affairs*, vol. 1, *January 1933–February 1934* (Cambridge, Mass.: Harvard University Press, 1969), 504–7.

119. Richard C. Keller, "Pennsylvania's Little New Deal," *Pennsylvania History* 29 (1962): 391–406 (see note 98 to this chapter for Keller's expanded work on this sub-

ject). See also Randolph H. Bates, "A Prophet Without Honor in His Own House: Governor George H. Earle III and Pennsylvania's Little New Deal" (thesis, Pennsylvania State University, 1996).

120. Lillian Erskine, "Report to Ralph M. Bashore," 21 March 1938, seven-page typescript, U.S. National Archives, Division of Labor Standards, Department of Labor, Classified General Files, Record Group 100, box 28, 1937–1941.

121. "Earle Taxes Assailed; Business Men Assert Industry Is Being Driven from State," *New York Times,* 9 April 1938.

122. "Earle Retraction Demanded by Foes," *New York Times,* 27 April 1938. Earle refers to the American Viscose Corporation as the "Viscose Company," the name by which the company had been known. The former had been set up as a holding company for the latter (along with the related Viscose Corporation of Virginia) as early as 1922. In 1937, however, all the assets were integrated into the American Viscose Corporation as the single remaining business entity (still owned by Courtaulds); see Coleman, *Courtaulds,* 302–3.

123. "Earle Retraction Demanded by Foes," *New York Times.*

124. Ibid.

125. Lillian Erskine to Verne Zimmer, 30 April 1938, two-page typescript, U.S. National Archives, Viscose Study (Rayon).

126. George Howard Earle, "Address of George H. Earle, Governor of Pennsylvania, from Station WHP, Harrisburg, and Over State-wide Radio Network, Wednesday Evening, October 5, 1938 at 6:45 P.M." (transcript), Democratic State Committee press release, Manuscript Group 342, George Howard Earle Papers, Speeches (series 342m.3), Pennsylvania State Archives, Harrisburg, Pennsylvania.

127. Keller, *Pennsylvania's Little New Deal;* see also "Lawrence Freed in Pennsylvania; Jury of Republicans Clears Him and 7 Other Democrats of Gift Assessing Plot," *New York Times,* 13 April 1940.

128. Fred G. Stuart to Frances Perkins, 6 June 1938, U.S. National Archives, Viscose Study (Rayon).

129. V. A. Zimmer to Fred G. Stewart, 12 June 1938, U.S. National Archives, Viscose Study (Rayon).

130. Samuel T. Gordy and Max Trumper, "Carbon Disulfide Poisoning," *Journal of the American Medical Association* 110 (7 May 1938): 1543–59.

131. Ibid.

132. Medical legal testimony by Gordy and Trumper was noted in Plaugher v. American Viscose Corp. (151 Pa. Super. 401; 30 A.2d 376, 1943). Thomas Plaugher was a deceased Marcus Hook employee, and the case was linked to testimony in the John Nichols case (see the following note). The court record refers to seven carbon disulfide cases, four heard by Referee Alessandroni (presumably from Marcus Hook) and three by Referee Patterson in Lewistown. The attorneys in the Plaugher case were Henry Temin and Todd Daniel. A letter dated 14 November 1934 from Estelle Lauder, secretary of the Consumers' League of Eastern Pennsylvania, to Clara Beyer of the Labor Standards Committee [sic] at the U.S. Department of Labor recommends both Trumper and Temper for consideration: "I am not suggesting their appointment to anything, you know, merely letting you know that they exist" (Department of Labor, Classified

General Files, Record Group 100, box 27, 1934–1937). One of Henry Temin's sons, Howard, went on to become a Nobel laureate for the discovery of reverse transcriptase.

133. "Medicine: CS2 [sic] Poisoning," *Time,* 18 March 1940, 56. This article recounts the case of a Marcus Hook American Viscose Company worker named John Nichols, which was heard by Alessandroni (see the preceding note). *Time* noted that Trumper and Gordy were at the hearing.

134. Although not directly connected with Trumper, Alice Hamilton did have ties to Estelle Lauder, who had written to the Division of Labor Standards on his behalf (see note 132 to this chapter). Hamilton was linked to Trumper also through Adele Cohn, whose report acknowledged the assistance of Max Trumper. Hamilton's *Exploring the Dangerous Trades* does not mention Trumper or Cohn by name.

135. Philip Drinker, "Ventilation of Churn Rooms in Viscose Rayon Manufacture," *Safe Practice Bulletin* 25 (Harrisburg, Pa.: Occupational Disease Prevention Bureau, Department of Labor and Industry, n.d. [c. 1939]), six pages with added illustrations, quoted passage on 4.

136. The American Standards Association was the initial arbiter of recommended workplace exposure limits. Another organization, the American Conference of Governmental Industrial Hygienists, was founded in 1938 (originally as the National Conference of Governmental Industrial Hygienists) and by the 1940s was recommending its own set of exposure limits. When OSHA was founded, it adopted the nonbinding limits proposed by both organizations as legal standards; see Liora Salter, *Mandated Science: Science and Scientists in the Making of Standards* (Dordrecht, Netherlands: Kluwer Academic, 1988), 36–37.

137. Drinker, "Ventilation of Churn Rooms," 4.

138. Hamilton, *Occupational Poisoning,* 61–62.

139. Bernard J. Alpers and Friedrich H. Lewy, "Changes in the Nervous System Following Carbon Disulfide Poisoning in Animals and in Man," *Archives of Neurology and Psychiatry* 44 (1940): 725–39. Additional details of experimental animal-exposure data, underscoring adverse vascular effects in the brain, appeared in Friedrich H. Lewey [sic] et al., "Experimental Chronic Carbon Disulfide Poisoning in Dogs: Clinical, Biochemical, and Pathological Study," *Journal of Industrial Hygiene and Toxicology* 23 (1941): 415–36. Alice Hamilton made a point of presenting Lewy's abstract on his work at the 1938 Congress of the Permanent Commission on Occupational Health held in Germany. She was one of the small U.S. delegation, which included Henry Kessler.

140. Alpers and Lewy, "Changes in the Nervous System," 738.

141. "Carbon Disulphide Poisoning—Dermatitis from Formaldehyde," *Journal of the American Medical Association* 103 (11 August 1934): 433–34.

142. Ibid.

143. Geraldine Strey, reference librarian, Wisconsin Historical Society, to the author, e-mail, 30 August 2010; data were derived from Madison city directories for years including 1927, 1929, 1931, and 1937.

144. Norman William Pettys, "Transparent Wrapping Materials," *Industry Report,* September 1932, Retail Credit Company, Atlanta, Georgia. The publication was noted in a synopsis published in *Industrial Medicine* 1 (1932): 128.

145. Samuel T. Gordy and Max Trumper, "Carbon Disulfide Poisoning: Report of 21 Cases," *Industrial Medicine* 9 (1940): 231–34.

146. Ibid., 234. Gordy and Trumper directly quote their own prior work (unpublished previously) presented at the Ninth Annual Convention of the Greater New York Safety Council, 20 April 1938.

147. Hamilton, *Exploring the Dangerous Trades,* 394.

Chapter 5. Rayon Goes to War

1. American Viscose Corporation, one-page printed letter, collection of the author. "Crown" refers to AVC's "Crown Rayon Yard" product line. By 1943, the AVC had plants in Marcus Hook, Lewistown, and Meadville, Pennsylvania (the last producing nonviscose cellulose acetate rayon); Roanoke and Front Royal, Virginia; and Parkersburg and Nitro, West Virginia.

2. American Viscose Corporation, *Rayon Goes to War* (Marcus Hook, Pa.: American Viscose Company, 1943), twelve-page pamphlet, unpaginated.

3. Ibid., 5.

4. DuPont (Cellophane), "The silent enemy in the steaming jungle" (advertisement), *Saturday Evening Post,* 19 June 1943.

5. Hutchins, *Labor and Silk,* 66.

6. Ibid., 67.

7. DuPont, "Our Company," History, 1900–26, "1917: Old Hickory," www.dupont.com/corporate-functions/our-company/dupont-history.html.

8. [DuPont], *Leonard A. Yerkes.*

9. U.S. Senate, *Hearings Before the Special Committee Investigating the Munitions Industry: Profiteering, Government Contracts, and Expenditures During World War, Including Early Negotiations for Old Hickory Contract,* 73rd Cong., pt. 13 (13 December 1934) and pt. 14 (14 December 1934) (Washington D.C.: Government Printing Office, 1935).

10. "Merchants of Death," 4 September 1934, U.S. Senate website, Senate History, www.senate.gov/artandhistory/history/minute/merchants_of_death.htm.

11. Royal Society, "Cross, Charles Frederick (1855–1935) Elected 1917," Certificate of a Candidate for Election, Archives, Royal Society, U.K., EC/1917/04.

12. Coleman, *Courtaulds,* 290–91.

13. Ibid., 419.

14. Wilkins, *History of Foreign Investment,* 151–52.

15. "To Make Artificial Silk; Girls Go from Here to Belgium to Learn Tubize Process," *New York Times,* 13 September 1920.

16. "Transformation of Munitions Plants to Needs of Peace," *Chemical Age* 28 (1920): 308.

17. Ferenc Horváth, *Sárvár Monográfiája* (Szombathely, Hungary, 1978), 491–94.

18. Jerzy Skoracki, "On Beginning of Chemical Fibres' Manufacturing in Poland," *Chemik* 65 (2011): 1307–18. A review of occupational substances toxic to the nervous system published in Poland just before World War II presented two clinical

case summaries of workers likely from the Tomoschaw factory; see M. Kalinska, "Zaburzenia psychiczne na skaieck zairnć zawadowych," *Polska Gazaeta Lekaska* 18 (1939): 164–68.

19. "Expansion and Competition in Artificial Silk Manufacture," *Chemical Age* 28 (1920): 320.

20. Kit Wiegel, "Since Tubize, Hopewell Has Never Been the Same," *Hopewell (Virginia) News,* 13 July 2012, reprinting the contents of an article that originally ran on 14 May 1976.

21. "Run Striker Gauntlet to Avert Blast; Fifty Guarded Men Enter Tubize Chatillon Factory to Remove Explosive Material," *New York Times,* 1 July 1934.

22. "Rayon Strike Shuts Mill Permanently; Tubize Chatillon to Abandon Hopewell, Va., Yarn Factory, Abolishing 1,500 Jobs; Union Attacks Proposal, Charges Report of Damage to Idle Machinery is a Ruse to 'Starve Out' Workers," *New York Times,* 25 July 1934.

23. "Rayon Company Formed; New Concern in Brazil to Get Equipment from Tubize," *New York Times,* 17 June 1935.

24. Wilkins, *History of Foreign Investment,* 328.

25. R. S. Borrows, "Plant Notice Number 106, Hopewell Works": "Agitated unemployed . . . working conditions in our Hopewell plant are generally good . . . [as] those at Rome . . . Without the influence of certain persons . . . there would be no strike talk in this community," Tubize Corporation, Hopewell, Va., 1934, collection of the Virginia Historical Society.

26. Michelle Brattain, *The Politics of Whiteness: Race, Workers, and Culture in the Modern South* (Princeton, N.J.: Princeton University Press, 2001), 61.

27. H. L. Barthelemy, "Ten Years' Experience with Industrial Hygiene in Connection with the Manufacture of Viscose Rayon," *Journal of Industrial Hygiene and Toxicology* 21 (1939): 141–51.

28. *Oxford English Dictionary,* s.v. "dope."

29. A. J. A. Wallace Barr, "Dope," in A. J. Swinton, ed., *The Aeroplane Handbook* (London: Aeroplane and General Publishing Company, 1920), 140–44. A footnote states, "This article was written just before the cessation of hostilities in November, 1918."

30. "Fatal Case of Poisoning by Tetrachloride of Ethane," *Lancet* 184 (26 December 1914): 1489–91. This report is a summary of the coroner's inquest in the death of Luxmore Drew, a thirty-six-year-old doping worker, including testimony by the forensic specialist Dr. William Henry Willcox. Willcox later provided additional information on this and another thirteen cases, three fatal; see William Henry Willcox, "Toxic Jaundice Due to Tetrachlor-Ethane Poisoning: A New Type Amongst Aeroplane Workers," *Lancet* 185 (13 March 1915): 544–47.

31. [Paul] Jungfer, "Tetrachloräthanvergiftungen in Flugzeugfabriken," *Zentralblatt für Gewerbehygiene* 2 (June 1914): 222–23. Jungfer reports four cases, one of which was fatal; the exposure occurred in the winter of 1913–14. Hamilton made this work known in the United States: Alice Hamilton, "Industrial Poisoning in Aircraft Manufacture," *Journal of the American Medical Association* 64 (15 December 1917): 2037–39.

32. "The King's Interest in Flying: Factory and Aerodrome Visited," *Times* (London), 1 June 1917.

33. Coleman, *Courtaulds*, 180–83.

34. "Scatter Acorns That Oaks May Grow," http://libraries.mit.edu/archives/exhibits/adlittle/history.html. See also Kahn, *Problem Solvers*.

35. Arthur D. Little, "Carbon Filament and Method of Manufacturing Same," U.S. Patent No. 532,568, 15 January 1895. The patent was filed on 18 June 1894.

36. Harry S. Mork, Arthur D. Little, and William W. Walker, "Artificial Silk," U.S. Patent 712,200, 28 October 1902.

37. *The Royal Little Story: Commemorating the Establishment of the Royal Little Professorship in Business Administration at the Harvard University Graduate School of Business Administration, March 1, 1966* (Cambridge, Mass: Harvard University, 1966).

38. *The War Record of the Fifth Company New England Regiment Second Plattsburg Training Camp* (Cambridge, Mass.: Harvard University Press, 1922), 34, listing Royal Little's business address as the Lustron Company, 44 K Street, South Boston. His service record was with the 42nd Division, Company K, 167th Regiment, which experienced some of the highest numbers of gas casualties among the American Expeditionary Force; see: Cory J. Hilmas, Jeffery K. Smart, and Benjamin A. Hill, "History of Chemical Warfare," in *Medical Aspects of Chemical Warfare*, ed. Shirley D. Tuornisky (Washington, D.C.: Borden Institute, Walter Reed Army Medical Center, 2008), 9–76.

39. Derrick C. Parmenter, "Tetrachlorethane Poisoning and Its Prevention," *Journal of Industrial Hygiene* 2 (1921): 456–65; George R. Minot and Lawrence W. Smith, "The Blood in Tetrachlorethane Poisoning," *Archives of Internal Medicine* 28 (1921): 687–702.

40. Parmenter, "Tetrachlorethane Poisoning," 457.

41. Dr. Legge visited Hendon on 4 December 1914; See William Henry Willcox, "Lettsomian Lectures (in Abridged Form) on Jaundice, with Special Reference to Types Occurring during the War," *Lancet* 193 (24 May 1919): 869–72. Willcox recounts the visit with Legge on 871.

42. [Department of Industrial Hygiene, Harvard School of Public Health], "Report upon the Activities of Industrial Hygiene with the Exception of the Industrial Clinic from 1918 to 1922" (22-page typescript), Archives of the Department of Industrial Hygiene, School of Public Health, Countway Medical Library, 1918–1934 (E72.5.A1). The discussion of Drs. Edsall and Legge's visit to Lustron appears on 6.

43. Harry S. Mork and Charles F. Coffin Jr., "Manufacture of Cellulose-Acetate Artificial Silk," U.S. Patent 1,551,112, filed 22 March 1923, awarded 25 August 1925.

44. "Artificial Silk Production Doubled in 1921," *Textile World* 61 (4 February 1922): 153, 285.

45. *Royal Little Story*. Eliot Farley was a lifelong associate, including serving as best man at Royal Little's wedding in 1932; see "Miss Ellis Bride of Royal Little," *New York Times*, 11 September 1932. The Special Yarns Corporation, with Harry S. Mork as its director, was listed in 1927 as being located at 60 K Street, South Boston, virtually next door to the 44 K Street address that had been Lustron's; see Callie Hull and Clarence J. West, comps., *Handbook of Scientific and Technical Societies and Institutions of the United States and Canada* (Washington D.C.: National Research Council, 1927), 99.

46. *Royal Little Story*, 12.

47. Ibid.

48. Valentin Wehefritz, *Wegbereiter der chemischen Technik: Prof. Dr. phil. Ernst Berl (1877–1946)—Ein deutsches Gelehrtenschicksal im 20. Jahrhundert* (Dortmund: Dortmund Universitätsbibliothek, 2010). For additional biographical details, see "Berl, Ernst," in *International Biographical Dictionary of Central European Emigrés, 1933–1945*, ed. Herbert A. Strauss and Werner Röder, vol. 2: *The Arts, Sciences, and Literature*, pt. 1: *A–K* (Munich: Saur, 1983), 93.

49. Ernst Berl, "Making Explosives Then and Now," *Chemical and Metallurgical Engineering* 46 (1939): 608–14.

50. Ibid., 608.

51. Ernst Berl, "Explosions That Weren't Planned," *Chemical and Metallurgical Engineering* 47 (1940): 236–39. For Sarvar, Hungary, see also note 17 to this chapter.

52. Wehefritz, *Wegbereiter der chemischen Technik Prof. Dr. phil. Ernst Berl.*

53. "Recipe for Fuel," *Time*, 23 September 1940, 56. Berl's American Chemical Society presentation was covered in greater depth by the *New York Times*: William T. Laurence, "Makes Coal or Oil of Grass in Hour," *New York Times*, 13 September 1940.

54. Baron Ralph S. von Kohorn, "Cellulose to Cell Phones," unpublished autobiographical manuscript, Wellington, New Zealand, 2008. Of the emigration from Germany, von Kohorn acknowledges that although not religiously observant, his parents were Jewish and his father began to transfer assets in anticipation of events. Baron Ralph von Kohorn, who was Oscar's younger son, died in 2010.

55. Robert N. Proctor, *The Nazi War on Cancer* (Princeton, N.J.: Princeton University Press, 1999), 13–15; see also Christopher Sellers, "Discovering Environmental Cancer: Wilhelm Hueper, Post-World War II Epidemiology, and the Vanishing Clinician's Eye," *American Journal of Public Health* 87 (1997): 1824–35.

56. David A. Hounshell and John Kelly Smith Jr., *Science and Corporate Strategy: Du Pont R&D, 1902–1980* (Cambridge: Cambridge University Press, 1988), 560.

57. G. H. Gehrmann, "Proposal for Scientific Medical Research," 28 November 1933, five-page typewritten memorandum, Willis Harrington Papers, Hagley Library and Museum, Accession No. 1813, box 16.

58. Ibid., 3.

59. John Parascandola, "The Public Health Service and Jamaica Ginger Paralysis in the 1930s," *Public Health Reports* 110 (1995): 381–83.

60. Maurice L. Smith, Elias Elvove, and W. H. Frazier, "The Pharmacological Action of Certain Phenol Ethers, with Special Reference to the Etiology of So-Called Ginger Paralysis," *Public Health Reports* 45 (17 October 1930): 2509–24.

61. John P. Morgan, "The Jamaica Ginger Paralysis," *Journal of the American Medical Association* 248 (15 October 1982): 1864–67. This highlights a 1932 FDA memorandum detailing the Celluloid Company as the ultimate source of Lindol (spelled as "Lyndol" in this report).

62. R. L. Shuman, superintendent, Chemical Specialties Dept., to P. Lorillard Company, attn. J. J. Driscoll, 27 August 1931, Truth Tobacco Industry Documents, University of California, San Francisco, https://industrydocuments.library.ucsf.edu/tobacco/docs/tfgx0169. An earlier letter from Celluloid to Reigel Paper (3 March 1931) re-

marked: "You will note that Lindol is used extensively by the Du Pont Cellophane Co.";
https://industrydocuments.library.ucsf.edu/tobacco/docs/fryho169.

63. Shuman to Lorillard, 27 August 1931.
64. Hounshell and Smith, *Science and Corporate Strategy,* 558.
65. Gehrmann, "Proposal for Scientific Medical Research," 3.
66. Hounshell and Smith, *Science and Corporate Strategy,* 560–63.
67. Wilhelm C. Hueper, "Etiologic Studies on the Formation of Skin Blisters in Viscose Workers," *Journal of Industrial Hygiene and Toxicology* 18 (1936): 432–47.
68. Frank C. Wiley, Wilhelm C. Hueper, and Wolfgang F. von Oettingen, "On the Toxic Effects of Low Concentrations of Carbon Disulfide," *Journal of Industrial Hygiene and Toxicology* 18 (1936): 733–40.
69. Ibid., 739. Given the prominent reputation of Dr. Karl B. Lehmann (1858–1940) as a leading toxicologist, it is not likely one would have been directly critical of his work. Only after his death did Alice Hamilton remember with dismay a 1933 visit to Lehmann in Würzburg and his "enthusiastic approval" of Nazi-led book burning (Hamilton, *Exploring the Dangerous Trades,* 377).
70. Hounshell and Smith, *Science and Corporate Strategy,* 206–9. Another driving force in the DuPont–I. G. Farben relationship was the commercial development of nylon of two types, DuPont's nylon-6,6 (the standard form in clothing textiles) and Farben's nylon-6, which still has major applications in furnishings (especially carpets). DuPont's counterpart on the Farben side during the negotiations related to synthetic rubber precursors was Dr. Fritz ter Meer, who came to Wilmington, Delaware, in 1935.
71. Hounshell and Smith, *Science and Corporate Strategy,* 560–63.
72. Pilar Barrera, "The Evolution of Corporate Technological Capabilities: Du Pont and IG Farben in Comparative Perspective," *Zeitschrift für Unternehmensgeschichte / Journal of Business History* 39 (1994): 31–45.
73. "Recent Developments in the German Rayon Industry," *Industrial and Engineering Chemistry* 18 (1926): 1356. The I. G. Farben takeover of Köln-Rottweil included the "Vistra" rayon staple plant in Premnitz; by 1932, Farben had gone on to establish a second major staple facility in Wolfen.
74. Witt, "Die deutsche Zellwolles-Industrie" 202, table 54. By 1938, I. G. Farben accounted for 32 percent of German rayon staple, and Glanzstoff approximately 30 percent (Glanzstoff's additional nonstaple rayon output was considerable).
75. Ibid., 92–95, 202–3 (tables 54, 55). The two new conglomerates accounted for 33 percent of German staple production in 1938. The factories in each group retained individual names, along with the overarching Phrix identity. The Wittenberg facility, for example, was Kürmarksiche Zellwolle- und Zellulose A.G.
76. Werner Knapp, "Die Grundlagen der Siedlungsgestaltung" and "Zellwolle Seidlung," in *Architektur Wettbewerbe I* (Stuttgart: Karl Krämer, 1938), 5–11. The "*Leben-Boden-Blut*" triangle appears on 5.
77. Universum Film AG (UFA), *Ein Phrix-Werk Entshtent,* twenty-nine-minute motion picture (c. 1938).
78. Robert Wilfer and Katja Beck, "Kalle: From Cellophane to Casings," *Fleiswirtschaft International* 1 (2011), www.kalle.de/fileadmin/user_upload/news_presse/pressespiegel/en/2011/Fleischwirtschaft_International_Ausgabe_Maerz_2011.pdf.

79. Ralf Foster and Jeanpaul Goergen, "Ozaphan: Home Cinema on Cellophane," trans. Anke Mebold, *Film History* 19 (18 February 2007): 372–83.

80. Although there was no strip in the Ozaphan series about the making the cellophane product (called Ozalid), there was a film documenting the process, but with no health hazard content: *Ozalid: Das Trocken-Lichtpaus-Verfahen,* produced by Döring-Film-Werke and directed by August Koch, c. 1935; it is film 101 in the archives of the Kalle company, Wiesbaden (Ralf Foster, e-mail to the author, 1 July 2013).

81. Coleman, *Courtaulds,* 186.

82. Fred Aftalion, *A History of the International Chemical Industry: From the Early Days to 2000,* 2nd ed. (Philadelphia: Chemical Heritage Press, 2001), 208. It has been argued that despite the priority on autarky, the staple industry was less a nationally coordinated effort (a so-called command economy) than might be presumed; see Jonas Scherner, "The Beginnings of Nazi Autarky Policy: The 'National Pulp Programme' and the Origin of Regional Staple Fibre Plants," *Economic History Review* 61 (2008): 867–95.

83. Hans Reiter, "Arbeitshygiene und Vierjahresplan" [Industrial Hygiene and the Four-Year Plan], in *Das Reichsgesundheitsamt 1933–1939: Sechs Jahre Nationalsozialistische Fuhrung* (Berlin: Julius Springer, 1939), 243–52. Reiter referred specifically to rayon staple factories (*zellwollefrabriken*) and exposure to hyrdogen sulfide (*schwefelwasserstoff*), but did not mention carbon disulfide (*schwefelkohlensstoff*).

84. Siegfried Seher, *Zellwolle: Ein Weg zur Freiheit* (Kelheim-Donau, Germany: Süddeutsche Zellwolle Aktiengesellschaft, 1938).

85. Paul G. Ehrhardt, *Zellwolle: Vom Wünder ihres Werdens* (Frankfurt am Main: Brönners Druckerei und Verlag, 1938). The book includes ninety-six photographic illustrations by Dr. Paul Wolff. The dust jacket only notes "Flox Zellwolle."

86. Anton Lübke, *Das deutsche Rohstoffwunder: Wandlungen der deutschen Rohstoffwirtschaft,* 4th ed. (Stuttgart: Forkel-verlag, 1939), 337–56. Plates 25–28 pertain to rayon, plate 27 being the Wolff image of the shirtless worker. This edition, with the Sudetenland supplement, is 572 pages long, as are all editions thereafter; the first through third editions are 556 pages.

87. The Library of Congress copy inscribed by Lübke to Adolf Hitler is in the Third Reich Collection, which includes Hitler's personal library; see https://catalog.loc.gov/vwebv/holdingsInfo?searchId=7990&recCount=25&recPointer=1&bibId=841388.

88. [Vereinigte Glanzstoff-Fabriken (VGF)], *Wir von Glanzstoff=Courtaulds,* issue 1 (January 1939). The St. Pölten facility was the Erste Östreichechische Glanzstoff-Fabrik. In 1939, A. G. Glanzstoff's main staple facility was in Kassel-Battenheim; staple production at Cologne was coming online as well.

89. [VGF], *Wir von Glanzstoff=Courtaulds,* issue 9 (September 1938), 225–52.

90. [Phrix-Gesellschaft M.B.H.], *Der Phrixer,* issue 7 (May 1940). "Zellwolle und Zucker," by "N.R.," appears on 30.

91. Jeffrey T. Schnapp, "The Fabric of Modern Times," *Critical Inquiry* 24 (1997): 191–245, cited passage on 195. A six-page appendix translates a previously

unpublished prose manuscript by Marinetti on the technical-chemical process of viscose synthesis.

92. Valerio Cerretano "The 'Benefits of Moderate Inflation': The Rayon Industry and Snia Viscosa in the Italy of the 1920s," *Journal of European History* 33 (2004): 233–84.

93. Felice Casorati, *Portrait of Riccardo Gualino,* 1922 (private collection); exhibited in Annalisa Scarpa, *Sguardi sul Novecento: Collezionismo privato tra gusto e tendenza* (Milan: Skira, 2012).

94. Cerretano, "'Moderate Inflation.'"

95. Riccardo Gualino, *Frammenti di vita* (Turin: Nino Aragno, 2007). Gualino began these memoirs while exiled on Lipari in 1931. A connoisseur and collector, Gualino worked closely with the art historian critic Lionello Venturi, a prominent antifascist who refused to sign a loyalty oath to Mussolini and was forced out of his chair at the University of Rome shortly afterward.

96. Schnapp, "Fabric of Modern Times."

97. [Ricerche e Studi S.p.A. (Mediobanca Milan)], *The Chemical Industry* [*L'industria chimica*] (Milan, 1970), 9, www.archiviostoricomediobanca.mbres.it/documenti/MONO GRAFIE_THE_CHEMICAL_INDUSTRY_1970_INTEGRAZIONE.pdf.

98. "Sniafiocco & Vistra," *Time,* 5 November 1934, 59. Vistra was the trade name for I. G. Farben staple.

99. Antonio Ferretti, "Process of the Manufacture of Artificial Textile Fibers," U.S. Patent 2,338,920, 11 January 1944, one of a series of Ferretti's casein-related patents.

100. "Lanital," *Time,* 6 December 1937, 100.

101. Federico Ferrari, "Torre Snia Viscosa, 1935–1937," Ordine degli Architetti, Pianificatori, Paesaggisti e Conservatori della provincia di Milano, www.ordinearchitetti .mi.it/it/mappe/itinerari/edificio/723-torre-snia-viscosa/17-milano-alta.

102. Jonathan D. Taylor, "Museum Review: A Future Woven in Rayon," *Chemical Heritage* 27, no. 3 (2009), www.chemheritage.org/discover/media/magazine/articles/27 -3-a-future-woven-in-rayon.aspx. This reviews a 2009 exhibition on the history of the SNIA plant held at Museo Territoriale Bassa Friulana in the present-day community of Torviscosa.

103. Schnapp, "Fabric of Modern Times."

104. Filippo Tommaso Marinetti, *Teoria e invensione Furturistica* (Verona: Arnoldo Mondadori, 1968), 1058. In 1940, the tetralogy of SNIA pieces were renamed in a series of "simultaneous" poems, *Il Poema non umano dei tecnicismi* (Schnapp translates this as *The Non-Human Poem of Technicisms*). Torre Viscosa became "Poesia simultanea dei canneti Arunda Donax," the latter being the Latin name of the reed harvested at Torviscosa.

105. Cinzia Sartini Blum, *F. T. Marinetti's Futurist Fiction of Power* (Berkeley: University of California Press, 1996), 142. Blum translates the final line of the poem as "High above traveling traveling endlessly the new constellation whose stars form the word Autarchy."

106. Schnapp, "Fabric of Modern Times," 243. I have substituted "xanthate" for "xanthogenate," his translation of *santogenato*.

107. Bruna Bianchi, "I tessili: Lavoro, salute, conflitte," *Annali Fondazione Giangiacomo Feltrinelli* 20 (1979–80): 973–1070. The fire suggests that the operation did involve inflammable carbon disulfide.

108. Ibid., 987.

109. Alice Sotgia, "Sul filo della pazzia: Produzione e malattie del lavoro alla Viscosa di Roma negli anni Venti e Trenta," *Dimensioni e Problemi della Ricerca Storica* 2003, no. 2: 195–210, quotation on 206.

110. Diego De Caro, "Le psicosi da solfocarbonismo," *L'Ospedale Psichiatrico* 9 (1941): 207–34, quotation on 214.

111. Luigi Tomassini, *La salute al lavoro: La Societa Italiana di Medicina del Lavoro e Igiene Industriale dale origini a oggi* (Piacenza: Nuova Editrice Berti, 2012), 55–57. Signatories to the founding resolution included Quarelli, Loriga, Vigliani, and Ranelletti.

112. Gustavo Quarelli, "L'intossicazione professionale da solfuro di carbonio," *Rassegna della Providenza Sociale* 21 (1934): 10–73. The Istituto nazionale fascista per l'assicurazione contro gli infortuni sul lavoro (INFAIL) also produced a separate reprint with a preface by Quarelli, 12 July 1934.

113. *Atti del congresso: XI° Congresso Nationale di Medicina del Lavoro*, 29–31 October 1934. Communications on carbon disulfide included an extensive report by Aristide Ranelletti on "*sulfocarbonismo*" (carbon disulfide illness).

114. Gustavo Quarelli, *L'impotenza sessuale nel solfocarbonismo professionale e la sua grande importanza nel problema razziale* (Rome: Edizioni Universitas, 1939). This fifteen-page tract begins, "Il Partito Nazionale Fascista alla continua lotta per la difesa ed il miglioramento della razza, ha efficacemente contribuito facilitando lo studio delle malattie professionali."

115. Enrico C. Vigliani, Fabio Visintini, and Paolo Emelio Maspes, "Prime osservazioni sulla miopatia da solfuro carbonio," *La Medicina del Lavoro* 35 (January–March 1944): 1–9. (The Fascist year had been dropped from the journal's masthead). There were limited German case reports of carbon disulfide after the mid-1930s. One notable exception is Hans Schramm, "Chronische Schwefelkohlenstoff Vergiftungen in der Kunsteide- und Zellwolle-Industrie," *Deutsche Medizinische Wochenscrift* 7 (1940): 80–182. A later article on combustion in rayon staple manufacturing appeared in a leading German medical forensic journal (one of a cluster of such fires in 1939–40 in German factories): Walter Specht, "Zur Frage der Selbstentzundüng von Zellwolle Untersnehnngen über die Ursache von Trocknerbründen," *Deutsche Zeitschrift für die gesamte gerichtliche Medizin* 36 (1942): 174–80.

116. S.K., *Agent in Italy* (Garden City, N.Y.: Doubleday, Doran, 1942). The book appeared in England in 1943 under the Hutchinson imprint.

117. Jacqueline M. Atkins, ed., *Wearing Propaganda: Textiles on the Home Front in Japan, Britain, and the United States, 1931–1945* (New Haven, Conn.: Yale University Press, 2005).

118. Brooklyn Museum, press release, 30 October 1942, https://www.brooklynmuseum.org/opencollection/exhibitions/854/Inventions_for_Victory.

119. Atkins, *Wearing Propaganda*, 286, 292.

120. Coleman, *Courtaulds*, 192–99.

121. U.S. Department of Defense, *The "Magic" Background of Pearl Harbor*, vol. 3, *August 5, 1941–October 17, 1941* (Washington D.C.: Government Printing Office, 1970), 173–74.

122. Atkins, *Wearing Propaganda*, 165–67.

123. "Clothing of Troops," *Barrier Miner* (Broken Hill, New South Wales), 24 January 1938.

124. Montserrat Llonch Casanovas, *Tejiendo en red: La industria del género de punto en Cataluña, 1891–1936* (Barcelona: Universitat de Barcelona, 2007), 105–7. A medical publication during the Spanish Civil War period reported on carbon disulfide poisoning in twenty-two workers in the industrial extraction of vegetable oils: Juan Dantín Gallego, "Erfahrungen über Schwefelkohlenstoffschädigungen bei der Olivenölbereitung in Andalusien mit einigen diesbezüglichen Tierexperimenten," *Archiv für Gewerbepathologie und Gewerbehygiene* 8 (6 November 1937): 124–38.

125. "La SNIACE, Empresa espanola de fribras textiles artificiales," *ABC* (Madrid), 16 April 1943. A second Spanish rayon concern also was established in Miranda de Ebro early in the Franco years: FEFASA (Fabricación Española de Fibras Textiles Artificiales S.A.). Its manufacturing process, like that of the Wittenberg Phrix plant, was meant to utilize straw; see "FEFASA en la Industria National," *ABC* (Madrid), 30 July 1957.

126. U.S. Office of Strategic Services, "Transfer of the Italia SNIA Viscosa Monopoly to Spain, 23 and 26 December 1943" (memorandum), Dissemination No. A 17849, 10452-1225A OSS Secret Stamp No. 52328, U.S. National Archives, declassified on 11 September 2010.

127. Coleman, *Courtaulds*, 463.

128. Ibid, 463–91. Lend-lease is also covered in Wilkins, *History of Foreign Investment*, 482–509. Wilkins details a number of British-held assets that were not sold but put up in part to collateralize U.S. loans to the British in 1941, including Celanese of America, Courtaulds' cellulose acetate competitor.

129. "Viscose Unveiled," *Time*, 26 May 1941, 89. This piece followed up on earlier coverage by *Time* on the impending deal: "Viscose Sale," *Time*, 24 March 1941, 76.

130. Steil, *Battle of Bretton Woods*, 104–13. Steil's cited source for Keynes's purpose to "sabotage the viscose deal" (113) is Lucius Thompson's diary entry for 22 May 1941 (Treasury Papers, National Archives, United Kingdom). The events from Morgenthau's perspective are covered in John Morton Blum, ed., *The Morgenthau Diaries: Years of Urgency, 1938–1941* (Boston: Houghton Mifflin, 1965), 235–41.

131. Charles P. Kindleberger, oral history interview, Truman Library, Independence, Missouri. In *The Battle of Bretton Woods*, Benn Steil notes that Harry Dexter White took a "hard line" on British reserves; speculating on the potential reasons for this, he does not allude to White's prior research on viscose and the tariff as a specific factor in the forced sale of AVC. Steil comments that Dean Acheson in the Department of State opposed White; Acheson's legal role in defending DuPont in viscose matters before he returned to State Department service is likely to have informed his views on AVC.

132. Coleman, *Courtaulds*, 473.

133. U.S. Department of Commerce, International Trade Administration, Office of Textiles and Apparel (OTEXA), "The Berry Amendment," http://web.ita.doc.gov/tacgi/eamain.nsf/d511529a12d016de852573930057380b/013cde07c51132d98525738c0074568e?OpenDocument&country=Berry.

134. Associated Press, "Jeffers Defies Cotton Block on Tire Rayon," 13 October 1942; see also Clifford Kennedy Berryman, "In the Lion's Den" (cartoon), 15 October 1942, Library of Congress, http://loc.gov/pictures/resource/acd.2a05901.

135. *Which Jobs for Young Workers, No. 4: Advisory Standards for Employment Involving Exposure to Carbon Disulfide* (Washington, D.C.: U.S. Department of Labor, Children's Bureau, December 1942); see also His Majesty's Factory Inspectorate, *Memorandum on Precautions Against Dangers of Poisoning, Fire, and Explosion in Connection with the Use of Carbon Bisulphide in Artificial Silk, India Rubber and Other Works*, Factory Department, form 836 (London: His Majesty's Stationery Office, 1943). Even the International Labour Office (a vestige of the League of Nations, with a wartime office in Montreal, Canada) put out an update on carbon disulfide: *Occupation and Health: Encyclopaedia of Hygiene, Pathology, and Social Welfare, Studied from the Point of View of Labour, Industry, and Trades. Special Supplement. Industrial Health in Wartime* (Montreal: International Labour Office, 1944), 19–21.

136. Rudolf Danner, *50 Jarhe Viscose Emmenbrücke, 1906–1956* (Emmenbrücke, Switzerland: Société de la Viscose Suisse, 1958), 25, 29–39.

137. From 1938 through 1942, only four cases of carbon disulfide poisoning received compensation in Switzerland; from 1943 to 1937 there were twenty-one such cases—data available only in four-year increments (Dr. David Meidinger, Swiss Accident Insurance Fund, to the author, e-mail, 21 August 2013).

138. Ragnar Magnusson, *En stråle av ljus: Svenska Rayon AB, 1943–1993; Fabriken, bygden, människor, händelser* (Edsvalla, Sweden: Ekens, 1993). For background on Svenskt Konstsilke (founded 1918), the Swedish private rayon concern predating Svenska Rayon, see Sylvia Danielsson, *Att förädla: En historia på 90 år; Konstsilke, ett högteknologiskt garn* (Borås: AB Svenskt Konstsilke, 2008).

139. "Consolidated statement on operational readiness of restored pulp and paper mills and previously hit viscose factory on Karelian Peninsula issued by the governmental commission on April 12, 1940; Statement issued by the governmental commission on the readiness of the restored pulp and paper mill in Keksholm for operation 1940/1941," reel 3.4641, 3.4644, file 27, 36; "Correspondence and other materials re logistical supplies to the USSR NKVD Gulag for restoration of a viscose-producing factory and pulp and paper enterprises in the Karelo-Finskaia SSR 1940," reel 3.6919, file 3910, Archives of the Soviet Communist Party and Soviet State Microfilm Collection, State Archives of the Russian Federation (Gosudarstvennyĭ arkhiv Rossiĭskoĭ Federatsii—GARF), 1903–90, Hoover Institution Archives, Stanford University. The 12 April 1940 memorandum asserts that the "quick and successful undertaking of work to restructure the factories of Karelia, given to the USSR by peaceful agreement with Finland . . . was only possible thanks to the actions and daily help of the Central Committee of the Party (Bolsheviks) and the Council of People's Commissars of the USSR"; those working on this included the deputy of the People's Commissariat of Cellulose Paper Production and the deputy head of the Main Administration of Labor Camp Production Construction of the

NKVD of the USSR. Georgii Mikhalovich Orlov, who also appears in the Jääski-related files, is a figure emblematic of the close links between the NKVD and the cellulose industry. He studied at the Leningrad Forestry Institute in the 1920s and then became an engineer in that industry. Arrested briefly in 1938 for "lack of political vigilance against enemies of the people" and then rehabilitated the same year, he was brought into the NKVD and became head of the "Cellulose and Paper Section" of the Gulag. By 1940–41 he had become deputy head of the entire Gulag operation and then the head of the Main Division of Camp Production and Building of the NKVD, before assuming the role in 1944 of commissar of Cellulose and Paper Production. He remained a powerful governmental bureaucrat for decades; see K. A. Zalesskii, *Imperiia Stalina: Biograficheskii entsiklopedicheskii slovar'* (Moscow: Veche, 2000).

140. Leo Noro was a leading figure in Finnish occupational medicine in the last decades of the twentieth century and was involved in later studies of the Finnish rayon industry (Henrik Nordman to the author, e-mail, 18 March 2009).

141. Leo Noro, "Svavelväte- och kolsvavlaförgiftningar," *Nordisk Medicin* 24 (October–December 1944): 2015–20.

142. Emil A. Paluch, "Two Outbreaks of Carbon Disulfide Poisoning in Rayon Staple Fiber Plants in Poland," *Journal of Industrial Hygiene and Toxicology* 30 (1948): 37–42. For additional information on Widzewska Manufaktura, see Arkadiusz Dąbrowski, "Widzew Manufacture," *Local Paper,* May 2007 (English-language edition of *Gazeta Lokalna,* Łódź).

143. Albert Langelez, "Viscose et sulfocarbonisme: Un enquête médicale," *Archives Belges de Médicine Sociale et d'Hygiène et Revue de Pathologie et de Physiologie de Travail* 4 (1946): 67–85, quotation on 84.

144. Jean Auffret, "L'Industrie des fibres artificielles et ses dangers," *Archives des Maladies Professionnelles de Médecine du Travail et de Sécurité Sociale* 7 (1946): 181–96.

145. Dominique Veillon, *Fashion Under the Occupation,* trans. Miriam Kochan (Oxford: Berg, 2002), 69–84.

146. Michelle Blondé, *Une usine sans la Guerre: La Société Nationale de la Viscose à Grenoble, 1939–1945* (Grenoble: Presse Universitaires de Grenoble, 2008); see also Patrice Ricard, Jean-Louis Pelon, and Michel Silho, *Memoires de Viscosiers: Ils filaient la soie artificielle à Genoble* (Grenoble: Presse Universitaires de Grenoble, 1992), 120–25. Text publications supported by the Musée de la Viscose, Isére, France.

147. Agnès Humbert, *Résistance: A Woman's Journal of Struggle and Defiance in Occupied France,* trans. Barbara Mellor (New York: Bloomsbury, 2009).

148. Ibid., 121–22.

149. Ibid., 142.

150. Ibid., 151–52.

151. Ibid., 157 ("nervous system is completely shattered"), 168 ("drugged and fuddled by the acid vapors"), 200 ("has just thrown herself out of the clothing store window").

152. In her descriptions of the plant's operations, Humbert refers either to "the viscose" or to "the acid." She footnotes the later term, explaining that the acid was carbon disulfide, but is deliberate in not using that word in the body of the text. Of note,

the single appearance of the footnote (and the only time in the book she reverts to this device) is in connection with the foul odor in the plant. Sulfur contaminants of carbon disulfide do indeed impart a rotten smell, even though the solvent, when pure, is supposedly rather sweet smelling. In the original, "l'odeur extrêment désagréable de l'acide," with the footnote "1. *Sulfure de carbone*" (Agnès Humbert, *Notre Guerre: Souvenirs de Resistance, Paris, 1940–41; Le Bagne; Occupation en Allemagne* [1946; repr., Paris: Tallandier, 2004], 250.). Her English translator, Barbara Mellor, concurs that Humbert was not likely to have understood the scientific aspects of rayon manufacturing at the time nor to have sought further information before rapidly writing her book and publishing it by May 1946 (Barbara Mellor to the author, e-mail, 3 January 2011).

153. Christopher Browning makes this point regarding slave labor in his cogent review of Wolf Gruner, *Jewish Forced Labor Under the Nazis: Economic Needs and Racial Aims, 1938–1944* (New York: Cambridge University Press, in association with the U.S. Holocaust Memorial Museum, 2006), in *Holocaust Genocide Studies* 21 (2007): 509–10.

154. German Federal Archives, Directory of Places of Detention, www.bundesarchiv.de/zwangsarbeit/haftstaetten/index.php.en?action=2.2&tab=4&id=100001104. Today the proper term is *Justizvollzugsanstalt* (justice enforcement facility).

155. Peter Zenker, *Zwangsarbeit in Siegburg*, www.peter-zenker.de/documents/Zwangsarbeit_SU_Langfassung.pdf. Rheinsiche Zellwolle came into operation shortly before it was absorbed into the Phrix combine, having taken over a defunct Bemberg operation and converted it to produce rayon staple; see the entry for 10 December 1936 in the Siegburg city archives, www.stadtarchiv-siegburg.de/web/stadtarchiv/01680.

156. The "Working Students" (*werkstudenten*) image of 1926 appeared in August Sander's *Face of Our Time* (1929). The image is held in the collection of the National Galleries of Scotland, https://www.nationalgalleries.org/object/AL00058.

157. The prison image of Erich Sander in the infirmary is held at the Tate Museum, London, www.tate.org.uk/research/publications/tate-papers/19/august-sanders-portraits-of-persecuted-jews. The Tate incorrectly attributes this image to August Sander, who had no possible access to take such a photograph. It may have been a triggered self-photograph or possibly one taken by Karl Hugo Schmölz (the son of another Cologne photographer, Hugo Schmölz, and a family friend). "Schmölz" is mentioned in a letter from Eric Sander as working with him on prison photographic documentation.

158. Erich Sander, letter from Siegburg Prison, 13 June 1942, typed transcription, August Sander Archive, GSA 164.19/42.42. I am indebted to Gerd Sander, Erich's nephew, for sharing these letters with me and allowing for their use.

159. Erich Sander, letter from Siegburg Prison, 27 June 1942, typed transcription, August Sander Archive, GSA 164.21/42.49.

160. Roger Repplinger, *Leg dich, Zigeuner: Die Geschichte von Johann Trollmann und Tull Harder* (Munich: Piper, 2008). For boxing statistics, see "Johann Trollmann," BoxRec, http://boxrec.com/list_bouts.php?human_id=062355&cat=boxer.

161. *Gibsy—Die Geschichte des Boxers Johann Rukeli Trollmann: Ein Film von Eike Besuden mit Hannes Wegener und Hannelore Elsner,* http://realfictionfilme.de/filme/gibsy/index.php.

162. U.S. Holocaust Memorial Museum, "Neuengamme," *Holocaust Encyclopedia*, www.ushmm.org/wlc/en/article.php?ModuleId=10005539.

163. Günter Rodegast, *Zwangsarbeiter und KZ-Häftlinge: Kurmärkische Zellwolle und Zellulose AG aus der Geschichte eines Wittenberger Phrix-Werkes* (Wittenberge: Prignitzer Heimatverein Wittenberge, 2000).

164. Gruner, *Jewish Forced Labor*, 208–13.

165. Hermann Kaienburg, "Zwangsarbeit fur das 'deutsche Rohstoffwunder': Das Phrix-Werk Wittenberge im zweiten Weltkrieg," *1999: Zeitschrift fur Sozialgeschichte des 20. und 21. Jahrhunderts* 9 (1994): 12–41.

166. Birgit Pelzer-Reith and Reinhold Reith, "Die 'Eiweißlücke' und die biotechnologische Eiweißsynthese," *Technikgeschichte* 79 (2012): 303–40. An innovation at Lenzing was "submerged fermentation," a technology that could have aided antibiotic development (and proved key in the United States), but was never applied by the Germans at the time; see Gilbert Shama and Jonathan Reinarz, "Allied Intelligence Reports on Wartime German Penicillin Research and Production," *Historical Studies in the Physical and Biological Sciences* 32 (2002): 347–67.

167. German Federal Archives, Directory of Places of Detention, "Zwangsarbeitslager für Juden Küstrin-Neustadt," www.bundesarchiv.de/zwangsarbeit/haftstaetten/index.php.en?action=2.2&tab=7&id=2397; see also Andreas Weigelt, "Küstrin," in *The United States Holocaust Memorial Museum Encyclopedia of Camps and Ghettos, 1933–1945: Early Camps, Youth Camps, and Concentration Camps and Subcamps under the SS-Business Administration Main Office (WVHA)*, ed. Geoffrey P. Megargee (Bloomington: Indiana University Press, 2009), 1:1321–23.

168. Gazyma Choptiany, "Hirschberg [Arbeitskommando]; Hirschberg [Arbeitslagen]," in Megargee, *Encyclopedia of Camps and Ghettos*, 1:748–49. The testimony of a survivor of one of the Hirschberg camps, Mordechai Schwimmer, thirteen at the time, recounts his experience unloading the logs; see "Testimony of Mordechai Schwimmer, born in Tirgu-Mures, Romania, 1931, regarding his experiences as a child in the Tirgu-Mures Ghetto and Bireknau, Hirschberg, Buchenwald and Theresienstadt camps," ID 3565405, Yad Vashem Archives, Jerusalem, transcript of video interview (in Hebrew). For another personal account, see Stanislaw Dziaduś, "'Hirschberg' i 'Treskau,' podobozy Gross-Rosen: Uurywki wspomień" [The "Hirschberg" and "Treskau" branches of the Gross-Rosen concentration camp: Excerpts from the memoirs], *Przegląd Lekarski* 25 (1969): 138–46. Dziaduś was imprisoned with a Polish work detail that was kept separate from Jewish slave laborers.

169. Pelzer-Reith and Reith, "Die 'Eiweißlücke' und die biotechnologische Eiweißsynthese."

170. Roman Sandgruber, *Lenzing: Anatomie einer Industriegründung im Dritten Reich* (Linz, Austria: Oberösterreichisches Landesarchiv, 2010), 87–100.

171. Gine Elsner, *Heilkräuter, "Volksernährung," Menschenversuche: Ernst Günther Schenck (1904—1988); Eine deutsche Arztkarriere* (Hamburg: VSA, 2010), 86–89.

172. Sandgruber, *Lenzing*, 287–312.

173. Margret Lehner, "Nebenlager Lenzing-Pettighofen," in Christian Hawle, Gerhard Kriechbaum, and Margret Lehner, eds., *Täter und Opfer: Nationalsozialistische*

Gewalt und Widerstand im Bezirk Vöcklabruck, 1938–1945 (Weitra, Austria: Bibliothek der Provinz, 1995), 33–57. The satellite camp at Lenzing is also covered in Sandgruber, *Lenzing*, 228–86.

174. Clare Parker, *Klara's Story* (London: Clare Parker, 1999), 63. The Yad Vashem archive contains the testimonies of a number of women who also survived the Lenzing camp. Elsa Kraus recounted how her right arm was caught between two machinery wheels in the plant, but because of the raw material in the process the arm was not crushed; a few weeks after that, her eyes and face became swollen from the fumes; see "Testimony of Elsa Kraus" (in English), file 069/40, Yad Vashem Archives, Jerusalem.

175. Annie Jacobsen, *Operation Paperclip: The Secret Intelligence Program That Brought Nazi Scientists to America* (New York: Little, Brown, 2014), 81–82, 302–4, 309. Jacobsen alludes to Scheiber's protein-substitute work, although mischaracterizing the product as being "made up of cellulose, or pieces of used clothing" (309).

176. Gerda Zorn, *Ostland geht unser Ritt: Deutsche Eroberungspolitik zwischen Germanisierung und Völkermord* (Berlin: Dietz, 1980, 58–64). For the takeover of the Nici thread factory in 1945, see the website of the Ariadna Thread Factory, www.ariadna.com.pl/index.php?id=53 (in Polish, with English translation available). Phrix took over the original Polish rayon plant in Tomaszów. Feliks Wiślicki, its founding director and a Polish Jew, had already gotten out by becoming an economic advisor to the Free Polish government in exile in London. For other sites of forced labor in the Thüringischen Zellwolle enterprises, and the role of the Schwarza factory, where more than two hundred Russian forced laborers worked, see Norbert Moczarski, Bernhard Post, and Katrin Weiß, *Zwangsarbeit in Thüringen, 1940–1945: Quellen aus den Staatsarchiven des Freistaates Thüringen* (Erfurt, Germany: Landeszentrale für politische Bildung Thüringen, 2002), 29, 54, 93–94, 181–82, 194, 196, 246, 248, 260.

177. Fragmentary evidence on *zellwolle* "recycling" operations in the Łódź Ghetto is contained in memoranda and letters in National and Provincial Archives of Poland, Zydowski Instytut Historyczny (Warsaw), held on microfilm by Yad Vashem, Jerusalem, specifically, "Getto-Verwaltunf-Lodz," Record Group 54, file 998 [Ko–K].

178. Material documenting I. G. Farben's use of slave labor in its rayon staple manufacturing is surprisingly fragmentary. For Farben's factory at Wolfen, which had a subcamp originally administered by the Ravensbrück concentration camp that later came under the aegis of Buchenwald, see Evelyn Zgenehagen, "Wolfen," in Megargee, *Encyclopedia of Camps and Ghettos*, 1:440.

179. Terezín survivors' testimonies that document the Glanzstoff work detail (*Arbeitskommando*) include those of František Soukop (testimony 878), Adolf Bureš (testimony 1281; it was he who reported that future Archbishop Josef Beran labored at Glanzstoff), František Kuno (testimony 2140), and Dr. Robert Bardfeld (testimony 2197); Miroslava Langhamerová, Historical Department, Terezin Memorial, Czech Republic, e-mail to the author, 16 June 2014.

180. Wolfgang E. Wicht, *Glanzstoff: Zur Geschichte der Chemiefaser, eines Unternehmens und seiner Arbeiterschaft*, Bergische Forschungen 22 (Neustadt/Aisch, Germany: Verlagsdruckerei Schmidt, 1992), 87–89.

181. Raimonod Finati, ed., *Allo straflager di Colonia: L'odissea di 369 giovanissimi ufficiali deportati nel campo di lavoro AK 96 alla Glanzstoff-Courtaulds nei racconti dei protagonist* (Cuneo, Italy: L'arciere, 1990), poem, 9; eye effects, 63; ersatz milk, 68.

182. Ervin O. Anderson, *Report on the International Synthetic Fiber Industry,* Economic Warfare Section, War Division, U.S. Department of Justice, confidential report, file 60-0-28, 1 August 1944, 129–30.

183. Božena Malovcová, ed., *História jednej myšlienky: Svit (1934–2009)* (Spišská Nová Ves, Slovakia: Bambow, 2009). As of 1937, there was a dispensary for the treatment of workers serviced two hours per week by a Dr. Pohl. See also Henrieta Moravčíková, "Architektúra koncernu Bat'a ako činitel' modernizácie: Príklad Slovensko" [The architecture of the Bata Company as a factor of modernization: The example of Slovakia], in L. Hornakova, ed., *The Bata Phenomenon: Zlin Architecture, 1910–1960* (Zlin, Czech Republic: Regional Gallery of Fine Arts in Zlin, 2009), 227–39.

184. Alice Jakubovic, oral history interview, U.S. Holocaust Memorial Museum, http://collections.ushmm.org/search/catalog/irn511521.

185. Some attribute the Cologne factory's escape to the stake that Courtaulds held. For this line of argument, see Claudio Sommaruga, *La "Glanzstoff & Courtaulds" di Colonia: Un'isola risparmiata nel ciclone della Guerra* (Naples: Guisco, 1996). Sommaruga was among the Italians imprisoned at Glanzstoff. In Coventry, at least one plant component of Courtaulds was destroyed. The story of George Hough, who had been a manager at Courtaulds in Coventry, was told by his son to the BBC as part of its "WW2 People's War" personal reminiscences project, www.bbc.co.uk/history/ww2peopleswar/stories/26/a5135726.shtml.

186. Elsner, *Heilkräuter,* "Volksernährung," *Menschenversuche.*

187. Harold Wickliffe Rose, *The Rayon and Synthetic Fiber Industry of Japan: Supplementary Material Gathered for the United States Department of State and War Department Textile Mission to Japan, January–March, 1946* (New York: Textile Research Institute, 1946).

188. Von Kohorn, "Cellulose to Cell Phones." Of the plants the von Kohorn memoir specifically mentions, however, most were scrapped during the latter part of the war, not bombed.

189. Armando Ferraro, George A. Jervis, and David J. Flicker, "Neuropathologic Changes in Experimental Carbon Disulfide Poisoning in Cats," *Archives of Pathology* 32 (1941): 723–38.

190. Richard B. Richter, "Degeneration of the Basal Ganglia in Monkeys from Chronic Carbon Disulfide Poisoning," *Journal of Neuropathology and Experimental Neurology* 4 (October 1945): 324–53, quotation on 338. Richter was chief of the neurology service at the University of Chicago. Among his many students was the future psychoanalyst Heinz Kohut, who was a trainee in neurology under Richter while he was carrying out the monkey experiments.

191. Deutsches Historisches Museum, "Gewebe mit Judensternen" [Textile with Jewish stars], inventory number KTe. 80/122, http://dhm.de/datenbank/dhm.php?seite

=5&fld_o=FM100434. As catalogued, the material is described as *"regeneratzellulose"* (regenerated cellulose), but is clearly *zellwolle* (rayon staple). The object's size is 78 × 103 cm (30.7×40.5 in).

192. American Viscose Corporation, *Rayon Goes to War,* final page (unnumbered).

Chapter 6. The Heart of the Matter

1. Samuel Courtauld to Lord Woolton, 19 June 1944, Office of the Minister of Reconstruction, Lord President of the Council and Minister for Science, CAB 124/701, The National Archives, United Kingdom.

2. Michael D. Kandiah, "Marquis, Frederick James, First Earl of Woolton (1883–1964)," *Oxford Dictionary of National Biography* (Oxford: Oxford University Press, 2004), www.oxforddnb.com/view/article/34885 (subscription required). For more on Woolton pie, see *Oxford English Dictionary,* s.v. "Woolton."

3. Samuel Courtauld to Lord Woolton, 29 June 1944, CAB 124/701, National Archives, United Kingdom.

4. Samuel Courtauld to Hugh Dalton, memorandum, 28 June 1944, CAB 124/701, National Archives, United Kingdom.

5. Charles Barr, *Ealing Studios,* 3rd ed. (Berkeley: University of California Press, 1998), 134–44; see also Amy Sargeant, "The Man in the White Suit: New Textiles and the Social Fabric," *Visual Culture in Britain* 9 (2008): 27–54.

6. Rose, *Synthetic Fiber Industry of Japan.*

7. The Textile Research Institute was founded in 1930 as the United States Institute for Textile Research, but at nearly the same time special enabling legislation funded the Textile Foundation. The foundation was legislated to have five directors, three from the textile industry, subject to presidential appointment. A main activity of the foundation was to spend down its assets through the Textile Research Institute; see John H. Dillon, *History of the Textile Research Institute, 1930–1965* (Princeton, N.J.: American Textile Institute, 1965).

8. Rose, *Synthetic Fiber Industry of Japan,* 326.

9. Ibid., 140–41. Another Hiroshima survivor named Hiroyuki Suzuki (a twenty-year-old soldier at the time) was "Bombed at Ujina-cho 3-chome, Hiroshima; Former site of Kinka Rayon (Daiwa Spinning Mill), Vessel Training Department of the Instructional Regiment"; see Memories of Hiroshima and Nagasaki, "Messages from Hiroshima," www.asahi.com/hibakusha/english/hiroshima/h01-00068-6e.html.

10. Masatane Takuhara and Tsuyoshi Okada, [Psychiatric diseases recently occurred in a viscose rayon silk factory], *Sangyo Fukuri* 8 (1933): 22–35. This describes four viscose workers confined to a psychiatric hospital over a two-month period; see also Masatane Takuhara, [Recent outbreak of carbon disulfide poisoning occurred in a viscose silk factory], *Sangyo Fukuri* 9 (1934): 36–59, detailing thirty-six workers with carbon disulfide poisoning.

11. J. Kubota, "Historical View of Carbon Disulphide Poisoning in the Japanese Viscose Rayon Industry," in Heinrich Brieger and Jaroslav Teisinger, eds., *Toxicology of Carbon Disulphide: Proceedings of a Symposium, Prague, Sept. 15th–17th, 1966* (Amsterdam: Excerpta Medica Foundation, 1967), 192–96.

12. The research appeared in a series of three publications by Yukio Suzuki: "Experimental research on carbon disulfide poisoning. 1. Animal experiments on the correlation between carbon disulfide gas concentration and exposure time. 2. Animal experiments on the accumulation effect from inhalation of carbon disulfide gas at various concentrations. 3. Effect of temperature on carbon disulfide gas poisoning" [in Japanese], *Rodo Kagaku Kenkyu* [Labor science research] 16 (May 1939): 49–55, and (September 1939): 11–15, 16–23. The three reports involved forty-five, twenty-six, and fifty-eight society (Bengalese) finches.

13. T. Toyama and H. Sakurai, "Ten-Year Changes in Exposure Level and Toxicological Manifestations in Carbon Disulfide Workers," in Brieger and Teisinger, *Toxicology of Carbon Disulphide*, 197–204, quotation on 197.

14. Rose, *Synthetic Fiber Industry of Japan*, 61. For general background on color doping see Avinash P. Manian, Hartmut Ruef, and Thomas Bechtold, "Mass Coloration of Regenerated Cellulosics: A Review," *Lenzinger Berichte* 85 (2006): 87–90.

15. Rose, *Synthetic Fiber Industry of Japan*, 257.

16. Ibid., 60. After the war, the Japanese industry retained a leadership position in crimped rayon technology; see M. Horio and T. Kondo, "Theory and Morphology of Crimped Rayon Staple," *Textile Research Journal* 23 (1953): 137–51.

17. LeRoy Henry Smith, *Synthetic Fiber Developments in Germany: Report Prepared by the Synthetic Fibers Team of the Technical Industrial Intelligence Committee* (New York: Textile Research Institute, 1946).

18. Ibid., 3.

19. Ibid.

20. Ibid., 3–4.

21. Ibid., 158–79. Mothwurf's name also appears as "Motwurf." The text lists four Mothwurf patents, two German and two American (179). The U.S. patents (2,345,622 and 2,365,096) were filed by Mothwurf in the summer of 1940, with mesne (that is, intermediate) assignment to the Industrial Rayon Company; the patents were awarded in 1944. The American team was not aware of a Canadian patent for his invention (CA 427212), issued on 1 May 1945, that lists Industrial Rayon, Edward G. Budd Manufacturing Company, and Budd International Corporation as co-owners of the patent. In 1938, Mothwurf began working as the agent of Budd International on contract to Thüringischen to install a continuous-process apparatus at its Schwarza plant. The machine was moved to Lenzing; Mothwurf left the employ of Budd to work directly for Schwarza in January 1942. Reports on Trusteeships, Sch 6/2 Lenzinger Zellwolle-und Papierfabrik AG: Correspondence (n.d., September 1945–September 1948), 120–21 (official translation), 127–29 (German-language original), Property Control Branch, U.S. Allied Commission for Austria, U.S. National Archives. For additional details on Mothwurf's history with Lenzing, see Sandgruber, *Lenzing*, 319–22.

22. Smith, *Synthetic Fiber Developments in Germany*, 3.

23. L. Hemsley et al., *The Viscose Continuous and Rayon Staple Fibre Plants of the British, American, and French Occupation Zones of Germany*, Final Report No. 290 (London: British Intelligence Objectives Sub-committee, [1946]), 2.

24. Ibid., 6. Detailed reports on specific plants appear as follows: Krefeld, 34–59; Glanzstoff Courtaulds Cologne, 87–102; Siegburg, 113–33; Kelheim, 255–72.

25. E. G. Locke, J. F. Saeman, and G. K. Dickerman, "The Production of Wood Sugar in Germany and Its Conversion to Yeast and Alcohol," in *Wood Yeast for Animal Feed*, Northeastern Wood Utilization Council (New Haven, Conn.), Bulletin 12, November 1946, 95–154, quotation on 139.

26. The information comes from a series of articles appearing in *Der Spiegel* over a period of more than a decade; see "Das Leben ist so bitter: Alles mit viel Liebe gekocht," 22 September 1949, 9–10; "Die Hellsherin befragt," 2 September 1953, 10–12; "Schweizer Touren," 17 February 1960, 20–21.

27. The United States of America vs. Carl Krauch et al., U.S. Military Tribunal Nuremberg, judgment of 30 July 1948, http://werle.rewi.hu-berlin.de/IGFarbenCase.pdf. One of the convicted defendants was Fritz ter Meer, the Farben negotiator who had visited DuPont (see note 70 to chapter 5).

28. Paul Macarius Hebert, "Dissenting Opinion on Count Three of the Indictment," *Nuremberg Trial Documents,* 28 December 1948, available at Louisiana State University Law Center, Digital Commons at LSU Law Center, http://digitalcommons.law.lsu.edu/cgi/viewcontent.cgi?article=1046&context=nuremberg_docs.

29. Jacobsen, *Operation Paperclip*, 302–14.

30. Elsner, *Heilkräuter*, "Volksernährung," *Menschenversuche,* 141–42.

31. Jacobsen, *Operation Paperclip*, 431–32. Jacobsen addresses the conjecture that Grünenthal's ties to concentration camp experimentalists may link wartime prisoners as the original source of the company's human-testing experience with thalidomide.

32. K.-J. Neumarkarker, "Karl Bonhoeffer and the Concept of Symptomatic Psychoses," *History of Psychiatry* 12 (2001): 213–26.

33. J. J. Martin, N. Partington, and S. Pearson, *The Manufacture of Carbon Bisulphide in Germany: With Notes on Sulphur Recovery and Thio-urea; B.I.O.S. Final Report 1702* (London: British Intelligence Objectives Sub-Committee, 1946), quotations on 21 and 28.

34. See Hedwig Wolff, *1946–1956: Zehn Jahre VEB Zellstoff- und Zellwollewerke Wittenberge* (Wittenberg: Betriebsparteiorganisation "Karl Liebknecht" im VEB Zellstoff- u. Zellwollewerke Wittenberge, 1956). A later, anonymous pamphlet is *Kämpfer: Garanten des sicheren Schutzes unserer sozialistischen Heimat; 30 Jahre Kampfgruppen der Arbeiterklasse im VEB Zellstoff- und Zellwollewerke Wittenberge* (Wittenberg: VEB Zellstoff- und Zellwollewerke, 1983). The forty-year anniversary publication referred to the plant as "an exemplar of the great struggles of our time."

35. Biographical details on Siegfried Rädel, after whom the Pirna plant was renamed, come from www.etg-ziegenhals.de/Siegfried_Raedel.html. Clara Zetkin, after whom the Glanzstoff facility in Elsterberg was renamed, was a radical feminist and comrade of Rosa Luxemburg.

36. V. A. Kritsman, V. I. Kuznetsov, and V. M. Pevzner, "Man-Made Fibres Research in the German Democratic Republic," *Fibre Chemistry* 3 (1971): 459–63. On the attractiveness of the East German viscose industry to West German investment, see "Die letze Börsenwoch," *Der Zeit,* 23 January 1947.

37. Johannes Karsch, "Die Keratoconjunctivitis chemicalis in den Sächsichen Kunstseidenwerken Pirna," *Klinische Monatsblätter für Augenheilkunde und für augenärztliche* 123 (1953): 440–49. It is unclear whether this is the Johannes Karsch after

whom Karsch-Neugebauer syndrome (a rare genetic combined eye and hand abnormality) was later named, based on authorship of a 1936 German case report. Another report from the DDR in this period details three cases of psychosis from carbon disulfide (classified as the exogenous Bonhoeffer type), including one in a rayon staple worker; see P. D. Krüger and E. Schilf, "3 Fälle von Schwefelkohlenstoffvergiftung des Nervensuystems nebst einer Erörterung der exogenen Reaktionstypen Bonhoeffers," *Psychiatrie Neurologie und Medizinische Psychologie* 4 (1952): 139–46.

38. Herbert Zenk, "Zur Symptomatik der Schwefelkohlenstoffeinwirkungen auf Grund von Reihenuntersuchungen Krankenstandsanalysenund Verdachtsmeldungen der Jahre 1956 bis 1965," *Das Deutsche Gesundheitswesen* 27 (1972): 518–20, quotation on 520.

39. Emil Paluch and Wanda Szamborska, "Badania toksykologiczne nad zatruciami dwurusiarczkiem węgla i siarkowodorem w polskim przemyśle wiskozowym," *Polski Tygodnik Lekarski* 1, no. 39 (1946): 1177–81; no. 40 (1946): 1217–22. The identification of Tomaszow and Chodakow is based on V. Vishnevskaya, "The Man-Made Fibre Industry of the Polish People's Republic," *Fibre Chemistry* 3 (1972): 168–70.

40. Feliks F. Sekuracki, "Occupational Medicine in Poland," *Industrial Medicine and Surgery* 27 (1958): 469–471, quotation on 470.

41. P. Bělin, V. Bencko, and J. Petráň, "Znečistenie oczdušia vo Svite v preibehu roku 1962: Zdroj znečistenia, jeho charakteristika, a poloha vzhľadóm k sídlisku" [Air pollution in Svit during 1962: Source of pollution, its characteristics, and degree in relation to the environment], *Československá Hygiena* 9 (1962): 73–77. For the soldier with the Czech army, see Ivan Brod, *From Auschwitz to Du Pont* (New Canaan, Conn.: Information Economics Press, 2008), 71–76.

42. Raimondo Finati, "Il rientro in Italia," in Raimondo Finati, ed., *Allo straflager di Colonia* (Naples: L'Arciere, 1990), 135–37.

43. Liliana Rimini Lagonigro, "Alessandro Rimini nei ricordi della figlia," in Giovanna D'Amia, ed., *Alessandro Rimini: Opere e silenzi di un architetto Milanese* (Milan: Maggioli, 2011), 121–28.

44. [Snia Viscosa], *Mezzo secolo di Snia Viscosa*, 40–42.

45. L'espresso, http://video.espresso.repubblica.it/tutti-i-video/torviscosa-il-corto-di-antonioni-/617/616. The credits list Antonioni as "Michelangiolo." Although not credited, the film score is by Giovanni Fusco, a close collaborator of Antonioni (*Red Desert*, among other films); see the biographical page of a website dedicated to the composer, www.giovannifusco.com/BiografiaENG.asp.

46. Sylwia Kuźma, "Popularizing Haute Couture: Acceptance and Resistance to the New Look in the Post-1945 United States," *Americanist: Warsaw Journal for the Study of the United States* 24 (2008): 143–57.

47. "Rose Pâle," ballad, c. 1947–48, words by Henri Contet, music by Paul Durand. The song was recorded by Simone Langlois and by Lucienne Delyle, each an important chanteuse in postliberation France. Contet is closely associated with Edith Piaf. "Two bits" is used to capture the casual feel of the French lyric, "*quatre sous.*"

48. Arthur Knight, "Courtaulds in Continental Europe in the 1950s and 1960s: Some Recollections and Reflections," *Business and Economic History* 16 (1987): 213–26.

Sir Arthur Knight coordinated overseas activities for Courtaulds starting in 1946, becoming its chairman in 1975.

49. Rowland Tennant, *A History of Holywell and Greenfield* (Wrexham, Wales: Bridge Books, 2007), 153.

50. Just before Courtauld's death, in October 1947, Anthony Blunt was appointed the institute's director; Courtauld Institute of Art, "History," www.courtauld.ac.uk /about/history.shtml, and "About the Gallery," www.courtauld.ac.uk/gallery/about/history .shtml.

51. R. and J. H-W, "Mr. Samuel Courtauld," *Times* (London), 17 December 1947. Although initialed only, "J. H-W." is almost certainly John Coldbrook Hanbury-Williams, who had become chairman of Courtaulds in 1946.

52. Tomassini, *La salute al lavoro*; Vigliani's editorial is covered on 75–76.

53. Enrico C. Vigliani, "L'intossicazione cronica da solfuro da carbonio: Una statistica di 100 casi," *La Medicina del Lavoro* 37 (1946): 165–93.

54. Enrico C. Vigliani, "Clinical Observations on Carbon Disulfide Intoxication in Italy," *Industrial Medicine and Surgery* 19 (1950): 240–42.

55. Vigliani, "L'intossicazione cronica da solfuro da carbonio," cases presented on 176–77, 179–80.

56. Enrico C. Vigliani and Benvenuto Pernis, "L'intossicazione cronica da solfuro da carbonio," in *Atti 11. Congresso Internazionale di Medicina del Lavoro, Napoli, 13–19 Settembre 1954* (Naples: Istituto di Medicina del Lavoro, 1954), 373–416. Vigliani had earlier reported on some of the initial cases in this series in Enrico C. Vigliani and C. L. Cazullo, "Alterazioni del sistema nervosa central di origine vascolare nel sulfocarbonismo," *Medicina del Lavoro* 41 (1950): 49–63.

57. Enrico C. Vigliani and Benvenuto Pernis, "Carbon Disulphide Poisoning in Viscose Rayon Factories," *British Journal of Industrial Medicine* 11 (1954): 234–44.

58. Enrico C. Vigliani and Benvenuto Pernis, "Klinische und experimentelle Untersuchungen über die durch Schwefelkohlenstoff bedingte Atherosklerose," *Archiv für Gewerbepathologie und Gewerbehygiene* 14 (1955): 190–202.

59. Vigliani and Pernis, "Carbon Disulphide Poisoning in Viscose Rayon Factories," 237.

60. Vigliani and Pernis, "L'intossicazione cronica da solfuro da carbonio," 381–88. The cases are summarized in a numerical listing; there is also a similar but shortened listing as an appendix to the *British Journal of Industrial Medicine* publication, also omitting the final eight cases (thus, forty-three as oppoed to the fifty-one cases in the original Italian).

61. Vigliani and Pernis, "Klinische und experimentelle Untersuchungen über die durch Schwefelkohlenstoff bedingte Atherosklerose." The early Swiss paper cited was E. Attinger, "Chronische Schwefelkohlenstoffvergiftung unter dem scheinbar ungewöhnlichen Bilde einer schweren Gefäßkrankheit," *Schweizerische medizinische Wochenschrift* 78 (10 July and 21 August 1948): 667–69, 815; see also Lewey [sic] et al., "Carbon Disulfide Poisoning in Dogs."

62. Among others, Vigliani cited the groundbreaking work of Dr. Gofman on cholesterol, which began appearing in the early 1950s. For the history of the "lipid hypothesis" of atherosclerosis in the first part of the twentieth century, see Daniel Steinberg, "An Inter-

pretative History of the Cholesterol Controversy: Part 1," *Journal of Lipid Research* 45 (2004): 1583–93.

63. In 1957, another publication based on further cases from Switzerland supported Vigliani's work; see Hans-Kaspar von Rechenberg, "Schwefelkohlenstoffvergiftung und das sulfocarbotoxische vasculäre Spätsyndrom," *Archiv für Gewerbepathologie und Gewerbehygiene* 15 (11 July 1957): 487–530.

64. [Morgan Stanley], *Memorandum on the Rayon Industry, Particularly American Viscose Corporation, Celanese Corporation of America, and Industrial Rayon Corporation* (New York: Morgan Stanley, 1949), 3.

65. DuPont, "Our Company," "History," "1941–69," "Orlon" http://www2.dupont.com/Phoenix_Heritage/en_US/1941_detail.html.

66. [Morgan Stanley], *Memorandum on the Rayon Industry*.

67. W. T. Astbury and C. J. Brown, "Structure of Terylene" (letter), *Nature* 158 (14 December 1946): 871.

68. J. R. Whinfield, "Chemistry of 'Terylene,'" *Nature* 158 (28 December 1946): 930–31, quotation on 931.

69. "Textile with a Great Stretch—Next Solar Eclipse," "Notes on Science," *New York Times*, 10 November 1946.

70. "'Dacron' Name Adopted; Du Pont Substitutes Trade-Mark for Former 'Amilar' Fiber," *New York Times*, 22 March 1951.

71. "New Dacron Plant Opened by Du Pont," *New York Times*, 24 May 1953.

72. One of the earliest uses of the term "polyester" in reference to the Terylene/Dacron class of fibers was a 1951 notice of DuPont's planned construction of its new facility; see "News and Notes—In the Laboratories," *Science* 115 (25 May 1951): 614.

73. "Coronation Spurs Rivalry in Fabrics," *New York Times*, 6 October 1952. At the coronation there was likely little synthetic, although the silk coronation banners, when they underwent conservation for the fiftieth jubilee, turned out to have a gold fringe made of rayon; see National Trust, Textile Conservation Studio, https://nttextileconservationstudio.wordpress.com/tag/queen-elizabeth-ii.

74. "Tire Cord Battle: Nylon vs Rayon; Fight to the Finish for $250 Million Market Pits Chemical Producers against Textile Industry," *Chemical and Engineering News* 38 (4 January 1960): 23–24.

75. [Morgan Stanley], *Memorandum on the Rayon Industry*.

76. Carlisle M. Thacker and Elmer Miller, "Carbon Disulfide Production: Effect of Catalysts on Reaction with Methane and Sulfur," *Industrial and Engineering Chemistry* 36 (1944): 182–84.

77. Coleman, *Courtaulds*, 309–10. The decision to establish a cellulose acetate facility (in Meadville, Pennsylvania) followed an extensive back-and-forth with DuPont, involving factors related to Courtaulds' European competitors as well.

78. [Clinton DuPont Cellophane] *Midwest Blend* 8, no. 1 (March–April 1950).

79. Crawford H. Greenewalt, *Address of Crawford H. Greenewalt, President, E.I. du Pont de Nemours & Company, Inc. before du Pont Employees and Business Leaders of Buffalo, New York Marking 25th Anniversary of Cellophane* (Wilmington, Del.: E. I. du Pont de Nemours, 1949).

80. [E. I. du Pont de Nemours], *Old Hickory Cellophane: A Quarter Century of Progress, 1929–1954* (Old Hickory, Tenn.: E. I. du Pont de Nemours, 1954).

81. *Spruance Cellophane News* 5 (25th anniversary issue), no. 20 (1 November 1955).

82. [Clinton DuPont Cellophane], *Midwest Blend* 8, no. 1 (March–April 1950). In 1952, Gordon K. Inskeep, an editor of *Industrial and Engineering Chemistry*, together with Prescott Van Horn, a member of DuPont's "film department," published a report on cellophane largely based on the operations of the Clinton plant: Inskeep and Van Horn, "Cellophane: A Staff-Industry Collaborative Report . . . ," *Industrial and Engineering Chemistry* 44 (1952): 2521–24, quotation on 2521 (title ellipsis in the original).

83. Du Pont Cellophane Co., Inc., v. Waxed Products Co., Inc., 85 F.2d 75 (2nd Cir. 1936).

84. "Dr. F. H. Reichel of Virginia Heads Giant Viscose Firm," *Washington Post*, 3 July 1948.

85. "DuPont Company Gives Fellowships in Science to Forty-Seven Schools," *Yale Daily News*, 13 December 1947.

86. United States v. E. I. du Pont de Nemours & Co., 118 F. Supp. 41 (D. Del. 1953); United States v. E. I. du Pont de Nemours & Co., 351 U.S. 377 (1956).

87. G. Edward White, *Earl Warren: A Public Life* (New York: Oxford University Press, 1982), 397.

88. "Arnold N. Nawrocki, Cheese Innovator, 78" (obituary), *New York Times*, 12 July 2003.

89. Although developed in the 1940s, Saran Wrap was first introduced as a household product in 1953, see *In: A Perspective on Global Packaging by Dow* (2013), http://storage.dow.com.edgesuite.net/dow.com/packaging/in_perspective/In_Perspective_1.pdf.

90. The Coventry rhyme was included in a discussion titled "Courtaulds (and its chimneys!)" on the Historic Coventry Forum, http://forum.historiccoventry.co.uk/main/forum-posts.php?s=0&id=4271&q=&member=&cat_id=3&show_cats=&var1.

91. "The 'Haff' Disease," *Journal of the American Medical Association* 83 (29 November 1924): 1783. Follow-up reports reiterated the suspected industrial source, citing arsine as the specific causative pollutant; see *Journal of the American Medical Association* 84 (17 January 1925): 216. Palytoxin was the biological factor eventually identified as the cause of Haff disease.

92. "Pollution Protests Halt Rayon Plant; Industry at Paw Paw, W. Va. Warmly Opposed," *Washington Post*, 12 April 1934.

93. Du Pont Rayon Co., Inc., v. Richmond Industries, Inc., et al., 85 F.2d 981 (4th Cir. 1936). Fifteen years later, DuPont was still focused on James River pollution, supporting a study reported as Academy of Natural Sciences of Philadelphia, Department of Limnology, *James River, Virginia Stream Survey Report, July–August 1951* (Philadelphia: Academy of Natural Sciences of Philadelphia, 1952).

94. Percy Harold McGauhey, H. F. Eich, William Herbert Jackson, and Croswell Henderson, "A Study of the Stream Pollution Problem in the Roanoke, Virginia, Metropolitan District," Engineering Experiment Station Series no. 51, *Bulletin of the Virginia Polytechnic Institute* 35 (May 1942), 1–120 (entire issue), quotation on 7–8.

95. Ibid., 9.

96. Arthur Evans, "Plant Waste, Sewage Add to TVA Problems; Some Sections Polluted, Engineer Says," *Chicago Daily Tribune,* 12 December 1945.

97. Max Forester, "Plan Advanced to End Stream Pollution in U.S.," *New York Herald Tribune,* 21 November 1948. The Water Pollution Control Act, sponsored by Senators Alben Barkley (D-Ky.) and Robert Taft (R-Ohio), passed on 30 June 1948.

98. Byron Filkins, "Rayon Fumes Blackened Her Porch," with the caption "Health Inspector Anthony Sidlow & Mrs. John C. Anderson, 3563 W. 100th Street," 1942, Cleveland Press Collection, Cleveland State University, Michael Schwartz Library, Special Collections, http://images.ulib.csuohio.edu/cdm/ref/collection/press/id/1663. Another image in this collection indicates that in 1927, a Cleveland grand jury investigated the question of air pollution from the local Industrial Rayon Corporation factory (formerly known as Industrial Fibre), "Witnesses before the grand jury in the probe of Industrial Fibre fumes nuisance," http://images.ulib.csuohio.edu/cdm/ref/collection/press/id/1654.

99. For the context of the Donora episode, see Devra Lee Davis, *When Smoke Ran Like Water: Tales of Environmental Deception and the Battle Against Pollution* (New York: Basic Books, 2004).

100. National Board of Fire Underwriters, *The Holland Tunnel Chemical Fire, New Jersey–New York, May 13, 1949* (New York: National Board of Fire Underwriters, 1949).

101. Ibid.

102. Sinclair Stone, *Saiccor: The First Fifty Years* (Pinegowrie, South Africa: Rollerbird, 2002).

103. Fighterworld, "About Port Stephens," "Post War Development," www.fighterworld.com.au/about-port-stephens/post-war-development.

104. Veronica Herrera-Moreno, Seong-Kyu Kang, and Aaron Sussell, *A Cross-Sectional Investigation of the Health Effects in Carbon Disulfide Exposed and Non-Exposed Workers in a Viscose Rayon Factory,* HETA 916-0114 (Cincinnati: National Institute for Occupational Safety and Health, [1996]), 2.

105. On the Continent, Courtaulds did not seem to have interests beyond Italy, Germany, and France. Courtaulds did extend its SNIA war-losses claim to French interests in its Spanish viscose venture, but was not otherwise engaged in the postwar rayon industry there.

106. [Morgan Stanley], *Memorandum on the Rayon Industry.* For Celanese operations in Mexico, see Celanese Mexicana, S. A. de C. V., History, www.fundinguniverse.com/company-histories/celanese-mexicana-s-a-de-c-v-history.

107. Neda Sherafat, "CUVZ Educational Park Pilot Project: In Search of Ecotourism Development and Opportunities in Zacapu, Mexico" (master's thesis, Chalmers University of Technology, Gothenburg, Sweden, 2013), 29.

108. Von Kohorn, *Cellulose to Cell Phones:* Egypt, 114–19; Peru, Chile, and Brazil, 119–22; India, 125–26; Palestine, 152–53; Ecuador, 178; Philippines, 184; Taiwan, 190; Pakistan, 206; Indonesia, 247–49. This source does not address the company's initial postwar foray back into the rayon business, which was a failed attempt to take control of the Lenzing, Austria, operation; see Sandgruber, *Lenzing,* 335–39.

109. "Rayon Company Formed; New Concern in Brazil to Get Equipment from Tubize," *New York Times*, 17 June 1935.

110. Robert B. Anderson to Walt W. Rostow, secret memorandum, 16 August 1961, released 20 September 1973, doc. CK3100352122, Declassified Documents Reference System (Framington Mills, Mich.: Gale, 2012).

111. For the competing scheme, see Leo D. Rosenstein, *Rayon: An Essential Industry for Palestine; A Survey for the Palestine Rayon Corporation* (New York: Palestine Rayon Corporation, 1948).

112. Hugh Thomas, *Cuba, or, the Pursuit of Freedom* (New York: Da Capo, 1998), 1165. See also "Ambassador of Fun," *Time Magazine*, 1 September 1958, 30. The eight-centavo stamp was issued on 30 January 1958 in a run of 100,00, along with a million four-centavo stamps also picturing Dayton and a textile-manufacturing plant, which may have been one of his other facilities (Alejandro Pascal, Cuban philatelic expert, e-mail to the author, 16 February 2013).

113. Paulo Roberto Ribeiro Fontes, *Trabalhadores e Cidadãos: Nitro Química; A Fábrica e as Lutas Operárias nos Anos 50* (São Paulo: Annablume [Sindicato Químicos e Plásticos-SP], 1997), 94–97, quotation on 95–96.

114. Antônio José de Arruda Rebouças, *Insalubridade: Morte Lenta no Trabalho* (São Paulo: Oboré Editorial, 1989), 161–79. Additional information came from Rudolfo Andrade de Gouveia Vileta (industrial hygienist) and Helio Neves (physician), e-mails to the author, 3 March 2011.

115. *The High Life* (1960), Internet Movie Database, www.imdb.com/title/tt0054007. There does not seem to have been any involvement of Lux Films in the international coproduction of *The High Life*. Lux Films was founded in Italy in 1935 by Riccardo Gaulino after he was forced out of SNIA; he became a major force in Italian filmmaking, including early collaborations with Carlo Ponti, Dino De Laurentiis, and Roberto Rossellini. Over time, the artistic caliber of his films fell off. In 1960, Lux Films produced movies such as *Cartagine in fiamme* [Carthage in Flames] and *Morgan il pirata* [Morgan the Pirate].

116. Richard S. F. Schilling, *A Challenging Life: Sixty Years in Occupational Health* (London: Canning, 1998). Schilling was apparently unaware of a contemporary doctoral dissertation on experimental carbon disulfide atherosclerosis, which begins by citing Vigliani: Josef Kösters, "Schwefelkohlenstoff-Vergiftung und Atherosklerose bei Meerschweinchen und Ratten" (Justus Liebig University, Giessen, Germany, 1960).

117. Howard B. Sprague, "Environmental Influences in Coronary Disease in the United States," *American Journal of Cardiology* 16 (1965): 106–13, quotation on 111.

118. P. Carmichael and J. Lieben, "Sudden Death in Explosives Workers," *Archives of Environmental Health* 7 (1963): 424–39. The earliest report on this subject appears to be Hans Joachim Symanski, "Schwere Gesundheitsschadigungen durch berufliche Nitroglykoleinwirkung," *Archiv für Hygiene und Bakteriologie* 136 (1952): 139–58.

119. John R. Tiller, "An Investigation of Byssinosis in a Rayon Mill," and John R. Tilling and Richard S. F. Schilling, "Respiratory Function During the Day in Rayon Workers: A Study in Byssinosis," *Transactions of the Association of Industrial Medical Officers* 7 (1957): 157–160, 161–62. Schilling was the president of the Association of Industrial Medical Officers at the time.

120. Jeremy Noah Morris et al., "Coronary Heart-Disease and Physical Activity of Work," *Lancet* 265 (28 November 1953): 1111–20.

121. "Risk of Coronary Thrombosis Among Rayon Workers Exposed to Carbon Bisulphide: Notes and Correspondence," *General Records of the Medical Research Committee and Medical Research Council, 1960–1962*, FD23/1277, National Archives, United Kingdom; J. N. Morris to H. Howard-Swaffield, 7 July 1960. I have used the numbers of cases as presented by Morris in this letter and later reiterated to Hinsworth, although these were somewhat modified after that in light of additional occupational data. Thus, the numbers that appear in the paper that was ultimately published (and in Schilling's memoir) differ, although the principal findings of elevated risk (overall, more than twice the expected proportion) remain the same.

122. Jeremy Noah Morris, *Uses of Epidemiology* (Edinburgh: Livingstone, 1957).

123. Schilling, *Challenging Life*. Schilling never identifies by name the specific Courtaulds factories he studied, but in fact, Aber, Castle, and Greenfield were the only three viscose plants in proximity to one another; Greenfield had been established in the 1930s.

124. "Risk of Coronary Thrombosis Among Rayon Workers." The 7 July letter (cited in note 121 to this chapter), in which Morris returned the check, refers to the exchange of letters that preceded this one. Howard-Swaffield, a former naval surgeon, joined Courtaulds in 1947. From shortly thereafter, he served as its chief medical officer until stepping down from that role in 1977; see W.H.L., "H. Howard-Swaffield MCRS, LCRP" (obituary), *British Medical Journal* 292 (22 February 1986): 567. "WHL" is, presumably, William Hugh Lyle, who succeeded Howard-Swaffield as chief medical officer at Courtaulds in 1977 and later testified in Washington, D.C., on behalf of the company in OSHA hearings (see note 154 to this chapter).

125. J. N. Morris to Sir Harold Hinsworth, 15 August 1960, General Records of the Medical Research Committee and Medical Research Council, 1960–62, FD23/1277, National Archives, United Kingdom.

126. Ibid., typed memorandum signed by Hinsworth, 18 August 1960.

127. Ibid., Joan Faulkner to J. N. Morris, 19 September 1960. Joan Faulkner, the wife of Richard Doll, was described in one reminiscence as having in her own right "considerable power and influence as MRC secretary"; see John Murray Last, *Last's Words* (blog), 14 May 2013, "Richard Doll," http://lastswords.blogspot.com/2013/05/richard-doll.html.

128. Ibid., J. N. Morris to Joan Faulkner, 6 October 1960.

129. Ibid., Joan Faulkner, typed memorandum, 18 October 1960, with handwritten addendum comment by Hinsworth, followed by Joan Faulkner to J. N. Morris, 24 October 1960.

130. Ibid., H. Howard-Swaffield to J. N. Morris, 26 November 1962.

131. Brieger and Teisinger, *Toxicology of Carbon Disulphide*.

132. Jana Pazderová-Vejlupková, *Profesor Jaroslav Teisinger a historie ceského pracovního lékarství* (Prague: Galén, 2005). Early work by Teisinger included human exposure studies at 50–100 ppm; see Jaroslav Teisinger and Bohumil Souĉek, "Absorption and Elimination of Carbon Disulfide in Man," *Journal of Industrial Hygiene and Toxicology* 31 (1949): 67–73.

133. Franz Goldstein, "Heinrich Brieger, M.D.—Eulogy," four-page typed manuscript [1972], AR 7034, box 5, folder 28, Ernst Hamburger Collection (Jewish Officials in Provincial and District Administrations, 1952–1970s), Leo Baeck Institute.

134. Heinrich Brieger, "Chronic Carbon Disulfide Poisoning," *Journal of Occupational Medicine* 3 (1961): 302–8, quotation on 306. Early on, Brieger had given specific attention to potential cardiac toxicity from carbon disulfide; see Heinrich Brieger, "On the Theory and Pathology of Carbon Disulfide Poisoning: IV. Effects of Carbon Disulfide on the Heart," *Journal of Industrial Hygiene and Toxicology* 31 (1949): 103–5 (same issue containing Teisinger and Souček's paper cited in note 132 to this chapter). Brieger's study was underwritten by the American Viscose Corporation; Brieger acknowledges AVC's medical director, Dr. J. A. Calhoun, "for his great personal interest in our work" (105).

135. P. G. Vertin, "Biochemical and Clinical Studies of the Pathogenesis of Carbon Disulphide," in Brieger and Teisinger, *Toxicology of Carbon Disulphide*, 94–98.

136. Discussion of P. G. Vertin paper cited immediately above, in Brieger and Teisinger, *Toxicology of Carbon Disulphide*, 99. Among participants who appear in these proceedings only as discussants is Dr. M. M. El-Attel, connected with the rayon factory in Egypt that was founded by the von Kohorn group. In one discussion, he makes the point that U.S. exposure limits may not be applicable elsewhere, "especially [in] developing countries where the majority of workers . . . suffer from different types of endemic diseases" (242). He specifically mentions bilharzia (the Egyptian endemic parasitic liver disease) and frequent work-related eye affections caused by "defective ventilation," but superimposed on prevalent eye disease, presumably trachoma, sometimes called "Egyptian ophthalmia."

137. John R. Tiller, Richard S. F. Schilling, and Jeremy Noah Morris, "Occupational Toxic Factor in Mortality from Coronary Artery Disease," *British Medical Journal*, 16 November 1968, 407–11.

138. "Sulphur and Heart Disease" (editorial), *British Medical Journal*, 16 November 1968, 405–6.

139. Richard S. F. Schilling, "Coronary Heart Disease in Viscose Rayon Workers" (editorial), *American Heart Journal* 80 (1970): 1–2.

140. T. Partanen et al., "Coronary Heart Disease Among Workers Exposed to Carbon Disulphide," *British Journal of Industrial Medicine* 27 (1970): 313–25. The Finnish factory studied was the rayon facility that had been in Karelia and that had to be relocated in 1941, Säteri Oy. The factory's smokestack was given the name "Noro's chimney" after the physician who studied the workers in the original facility in Karelia (Henrik Nordman, e-mail to the author, 18 March 2009).

141. National Institute for Occupational Safety and Health, *Criteria for a Recommended Standard . . . Occupational Exposure to Carbon Disulfide* (Washington D.C.: U.S. Department of Health, Education, and Welfare, National Institute for Occupational Safety and Health, 1977). By 1980, the cause-and-effect relationship between carbon disulfide and heart disease was well enough accepted in the United States that the "Medical News" column of the *Journal of the American Medical Association*, under the banner "No Proof of Environmental Ill Effects on the Heart," identified two occupational groups clearly at risk of developing work-related heart disease: explosives workers and "viscose

rayon workers, who inhale carbon disulfide"; see E.R.G., "But with Regard to Two Occupational Inhalants," within "No Proof of Environmental Ill Effects on Heart," *Journal of the American Medical Association* 243 (28 March 1980): 1220.

142. Thomas F. Mancuso and Ben Z. Locke, "Carbon Disulfide as a Cause of Suicide: Epidemiological Study of Viscose Rayon Workers," *Journal of Occupational Medicine* 14 (1972): 595–606. Dr. David A. Savitz, later a leading public health researcher, based his 1982 doctoral thesis, "Behavioral Effects of Carbon Disulfide Exposure in Viscose Rayon Workers," on the Mancuso cohort. He remembered the factory as being in Painesville, Ohio, which was consistent with both the Industrial Rayon employer and the 1938 start date of personnel records (the year the Painseville factory was established) as stated in the paper (David Savitz, e-mail to the author, 1 March 2010).

143. Thomas F. Mancuso, *Help for the Working Wounded* (Washington, D.C.: International Association of Machinists and Aerospace Workers, 1976), 75–77, quotation on 77.

144. Ibid.

145. "Va. 'Triangle' Case Slayer Dies in Chair," *Washington Post*, 3 March 1945.

146. Thomas F. Mancuso, "Epidemiological Study of Workers Employed in the Viscose Rayon Industry," study no. 210-76-0186 (Cincinnati, Ohio: National Institute for Occupational Safety and Health, 1981).

147. Matti Tolonen, Markku Nurminen, and Sven Hernberg, "Ten-Year Mortality of Workers Exposed to Carbon Disulfide," *Scandinavian Journal of Work and Health* 5 (1979): 108–14. Follow-ups at five and eight years had been previously published.

148. Michel Vanhoorne, "Epidemiological and Medico-Social Study of the Toxic Effects of Occupational Exposure to Carbon Disulphide" (thesis, University of Ghent, Belgium, 1992).

149. Michel Vanhoorne, D. De Bacouer, and G. De Backer, "Epidemiological Study of the Cardiovascular Effects of Carbon Disulphide," *International Journal of Epidemiology* 21 (1992): 745–52, quotation on 746.

150. The study's author, Dr. Michel Vanhoorne, was the source of information on the investigation's being delayed because of company resistance; he also confirmed the location of the factory in Belgium (which is not specified in his thesis or in the published paper cited immediately above) as Ninove, site of the manufacturer Fabelta Ninove (Michel Vanhoorne, e-mail to the author, 6 January 2013).

151. P. G. Vertin, "Über das Vorkommen von Herz- und Gefäßkrankheiten in einer Rayon-Fabrik," *International Archives of Occupational and Environmental Health* 35 (1975): 279–90.

152. P. G. Vertin, "Incidence of Cardiovascular Diseases in the Dutch Viscose Rayon Industry," *Journal of Occupational Medicine* 20 (1978): 346–50. He does not refer to the analysis that he presented in Prague.

153. J. Lieben et al., "Cardiovascular Effects of CS_2 Exposure," *Journal of Occupational Medicine* 16 (1974): 449–53, quotation on 449. Aside from the link to suicide, the statement that there had been no cases of severe illness in the U.S. industry was false. A report appeared in *JAMA* in the mid-1950s of two cases of severe carbon disulfide poisoning in an unnamed rayon factory in New York. The treating physicians were in Utica and Whitesboro, New York. See Morris Kleinfeld and Irving R. Tabershaw, "Carbon

Disulfide Poisoning: Report of Two Cases," *Journal of the American Medical Association* 159 (15 October 1955): 677–79.

154. Brian MacMahon and Richard R. Monson, "Mortality in the US Rayon Industry," *Journal of Occupational Medicine* 30 (1988): 698–705, quotation on 703.

155. American Conference of Governmental Industrial Hygienists, "Carbon Disulfide," in *Documentation of the TLVs and BEIs, 7th ed., 2006 Supplement* (Cincinnati: American Conference of Governmental Industrial Hygienists, 2006), 1–10 and chap. 4, n. 136.

156. U.S. Department of Labor, Occupational Safety and Health Administration, *Informal Pubic Hearing: Proposed Rule on Air Contaminants, August 2, 1988*, hearing transcript, 4:37–140. (Additional question-and-answer interchanges follow through page 200.)

157. National Institute for Occupational Safety and Health, "Carbon Disulfid" [*sic*], 1988 OSHA PEL Project Documentation, www.cdc.gov/niosh/pel88/75-15.html.

158. Anne T. Fidler and Michael S. Crandall, *Teepak, Inc.*, Hazard Evaluation and Technical Report HETA 95-098-L1959 (Cincinnati: National Institute for Occupational Safety and Health, April 1989).

159. AFL-CIO v. OSHA, 965 F.2d 962 (11th Cir. 1992).

160. P. M. Sweetnam, S. W. C. Taylor, and P. C. Elwood, "Exposure to Carbon Disulfide and Ischaemic Heart Disease in a Viscose Rayon Factory," *British Journal of Industrial Medicine* 44 (1987): 220–27.

161. Schilling, *Challenging Life*, 153.

162. Ibid.

163. "Coronary Heart Disease Among Viscose Rayon Workers," Ministry of Pensions and National Insurance, Industrial Injuries Advisory Council, Minutes and Papers, June 1971–31 December 1981, PIN 65/96 (former ref. no. II AC 133), minutes of meeting, 17 March 1978, National Archives, United Kingdom.

164. W. R. Henwood to A. J. Collins, memorandum, 28 November 1977, Ministry of Pensions and National Insurance, Industrial Injuries Advisory Council, minutes and papers.

165. Chris Schilling, Richard Schilling's son, recalled of his father, "He was very angry. It was Courtaulds and he was disgusted with what they did in regards to this work. And I do believe they sacked Dr. Tiller. I can remember them working together at home on the topic and him introducing me to John Tiller and him saying afterwards that he is a very jolly decent sort and they had given him the sack" (Chris Schilling to the author, 15 January 2009).

Chapter 7. Rayon Will Be with Us

1. Kathryn Rudie Harrigan, *Declining Demand, Divestiture, and Corporate Strategy* (Washington, D.C.: Beard Books, 1980), 280–85.

2. Geoffrey Owen, *The Rise and Fall of Great Companies: Courtaulds and the Reshaping of the Man-Made Fibres Industry* (Oxford: Oxford University Press, 2010), 89–92.

3. European Communities Press and Information Office, "International Trade Union Action Prevents AKZO Dismissals," *Trade Union News from the European Community* 10 (Winter 1972–73): 24.

4. Eddy Posthuma de Boer, "Fabriekshal van Enka-Glanzstoff," from "Werkloosheid in Nederland," Rijksmuseum, Amsterdam, NG-1977-276-69.

5. AkzoNobel, *Tomorrow's Answers Today: The History of AkzoNobel since 1646* (Amsterdam: AkzoNobel, 2008), 19–20.

6. Hansard, *Parliamentary Debates*, Commons, 6th ser., vol. 78 (2 May 1985), cols. 526–34, Keith Raffan (Delwyb) and response by the Parliamentary Under-Secretary of State for Wales (Wyn Roberts).

7. Ken Davies, e-mail to the author, 5 April 2011.

8. Nina Pauer, "Zum Beispiel Wittenberge," *Der Zeit ZEITmagazin*, 4 March 2010; see also Derek Scally, "Since the Wall Came Tumbling Down," *Irish Times*, 23 August 2014. Poland's rayon production took a similar path. Images of the abandoned Polish facility at Tomascow by Jakob Ehrensvärd can be seen on his "Ruins of Despair," a Web-based series, www.pbase.com/jakobe/another_viscose_factory.

9. Magnus Linnarsson, "Rayon för rikets försörjning: Kris, korporatism och beredskapspolitik, 1972–1983," *Historisk Tidskrift* 133, no. 1 (2013).

10. Ragnar Magnusson, e-mail to the author, 16 June 2012. See also Magnusson's book of recollections of the factory, *En stråle av ljus: Svenska Rayon AB 1943–1993; Fabriken, bygden, människor, händelser* (Edsvalla, Sweden: Ekens, 1993).

11. "Tio år sedan Svenska Rayons konkurs," 5 January 2014, http://sverigesradio.se/sida/artikel.aspx?programid=93&artikel=5747593.

12. *Tencel at Courtaulds: From Genesis to Exodus and Beyond* ... (blog maintained by former Courtaulds employees involved in Tencel research and development), www.lyocell-development.com/p/introducing-tencel.html.

13. James H. Jones and Sherry G. Selevan, *Walk-Through Survey Report No. 75-16* (Cincinnati: Industry-Wide Studies, NIOSH, 1975). A later, more detailed study by NIOSH appeared as John Fagen, Bruce Albright, and Sanford S. Leffingwell, "A Cross-Sectional Medical and Industrial Hygiene Survey of Workers Exposed to Carbon Disulfide," *Scandinavian Journal of Work Environment and Health* 7, supp. 4 (1981): 20–27. The factory studied is not identified.

14. Veronica Herrera-Moreno, Seong-Kyu Kang, and Aaron Sussell, *A Cross-Sectional Investigation of Health Effects in Carbon Disulfide Exposed and Non-Exposed Workers at a Viscose Rayon Factory*, NIOSH report no. HETA 96-0114. This report, which documents the participants and methods used, was sent to Courtaulds with a cover letter from Aaron Sussell, dated 2 June 1997, that began, "As we promised last year, enclosed is a draft protocol for the National Institute for Occupational Safety and Health (NIOSH) evaluation of health effects related to carbon disulfide at Courtalds Fibers, Inc., Mobile, Alabama."

15. Aaron Sussell, NIOSH, to the vice president for operations of Courtauld Fibers, 25 August 1997, NIOSH records related to Herrera-Moreno, Kang, and Sussell, *Cross-Sectional Investigation of Health Effects*. The letter addresses the eight-page response and the twenty-seven pages of comments from consultants that Courtaulds had sent in

response to Sussell's June letter and attached protocol for the proposed but delayed investigation.

16. David S Sundin, NIOSH, to Brian Hughes, Alabama Department of Public Health, 29 December 1997, NIOSH records related to Herrera-Moreno, Kang, and Sussell, *Cross-Sectional Investigation of Health Effects*. This letter closes out the proposed investigation and is copied to Courtaulds Fibers and others.

17. Bertram Price et al., "A Benchmark Concentration for Carbon Disulfide: Analysis of the NIOSH Carbon Disulfide Exposure Database," *Regulatory Toxicology and Pharmacology* 24 (1996): 171–76. This industry-sponsored reanalysis critiqued G. M. Egeland et al., "Effects of Exposure to Carbon Disulfide on Low Density Cholesterol Concentration and Diastolic Blood Pressure," *British Journal of Industrial Medicine* 49 (1992): 287–93.

18. Howard Frumkin, "Multiple System Atrophy Following Chronic Carbon Disulfide Exposure," *Environmental Health Perspectives* 106 (1998): 611–13.

19. Doyle G. Graham, "Carbon Disulfide," *Environmental Health Perspectives* 108 (2000): A110-12. Graham cites the decision against the heir to the patient from the U.S. District Court, Southern District of Texas, Houston Division, in August 1998 (civil action no. H-96-1904). The Supreme Court of Alabama had ruled that the plaintiff could not bring action under the hazardous waste provisions of federal law: Supreme Court of Alabama, West Berry Becton, Sr., and Mary L. Becton v. Rhone-Poulenc, Inc., as successor to Stauffer Chemical Company, Inc.; and Courtaulds PLC, 1960276 (7 November 1997).

20. Courtaulds Fibers v. Long, 779 So.2d 198 (15 September 2000) (Alabama Supreme Court, Summary of Opinion).

21. U.S. Environmental Protection Agency, Minutes of the December 21, 1998 Meeting with Acordis Cellulosic Fibers Inc., EPA Contract No. 68-D6-0012, Task Order No. 0021, ESD Project No. 97/06, MRI Project No. 4801-21, www.epa.gov/ttnatw01/cellulose/minacor2.pdf.

22. Courtaulds Fibers v. Long, 779 So.2d 198 (Summary of Opinion).

23. The Finnish viscose industry closed down late in 2008 after changing from Säteri Oy to Kuitu Finland Oy in 1977. An exception to the closures of the period was the reopening of a shuttered plant in Lovosice, Czechoslovakia, an operation that took back the defunct name of Glanzstoff.

24. EPA, Site Visit—Lenzing Fibers Corporation, Lowland, Tennessee Miscellaneous Cellulose Manufacturing Industry, NESHAP, EPA Contract No. 68-D6-0012, Task Order No. 0011, EPA Project No. 97/06, MRI Project No. 4800-11, 22 May 1998, www.epa.gov/ttnatw01/cellulose/lenzing.pdf.

25. Samuel Gottscho, *American Enka Corp., Morristown, Tennessee. Saw teeth*, 21 September 1948, LC-G613-T-53858, Gottscho-Schleisner Collection, Library of Congress, Prints and Photographs Division; see also *Lenzing-Lowland,* a photo blog of the former plant, http://lenzinglowland.blogspot.com.

26. Roger Mackey, e-mail to the author, 24 December 2012. Mackey's report of workers' concerns over a carbon disulfide–caused conflagration is consistent with the known risks of explosion, fire, and asphyxiation associated with the chemical. Not long after FMC changed to Avtex, OSHA was called into Front Royal, a sister facility of

Parkersburg, concerning the death of a worker who had "stuck his head into the manhole while trying to thaw the line and was overcome by hydrogren sulfide and carbon disulfide." Avtex was fined $2,160 for multiple serious violations; see OSHA Inspection 3355112, Avtex Fibers, Inc., 4 December 1986, https://www.osha.gov/pls/imis/establishment.inspection_detail?id=3355112. Two years before that, just before Christmas 1984, the *New York Times* reported on a damaging sewer explosion in downtown Yonkers, New York, believed to have been caused by carbon disulfide waste from a chemical factory: Edward Hudson, "Aftermath of Blast: Repairs and Mystery," *New York Times*, 23 December 1984.

27. Robert E. Rosensteel, Steven K. Shama, and Jerome P. Flesch, *American Viscose Division, FMC Corporation, Nitro, West Virginia*, Health Hazard Evaluation Determination Report 72-21-91, November 1973 (Cincinnati: National Institute for Occupational Safety and Health, 1973). The evaluation was carried out in the summer and fall of 1972.

28. Rachel Scott, *Muscle and Blood: The Massive, Hidden Agony of Industrial Slaughter in America* (New York: Dutton, 1974), 97–105, quotation on 100. This is also the source for the denial of Sayre's workers' compensation claim.

29. B. Drummond Ayres Jr., "Jobs Are Lost in Plant Shutdown, but So Is Foul Smelling Air," *New York Times*, 21 November 1989. Early plant closures are documented in U.S. Trade Commission, *Rayon Staple Fiber from Sweden*, USTC Publication 1360 (Washington, D.C.: U.S. Trade Commission, March 1983).

30. EPA, "Fact Sheet: Partners Mend Site, Sew Quilt of Future Uses—Avtex Fibers," fact sheet no. 900098 (undated), https://semspub.epa.gov/src/collection/HQ/SC31206.

31. "FMC Slates Closedown of Cellophane Plant at Marcus Hook, Pa.," *Wall Street Journal*, 4 January 1977; see also Mark Hedlund, "Marcus Hook," *Encyclopedia of Forlorn Places*, http://eofp.net/marcushook.html.

32. "FMC to Stop Making Cellophane, Plans to Close Facility," *Wall Street Journal*, 14 February 1978.

33. Chip Hughes and Len Stanley, "OSHA: Dynamite for Workers," in "'Here Comes a Wind': Labor on the Move," ed. Bill Finfer, special issue, *Southern Exposure* 4, nos. 1–2 (Spring–Summer 1976), 75–82, quotation on 77.

34. Ibid., 80.

35. Jerome P. Flesch and James B. Lucas, *Olin Corporaton Film Division, Pisgah Forest, North Carolina*, Health Hazard Evaluation Determination Report 73-8-132, April 1974 (Cincinnati: National Institute for Occupational Safety and Health, 1974).

36. "EPA Superfund Program: Ecusta Mill, Pisgah Forest, N.C.," www.epa.gov/region4/superfund/sites/npl/northcarolina/ecumlnc.html.

37. "Du Pont Cellophane Unit in Tennessee to Close," *Wall Street Journal*, 18 May 1964.

38. "Du Pont Co. to Stop Making Cellophane at Plant in Virginia," *Wall Street Journal*, 31 December 1975.

39. "Flex (Atlanta, Georgia) Has Acquired the Assets of E. I. DuPont de Nemours and Company's U.S. Cellophane Business" (paid advertisement), *Wall Street Journal*, 9 October 1986.

40. Morgan Chilson, "Belgian Firm Aims to Reopen Flexel plant," *Topeka Capital-Journal*, 12 June 1997; see also Robert H. Beach et al., *Economic Analysis of Air Pollution Regulations: Miscellaneous Cellulose Manufacturing Industry*, EPA Contract Number 68-D-99-024 (Research Triangle Park, N.C.: Research Triangle Institute Center for Economics Research, 2000), 11–13.

41. Innovia Films, www.innoviafilms.com.

42. Viskase, http://viskase.com/about-us.

43. Viscofan, www.viscofan.com/EN/Pages/default.aspx. For the NIOSH investigation of Teepak, see note 158 to chapter 6.

44. "Spartan Personality," *Record: Spartan Alumni Magazine* [Michigan State University], 15 September 1954, 14. This profiled Chester Hardt, one of O-Cel-O's three founders.

45. 3M.com, "Locations," http://solutions.3m.com/wps/portal/3M/en_US/3M-Company/Information/Resources/Locations. Alfred Politzer, the founder of the company, was a chemical engineer who fled Czechoslovakia in 1939; he was working at the time for the Bata facility in Svit (Peter Politzer, e-mail to the author, 11 August 2015).

46. Spontex, www.spontex.com.

47. Coloribus Advertising Archive, "Spontex Swing: 'The Hedgehog (French)' TV commercial," December 1999, www.coloribus.com/adsarchive/tv-commercials/spontex-swing-the-hedgehog-french-1874055.

48. André Cicolella and Raymond Vincent, "Workplace Pollution in Two Viscose Plants," *Giornale Italiano di Medicina del Lavoro* 6 (1984): 101–6.

49. Jungsun Par, Naomi Hisanaga, and Yangho Kim. "Transfer of Occupational Health Problems from a Developed to a Developing Country: Lessons from the Japan–South Korea Experience," *American Journal of Industrial Medicine* 52 (2009): 625–32.

50. For Dr. Rok Ho Kim, see the Collegium Ramazzini website, www.collegiumramazzini.org/fellows1.asp?id=300.

51. S. K. Cho et al., "Long-Term Neuropsychological Effects and MRI Findings in Patients with CS_2 Poisoning," *Acta Neurologica Scandinavica* 106 (2002): 269–75; see also H. J. Jhun et al., "Electrocardiographic Features of Korean Carbon Disulfide Poisoned Subjects After Discontinuation of Exposure," *International Archives of Occupational Environmental Health* 80 (2007): 547–51.

52. Special Committee for Wonjin Rayon Measures, *The Sullied History of Wonjin Rayon* [in Korean] (Seoul: Special Committee for Wonjin Rayon Measures, 1994), 59–62.

53. Ing. A. Maurer, S.A., client list (projects reference list) through 2012, www.maurer-sa.ch/Maurer2/projects.htm.

54. The Taiwanese rayon industry has been the subject of a number of biomedical publications, the most recent of which appeared in 2011: Jiin-Chyuan John Luo et al., "Blood Oxidative Stress in Taiwan Workers Exposed to Carbon Disulfide," *American Journal of Industrial Medicine* 54 (2011): 637–45.

55. Nick Bywater, "The Global Viscose Fibre Industry in the 21st Century: The First 10 Years," *Lenzinger Berichte* 89 (2011): 22–29.

56. H. Y. Song et al., "Studying the Health Status of Workers Occupationally Exposed to Carbon Disulfide" [in Chinese], *Zhonghua Lao Dong Wei Sheng Zhi Ye Bing*

Za Zhi 30 (2012): 443–47. Although not identified, the factories were in Xinxiang City, Henan Province (Dr. Yu Shanfa, e-mail to the author, 27 February 2012). In 2010 the Lenzing (Nanjing) Fibers Co. opened (70 percent Austrian owned). An early report focusing on the Shanghai Number 12 Viscose Rayon Factory, in operation since 1961, found that more than one in ten workers suffered from carbon disulfide poisoning; exposures were up to nine times as high as the levels legally allowed. See Liang Youxin and Qu De-zhen, "Cost-Benefit Analysis of the Recovery of Carbon Disulfide in the Manufacturing of Viscose Rayon," *Scandinavian Journal of Work Environment and Health* 11, supp. 4 (1985): 60–63. Between 2000 and 2004, other publications on Chinese viscose manufacturing appeared that were carried out in collaboration with Belgian investigators.

57. Grasim Industries Limited, "Another Name for Comfort," www.grasim.com/products/birla_viscose.htm, and Aditya Birla Group, http://adityabirla.com.

58. Bywater, "Global Viscose Fibre Industry."

59. Christopher Pinney, "On Living in the Kal(i)yug: Notes from Nagda, Madhya Pradesh," *Contributions to Indian Sociology* 22 (1999): 77–106.

60. Ibid., 91. Pinney cites an undated article from the local press for these complaints and inserts parenthetically *"napusanktva"* as the Hindi word for impotence, apparently a typographical error for *"napunsakta."* A more recent publication from India on the human health effects of carbon disulfide reports a threefold increase in miscarriage rates linked to husbands' exposures at an unnamed rayon factory. According to this paper, "It was also found that in the spinning department the exposure exceeds many times the Threshold Limit Values"; see K. G. Patel et al., "Male Exposure Mediated Adverse Reproductive Outcomes in Carbon Disulphide Exposed Rayon Workers," *Journal of Environmental Biology* 25 (2004): 413–18, quotation on 413. A nonbiomedical work that touches briefly on carbon disulfide exposure is E. A. Ramaswamy, *The Rayon Spinners: The Industrial Management of Industrial Relations* (Delhi: Oxford University Press, 1994): "The [rayon] cake removers' complaint about fumes emitted by spinning machines was the oldest industrial relations problem, having surfaced almost immediately after the plant was commissioned" (124).

61. Pinney, "On Living in the Kal(i)yug," 98. A particular foil of Pinney's is a figure named V. T. Padmanabhan, who wrote a small booklet, *The Gas Chamber on the Chambal: A Study of Job Health Hazards and Environmental Pollution at Nagda, Madhya Pradesh* (New Delhi: People's Union for Civil Liberties, Madhya Pradesh, 1983). Pinney is particularly skeptical of Padmanabhan's presumed upper-caste perspective. The Chaliyar River in Kerala, where another Birla Group facility was located, was also the site of contentious battles over pollution; see Asima Nusrath and A. M. Shabeer, "Impact of Industrial Shut Down and Land Use Change in Chaliyar Basin," *Journal of Geography and Geology* 3 (2011): 247–57.

62. Bill Weir, "A Trip to the iFactory: 'Nightline' Gets an Unprecedented Glimpse Inside Apple's Chinese Core," *Nightline*, 20 February 2012, http://abcnews.go.com/International/trip-ifactory-nightline-unprecedented-glimpse-inside-apples-chinese/story?id=15748745.

63. Tania Carreón et al., "Coronary Artery Disease and Cancer Mortality in a Cohort of Workers Exposed to Vinyl Chloride, Carbon Disulfide, Rotating Shift Work, and

o-Toluidine at a Chemical Manufacturing Plant," *American Journal of Industrial Medicine* 57 (2014): 398–411.

64. Jim Morris, "High Bladder Cancer Rate Shrouds New York Plant, Exposing Chemical Hazards in the Workplace," Center for Public Integrity, 16 December 2013, www.publicintegrity.org/2013/12/16/13972/high-bladder-cancer-rate-shrouds-new-york-plant-exposing-chemical-hazards-workplace.

65. E. E. Williams, "Effects of Alcohol on Workers with Carbon Disulfide" [Letter, Queries and Minor Notes], *Journal of the American Medical Association* 109 (30 October 1937): 1472–73.

66. Satyakam Mohapatra, Manas Ranjan Sahoo, and Neelmadhav Rath, "Disulfiram-Induced Neuropathy: A Case Report," *General Hospital Psychiatry* 37 (2015): 97.e5–6. This case report also reviews the literature, including eight other case reports and case series. On atherosclerosis, see J. M. Rainey and R. A. Neal, "Disulfiram, Carbon Disulphide, and Atherosclerosis" (letter), *Lancet*, 1 February 1975, 284–85.

67. Keron Fletcher et al., "A Breath Test to Assess Compliance with Disulfiram," *Addiction* 101 (2006): 1705–10; also see Keron Fletcher, "Disulfiram and the Zenalyser: Teaching an Old Dog New Tricks" (letter), *Alcohol and Alcoholism* 50 (2015): 255–56.

68. Daniel E Jonas et al., *Pharmacotherapy for Adults with Alcohol-Use Disorders in Outpatient Settings: Comparative Effectiveness Reviews*, no. 134 (Rockville, Md.: Agency for Healthcare Research and Quality, May 2014).

69. M. W. Glenn and W. M. Burr, "Toxicity of a Piperazine–Carbon Disulfide–Phenothiazine Preparation in the Horse," *Journal of the American Veterinary Medical Association* 160 (1 April 1972): 988–92.

70. Henry A. Peters et al., "Extrapyramidal and Other Neurologic Manifestations associated with Carbon Disulfide Fumigant Exposure," *Archives of Neurology* 45 (1988): 537–40, quotation on 537.

71. Guy Darst, "Manufacturers Take Grain Fumigant Off Market in Face of EPA Testing," Associated Press, 12 February 1985, www.apnewsarchive.com/1985/Manufacturers-Take-Grain-Fumigant-Off-Market-in-Face-of-EPA-Testing/id-baee5a0b83efd39fc92f08db6b4a58ee. The EPA banned 80/20 a few months later.

72. James F. Cone et al., "Persistent Respiratory Health Effects After a Metam Sodium Pesticide Spill," *Chest* 106 (1994): 500–508. Railroad mishaps led to carbon disulfide fumigant release as well; for example, a 1973 explosion involved a truck siphoning off the fumigant from a railroad tank car situated near a veterans' retirement home: "Truck Blast Spreads Lethal Fumes," *Los Angeles Times*, 17 November 1973.

73. EPA, "Soil Fumigant Labels: Metam Sodium/Potassium," EPA registration no. 5481-483, AMV 540, AMVAC Chemical Corporation, 11 June 2014, http://www2.epa.gov/soil-fumigants/soil-fumigant-labels-metam-sodiumpotassium.

74. Marie Bretaudeau Deguigne et al., "Metam Sodium Intoxication: The Specific Role of Degradation Products—Methyl Isothiocyanate and Carbon Disulphide—as a Function of Exposure," *Clinical Toxicology* 49 (2011): 416–22. An explosion in Akron, Ohio, in 1982 sent manhole covers into the air after a chemical spill at Goodyear re-

leased hundreds of gallons of carbon disulfide into the city's sewer line; "Akron Sewer Explosions Follow Factory Spill," *New York Times*, 16 November 1982.

75. Arysta LifeScience, "Arysta Expands Global Specialty Crops Portfolio: Buys Enzone, a Soil Fumigant, from DuPont" (press release), 27 October 2004, www.arystalifescience.com/eng-us/news/2004-news-media/arysta-expands-global-specialty-crops-portfolio-buys-enzone%C2%AE-a-soil-fumigant-from-dupont.html.

76. EPA, "Fenoxycarb, Sodium Tetrathiocarbonate, and Temephos: Registration Review Proposed Decisions; Notice of Availability Sodium Tetrathiocarbonate Registration Review Final Decision," EPA-HQ-OPP-2011-0381, FRL-8872-1, September 2011, *Federal Register* 76, no. 86 (4 May 2011): 25340–42. This documents the voluntary withdrawal as well as final EPA cessation of its use.

77. Arysta LifeScience, "Arysta LifeScience Suspends MIDAS in United States" (press release), 20 March 2012, www.arystalifescience.com/release/MIDASPress3-20-12FINAL.pdf. See also Sarah Rubin, "Arysta Would've Lost in Court—And Still Might, Even in Seeking Dismissal on Methyl Iodide Case," *Monterey (California) County Weekly*, 22 March 2012.

78. Arysta LifeScience, "Arysta LifeScience Acquired by Platform Specialty Products Corporation" (press release), 17 February 2015, www.arystalifescience.com/2015release/ArystaLifeScienceacquiredbyPlatform.pdf.

79. "Meet the Locals: More Styles from the Streets of Miami," *Hemispheres Magazine*, June 2012, 59.

80. Rayon episode in *Legally Blonde* (2001).

81. *Daily Show*, Stephen Colbert segment "Nuclear Secrets," 10 February 2004, www.cc.com/video-clips/7qf7ll/the-daily-show-with-jon-stewart-nuclear-secrets.

82. Jeff Lindsay, *Darkly Dreaming Dexter* (New York: Vintage, 2009), 31.

83. "I.D.," *Law & Order*, season 7, episode 2 (first aired 25 September 1996).

84. Donald Allen, ed., *Collected Poems of Frank O'Hara* (Los Angeles: University of California Press, 1995), 150. A more recent symbolist use of cellophane occurs in an opera about Walter Benjamin from 2005, which includes the recitative "The answer comes in the form of a question, an echo inside a shadow wrapped in cellophane"; Charles Bernstein (libretto) and Brian Ferneyhough (music), *Shadowtime* (Copenhagen: Green Integer, 2005), 79, act 4.

85. Timothy Noah, "Bush's Mr. Cellophane: Goodbye to Mel Martinez," *Slate*, 10 December 2003, www.slate.com/articles/news_and_politics/chatterbox/2003/12/bushs_mr_cellophane.html. See also Michael Jonas, "Mr. Cellophane," *CommonWealth*, 22 July 2008, addressing the issue of mandated governmental transparency.

86. Erin Dooley, "Sarah Palin Talks About Killing Her Own Dinner in New Reality Show," ABCNews.com, 4 April 2014, http://abcnews.go.com/blogs/politics/2014/04/sarah-palin-kills-her-own-dinner-in-new-reality-show.

87. Francis Spufford, *Red Plenty* (London: Faber and Faber, 2010), 217, 397n.

88. Synopses of the convoluted plot twists of *General Hospital* are available on the Web, along with profiles of the characters, such as Lord Larry Ashton, on SoapCentral.com; see http://soapcentral.com/gh/whoswho/larry.php.

89. Bywater, "Global Viscose Fibre Industry."

90. Outlast, "Technology," www.outlast.com/en/technology. Outlast is manufactured by Kelheim Fibres, originally Süddeutsche Zellwolle. A takeover by Lenzing was blocked by the German courts because it would interfere with competition in the tampon market. Cellophane also has a high-tech niche through applications in dialysis membranes, one made by the cellophane-sausage-casing maker Viskase; see http://viskase.com/products/applied-technologies/membra-cel.

91. Kenneth P. Wilson, *NARC Rayon Replacement Program,* NASA Technical Reports Server, 3 April 2002; see also "NASA Hopes to Weave a Plan to Keep the Avtex Rayon Coming," *Journal of Commerce,* 6 November 1988.

92. Heli Lagusa et al., "Prospective Study on Burns Treated with Integra, a Cellulose Sponge and Split Thickness Skin Graft: Comparative Clinical and Histological Study—Randomized Controlled Trial," *Burns* 39 (2013): 1577–87.

93. Vinicius C. S. Antao et al., "Rayon Flock: A New Cause of Respiratory Morbidity in a Card Processing Plant," *American Journal of Industrial Medicine* 50 (2007): 274–84; the indentity of the facility comes from Chris Piacitelli and Vinicius Antao, *Hallmark Cards, Inc., Lawrence, Kansas,* National Health Hazard Evaluation Report HETA #2004-0013-2990, January 2006 (Cincinnati, Ohio: National Institute for Occupational Safety and Health, 2006).

94. "As Pure as Edelweiss," *Lenzing Inside: The Global Magazine of the Lenzing Group* 1, no. 12 (2012): 40–43.

95. For a consideration of the "precautionary principle" asserting that potentially toxic chemicals should be proved, not presumed, safe, see Marco Martuzzi, "The Precautionary Principle: In Action for Public Health," *Occupational Environmental Medicine* 64 (2007): 569–70.

96. Smartfiber AG, "SeaCell—The Power of Seaweed in a Fiber," www.smartfiber.info/seacell.

97. Asahi Kasei Fibers Corporation, "More comfortable. More friendly. Our Bemberg Business is gentle on people and the environment," www.asahi-kasei.co.jp/fibers/en. See earlier text on the chemical composition of Bemberg silk.

98. David Coleman, "What's New in Suits? Look Closely," *New York Times,* 11 April 2004.

99. FTC, "FTC Charges Companies with 'Bamboo-zling' Consumers with False Product Claims" (press release), 11 August 2009, www.ftc.gov/news-events/press-releases/2009/08/ftc-charges-companies-bamboo-zling-consumers-false-product-claims.

100. Greener Ideal, "Is the FTC 'Bamboo-zling' Consumers?," 20 June 2013, www.greenerideal.com/science/0620-is-the-ftc-bamboo-zling-consumers.

101. Innovia Films, "NatureFlex Films Use Only FSC and PEFC Certified Wood Pulp" (press release), 24 July 2015, www.innoviafilms.com/NatureFlex/News/Media-Centre.aspx?id=111.

102. OSHA, Citation and Notification of Penalty, Inspection Number 924097, 28 January 2014, https://www.osha.gov/ooc/citations/InnoviaFilmsInc_924097_0128_14.pdf.

103. Viscose, "Celons/Viskrings," www.viscoseclosures.com/celons-viskrings. The manufacturing company, Viscose, is based in Swansea, Wales.

104. "Use of Atomic Liquids by Industry Predicted," *Los Angeles Times*, 23 February 1948.

105. [William Darton], *Little Jack of All Trades, or, the Mechanical Arts Considered in Prose and Verse Suited to the Capacities of Children* (London: Harvey and Darton, 1823), 63–64.

106. Philip J. Klemmer and Alexis A. Harris, "Carbon Disulfide Nephropathy," with accompanying editorial, Richard P. Wedeen, "Occupational Renal Diseases," *American Journal of Kidney Diseases* 36 (2000): 626–29, 644–45.

107. American Conference of Governmental Industrial Hygienists, "Carbon Disulfide."

108. Heinz-Peter Gelbke et al., "A Review of Health Effects of Carbon Disulfide in Viscose Industry and a Proposal for an Occupational Exposure Limit," *Critical Reviews in Toxicology* 39, supp. 2 (2009): 1–126; this research was funded by the Industrievereinigung Chemiefaser e.V., a viscose fiber trade association.

109. Institute for Occupational Safety and Health of the German Social Accident Insurance, list of international exposure limits (GESTIS database), www.dguv.de/ifa/Fachinfos/Occupational-exposure-limit-values/index.jsp.

110. Marilyn Silva, "A Review of Developmental and Reproductive Toxicity of CS_2 and H_2S Generated by the Pesticide Sodium Tetrathiocarbonate," *Birth Defects Research, Part B* 98 (2013): 119–38.

111. Y. Nishiwaki et al., "Six Year Observational Cohort Study of the Effect of Carbon Disulphide on Brain MRI in Rayon Manufacturing Workers," *Occupational and Environmental Medicine* 61 (2004): 225–32.

112. Professor Toru Takebayashi, Department of Preventive Medicine and Public Health, Keio University School of Medicine, Tokyo, e-mail to the author, 23 May 2015.

113. Bradly Haran, "Barking Dog (Slow Motion)," *Periodic Table of Videos*, University of Nottingham, 13 June 2013, https://www.youtube.com/watch?v=c3fJRRCAIdk. This can be contextualized with an older news item, "35 Hurt in Toxic Leak," *Los Angeles Times*, 24 April 1986, reporting on those made sick by carbon disulfide fumes at a local high school "after a chemistry class experiment went awry."

114. For EPA Toxic Release Inventory data for carbon disulfide: http://iaspub.epa.gov/enviro/efsystemquery.tri?fac_search=primary_name&fac_value=&fac_search_type=Beginning+With&postal_code=&location_address=&add_search_type=Beginning+With&city_name=&county_name=&state_code=&selecttribe=&triballand=&tribedistance=fac_beginning&sic_type=Equal+to&sic_code_to=&naics_type=Equal+to&naics_to=&chem_name=carbon+disulfide&chem_search=Beginning+With&cas_num=&program_search=2&page_no=1&output_sql_switch=TRUE&report=1&database_type=TRIS. The EPA site identifications are as follows: 14150GNRLM305SA (3M Co.—Tonawanda); 37774VSKSCEASTL (Vikase Corp); 61832TPKNC915NM (Viscofan USA Inc.); 66542FLXLN6000S (Innovia Films).

115. Xin Zhou, "A Leak at a Chemical Factory in China's Northern Shanxi Province Killed at Least Eight People over the Weekend, Underscoring Concerns About the Safety of Industrial Projects in the Country," *Bloomberg*, 18 May 2015, available at HydrocarbonProcessing.com, www.hydrocarbonprocessing.com/Article/3454515/Carbon-disulfide-leak-kills-eight-at-Chinese-factory.html.

116. Booth Moore, "'Hunger Games' Fashion Comes to Fiery Life," *Los Angeles Times,* 18 March 2012.

117. Eva Bellakova, "Seven Storey Storage House No. 11: Convervsion to Museum of Industrial Architecture," www.archiprix.org/2015/index.php?project=3436.

118. Ultrasuede LLC and Vainglorious Pictures, *Ultrasuede: In Search of Halston* (dir. Whitney Sudler-Smith), 2010.

Acknowledgments

The research for *Fake Silk* was supported in part by a grant from the National Library of Medicine (NIH 13LM010076). This book grew out of earlier work for "Going Crazy at Work" in *How Everyday Products Make People Sick* (University of California Press, 2006), after I realized that the full story of carbon disulfide had not yet been told. I originally developed some of the material on carbon disulfide in the nineteenth century in "From Balloons to Artificial Silk: The History of Carbon Disulfide Toxicity," in *Occupation Health and Public Health: Lessons from the Past, Challenges for the Future,* ed. N. D. Nelson (Stockholm: Arbete och Hälse, Ventenskaplig Skriftserie, National Institute for Working Life, 2006), 87–97. Chapter 2 builds on "Rayon, Carbon Disulfide, and the Emergence of the Multinational Corporation in Occupational Disease," in *Dangerous Trade: Histories of Industrial Hazard Across a Globalizing World,* ed. Joseph Melling and Christopher Sellers (Philadelphia: Temple University Press, 2011), 73–84. A major portion of the writing of *Fake Silk* was done while I was a Mellon Fellow at the Center for Advanced Studies for the Behavioral Sciences at Stanford University during the 2013–14 academic year; I benefited especially from the assistance of its research librarians, Amanda Thomas and Tricia Soto. Azar Khatibi, the interloan librarian at the University of California, San Francisco, provided ongoing assistance thoughout my research.

I am indebted to many colleagues and friends around the world who assisted me in visiting current and former rayon production facilities, helped me gain access to archival and other research materials, and provided translations and contexts for selected texts. Dr. Ethan Pollock was key in accessing and translating Mendeleev's text on viscose and in translating and placing in political context the Soviet archival memoranda on the strategic Finnish rayon works captured by the USSR in Jääski, Keralia. Dr. Henrik Nordman shared with me his personal knowledge of the Finnish physician who reported illnesses from the factory. Dr. Denis Vinnikov called to my attention and translated the early Russian medical report on viscose illness. Dr. Marcin Skrzypski and Dr. Anna Wald assisted me with Polish medical materials. Dr. Daniela Pelclová made possible my visit to the Glanzstoff Bohemia facility in the Czech Republic and assisted me with details regarding the work of Dr. Jaroslav Teisinger. Miroslava Langhamerová of the Historical Department, Terezín Memorial, assisted me with sources on forced labor at the Glanzstoff facility during the war years. Eva Bellakova and Henrieta Moravčíková provided access to and translation of texts related to the viscose facility at Svit, Slovakia; Dr. Peter A. Politzer shared reminiscences of his father, who worked as a chemist at Svit before fleeing Slovakia ahead of World War II.

Dr. Heiner Ruschulte (Hannover, Germany) graciously hosted me and made possible my visit to Wittenberge Elbe, where Günter Rodegast, a local historian, and Helmut Worbs, a retired engineer at Wittenberge's closed viscose factory, gave generously of their time and provided me with source material. Dr. Reimer Moller of the Neuengamme Archives assisted me with additional material regarding the satellite concentration camp for Phrix at Wittenberge Elbe. Shaul Ferrero at Yad Vashem, Jerusalem, provided me assistance in my research of Holocaust witness testimonies. Gerd Sander shared with me the letters of his uncle Erich Sander, who perished at Siegburg; the late Dr. Susan Groag Bell assisted me with translating the letters. Material related to the satellite camp of Mauthausen that supplied workers to Lenzing was provided by Helga Amesberger of the Institute of Conflict Research and Katherine Czachor of the Mauthausen Archives, both in Vienna, Austria. Clare Parker of London, England, a survivor the Mauthausen satellite at Lenzing, kindly agreed to meet with me and share her personal memories. Angelika Guldt, head of corporate communications for Lenzing Aktiengesellschaft, made possible my visit to the modern facility at Lenzing. Dr. Gine Elsner shared with me her work on Ernst Günther Schenck and Biosyn-Vegetabil-Wurst (synthetic sausage) in relation to viscose production. Dr. Petra Moser, an expert on the contributions of exiled European scientists after emigration, assisted me on the work of Dr. Ernst Berl. Jeanpaul Goergen and Ralf Forster shared with me details of their research on Kalle and its "ozaphan" cellophane-based film strips.

Barbara Mellor, the English translator of Angès Humbert's memoir of her experiences as a forced laborer, kindly provided me with her insights. Dr. Paul-André Rosental provided information on viscose manufacturing in Strasbourg, France, during the interwar period. Dr. Alfredo Mendez Navarro assisted me with sources on the history of the viscose industry in Franco's Spain, as did Dr. Francesco Carnevale and Dr. Valerio Cerretano in regard to viscose manufacturing in the Fascist period in Italy. Workers' compensation data on industrial illness in Switzerland was provided by Dr. David Miedinger. Sara Catella and Alexander Nikles of Ing. A. Maurer were kind enough to meet with me at the Bern offices of this Swiss rayon manufacturing engineering firm and provided me with a list of their clients going back to the 1930s.

My friend and colleague Dr. Kjell Torén (Gothenburg, Sweden) made possible my visit to meet with the former workers at Svenska Rayon, especially Ragnar Magnusson, and helped with the translation of material from Swedish and Norwegian. In Flintshire, North Wales, Brian Davies, his brother Ken Davies, and his sister-in-law, Debbie Davies, not only hosted me on a tour of the area where Courtaulds had once dominated, but also shared the story of Robert William Davies, Brian and Ken's late father, who worked in the factory and died prematurely from vascular disease. Dr. Bernard Thomann identified multiple Japanese experimental publications on carbon disulfide from the 1930s, and Dr. Seichi Horie (Fukuoka, Japan) translated material for me. Dr. Yangho Kim (Ulsan, Korea) provided me with the original report on the Korean victims of carbon disulfide; my UCSF colleague in occupational health nursing, Dr. Soo-Jeong Lee, translated the key text. Dr. Leon Guo of Taipei made possible my visit to the only currently operating rayon production facility in Taiwan. Dr. Shanfa Yu (Henan, China) provided information on the rayon industry there. Rudolfo Andrade de Gouveia Vileta (industrial hygienist) and Dr. Helio Neves (physician) of the workers' union at Nitro Química in the 1980s, provided me with information on exposure conditions at the Brazilian factory.

I am deeply indebted to Dr. Barbara Sicherman, the biographer of Alice Hamilton, who called to my attention a number of key materials on the U.S. viscose industry. I was also assisted by Max Wright, who provided me with materials related to his mother, Dr. Adele Cohn, who carried out the first study of rayon workers for Dr. Alice Hamilton at Bryn Mawr. Sarah Hambleton of the Schlesinger Library, Radcliffe, was diligent in tracking down Alice Hamilton's pocket diaries for me; Jessica Murphy at the Harvard Countway Library was pivotal in my gaining access to records there, particularly in regard to Philip Drinker. Amber Dushman, senior archivist at the American Medical Association, assisted me in accessing material related to the "viscose treatment." Dr. Charles B. Mundy, a former plant physician for the American Viscose Corporation, shared his reminiscences with me. Ralph von Kohorn's son, Michael von Kohorn, gave me invaluable access to his father's unpublished memoirs. Eli Reinhard shared memories of his father, Isidor, and the Arcadia Mills. Dr. Bruce P. Bernard of the National Institute for Occupational Safety and Health assisted me greatly with access to its reports and investigations. My late cousin, Joan Pinkham, called to my attention the French postwar ballad "Rose Pâle" and left me a vinyl recording of the song. The rayon episode in the film *Legally Blonde* was brought to my attention by Hannah Kohrman, as was the scene in *Animal House*.

Index

acetate silk, 58, 61, 74, 160, 175. *See also* cellulose acetate
Acheson, Dean, 68, 70, 259n131
Acordis, 197, 200
acrylonitrile, 168
Addams, Jane, 82
Aditya Birla Group, 207–8
aeropoetry, 131
AFL-CIO, 191
Agent in Italy (novel), 213
Agfa, 126
agribusiness, 210–12
Agriphar Group, 212
airplane doping, 115–16
air pollution, 47, 51, 163, 172–74, 199–200, 219, 273n98
AKU, 164, 194
Akzo, 194
AkzoNobel, 194, 197–98, 200
Alabama, health and pollution issues in, 197–200
Albert, Prince Consort, 8

alcohols, 6
Alfred, Roy, 186
alkahest, 6
Alkali Acts (Great Britain), 47
Alsberg, Carl L., 72–73, 239n80
amblyopia, 16, 18–19, 49, 51, 80. *See also* eye damage
American Bemberg, 87–91
American Cellophane League, 64
American Cellulose & Chemical Company, 118
American Chatillon Corporation, 114
American Conference of Governmental Industrial Hygienists, 189, 218, 250n136
American Dry Goods Association, 73
American Glanzstoff, 87–91
American Heart Journal, 185
American Liberty League, 63–64
American Medical Association, 75
American Standards Association, 105–6, 250n136

American Viscose Company, 26, 31, 170–71. *See also* Viscose Company
American Viscose Corporation (AVC), 44, 69, 94, 99, 102–3, 105, 113, 137–38, 151, 154, 156, 168, 173, 188, 193, 202, 249n122
Anderson, Mrs. John C., 173
Anderson, Sherwood, 90, 91
anesthesia, 5, 12, 17, 59
aniline, 79–80
animal experiments, 5, 11, 36, 39, 106, 123–24, 150–51, 155
Animal House (film), 213
The Annual Report of the Chief Inspector of Factories and Workshops, 33–34, 47–48, 51
Antabuse, 209–10
antibiotic synthesis, 263n166
antitrust charges, 171
Antonioni, Michelangelo, 164
Arcadia Development Company, 70
Arcadia Knitting Mills, 67–70
Arezzi, M., 39–40
Arliss, George, 30
Aron, Moses (Israel), 149
"artificial," 72–74
artificial leather, 73, 240n91
artificial silk, 23, 30–31, 59, 73–74, 87. *See also* viscose rayon
artificial wool, 130
artsilk, 73
Arysta LifeSciences, 212
Asahi-Kasei, 216
asbestos, 199
Ashton, Frederick, 76
Associated Rayon Corporation, 58
Astraline, 141
atherosclerosis, 106, 166–67, 184, 210
Atkinson, C. P., 62
Atlantic Rayon Corporation, 118
Auschwitz concentration camp, 149, 160, 163
Australia, 19, 63, 175
autarky, 130–31, 136, 140, 256n82
AVC. *See* American Viscose Corporation

Avtex Fibers, 202, 214
Avtex Fibers Superfund site, 202–3
Axis, Alabama, 175, 197–200

bacteria, 3
bagasse, 176
Bahin, Arthur, 13
Bakker, C., 49–50
bamboo, viii, 216
Bangladesh, 176
Banque Misr, 176
Bard, C. L., 20–21
barking dog experiment, 219
Barkley-Taft Water Pollution Control Act, 173
Barthelemy, H. L., 114–15
BASF, 194, 200
Bashore, Ralph, 97–98, 102, 104
Bata, 149, 163, 265n183
Bat'a, Jan, 149
Batista, Fulgencio, 177
Beaunit, 193
Becton, West Berry, 199
Bedford Dyers' Association, 152
beech pulp, 140, 215
Beer, Edwin, 25–26
Belgium, 140, 164
Bellakova, Eva, 220
Bemberg (company), 24, 125. *See also* American Bemberg
Bemberg silk, 24, 71, 216
Ben-Gurion, David, 177
benzene, 92
Beran, Josef, 148
Bergk (works manager), 158
Berl, Ernst, 119–20
Berry Amendment, 138
Berzelius, Jöns Jacob, 2, 3, 4
Bethlehem Steel, 84
Bevan, Edward, 24–26, 28
Bianchi, Bruna, 132
bilharzia, 276n136
BIOS. *See* British Intelligence Objectives Subcommittee
Biosyn-Vegetabil-Wurst, 147–48

Birla conglomerate, 176
bladder cancer, 121–22
blends, of rayon staple and natural fibers, 42, 63, 126, 136, 155
Blüthgen, Fritz, 58, 91
Bollettino della Società Medico-Chirurgica, Pavia (journal), 39–40
Bonhoeffer, Karl, 53, 160
bottle closures, viscose, 75, 107, 217
Bowen, Margaret, 88
Bowing, Emily, 84–85
BPD, 194
brain damage, ix, 101, 106, 123–24, 150, 166, 219
Breda Visada, 47
Brentford, Viscount. *See* Joynson-Hicks, William
Bretton Woods Conference, 67
Bridge, John C., 35, 48, 50–51
Bridge Hall Mills, 61–62
Brieger, Heinrich, 183–84
Brigo, Bartolomeo, 132
British Cellophane, 62, 204
British Cellulose, 117
British Intelligence Objectives Subcommittee (BIOS), 156–58, 160–61
British Journal of Industrial Medicine, 165, 179
British Medical Journal, 33, 36, 47, 51, 52, 185
Brod, Ivan, 163
Brooklyn Museum, 135
Brooks Paper Company, 77
Bryn Mawr College, 94–95
Bubbifil, 135
Buchenwald concentration camp, 264n178
Bulganin, Nikolai Alexandrovich, 139
Bulletin of the Virginia Polytechnic Institute (journal), 173
byssinosis, 180

Caetani, Princess, 130
Calhoun, J. Alfred, 99, 248n110, 276n134
Cameron, Richard, 45–46
Canada, 57, 175, 200
carbon disulfide: early uses of, 4–5; extinction treatment, 19–20; manufacturing uses of, viii–ix; molecular properties of, 1–2; names for, 2; in nature, 3; other industrial uses of, ix, 19–21; in pesticide industry, 19–21; poisoning, case examples of, 18, 20–21, 26, 45, 96, 100, 133; in rubber trade, 8–12, 16–19, 21–22; solvent capabilities of, 5–9, 19; synthesis of, 1–2, 170; in viscose manufacturing, viii–ix, 25, 42, 106–7, 169–70; volatility of, 4–5
—maladies caused by, 219; amblyopia, 16, 18–19; brain damage, ix, 101, 106, 123–24, 150, 166, 219; conjunctivitis, 48, 51, 53, 115; eye damage, 34, 47–51, 115, 123, 161, 196, 276n136; heart disease, 179–92, 276n141; keratitis, 49, 115, 161; mental disturbance, ix, 10–12, 17–18, 20–22, 40, 53, 78–79, 95–97, 102–3, 132–33, 144–45, 155; neuropathy, 34, 38; Parkinsonism, 52–54, 106, 133, 150, 211, 219; sexual disturbance, 12, 17, 133–34, 218–19; suicide, 17, 34, 41, 97, 143, 186, 201, 204–7
—toxicity of: awareness of, ix, x, 214, 218–19; British handling of, in early twentieth century, 32–36, 46–56; early scientific studies of, 5, 10–19, 28–29; exposure levels relating to, 99, 105–6, 124, 155, 177, 185, 189–90, 202, 208, 218, 250n136; Hamilton's expertise in, 78–80, 91–93; industrial studies of, 33–41, 45–46, 93–109, 123–24, 155, 165–66, 177, 180–92, 208–9, 230n50; official response to, x–xi, 187–92 (*see also* viscose rayon industry: secrecy and obstruction concerning operations of); recent non-viscose sources of, 209–12; symptoms of, ix, 10–12, 14–17, 26, 34, 99, 139; in viscose

carbon disulfide (*continued*)
  production, vii–viii, 25–27, 42, 51–52, 60, 78, 84–109, 114–15; in vulcanized vegetable oil process, 73; wartime studies of, 139–40; in World War II, 132–33, 140, 142–46
Carbon Disulfide Panel of Chemical Manufacturers Association, 198
carbon tetrachloride, 92, 210
casein, 130
casings. *See* sausage casings
Casorati, Felice, 129
Ceconi, Angelo, 38–39
Celanese, 74, 117, 118
Celanese Corporation of America, 170, 175, 259n128
Cellini, Benvenuto, 170
Cellon Company, 107
Cellonite Company, 117
cellons, 107, 217
cellophane: development and spread of, 60–62; dialysis membranes made from, 286n90; in Germany, 126, 256n80; in Great Britain, 61–62; green marketing of, 216–17; health hazards linked to, 107–8, 122–23; manufacturing process for, viii–ix, 107; marketing of, 63; name of, 61, 236n20; popular perceptions of, 70–71, 75–77, 213–14, 285n84; Sylvania and, 70; trademarking of, 70; in United States, 61, 170–71, 203–5; viscose rayon vs., 70–71
Cellophane League, 64
"Cellophane Symphony" (song), 214
Celluloid Corporation, 122–23
cellulose, 10, 23–25, 130–31, 197. *See also* wood pulp
cellulose acetate, 24, 74, 114, 115–17, 118, 160, 170, 175, 251n1
Century Rayon, 176
cerebral atherosclerosis, 106, 166–67
Challenger (spacecraft), 214
Chalmers, Noble R., 106
Chamberlain, Arthur Neville, 47, 152

Chambers, Whittaker, 60
Charcot, Jean-Martin, 13–15
Chardonnet silk, 23
Charles Macintosh and Company, 8, 16
chemical weapons, 117, 147, 160
Chemsvit, 163
Chemtura, 212
Chevalier (inspector), 230n50
Chicago (musical), 214
child labor, 40, 83, 138
China, 207, 218, 220
China Man-Made Fiber Corporation, 176
chlorinated hydrocarbons, 116
Churchill, Winston, 31
churn rooms, viii, 35, 42, 48, 51, 54, 92, 95–97, 105, 108
cigarette smoking, 179
"Clara Zetkin" plant, 161
Cleveland Press (newspaper), 173
Clynes, John Robert, 53
coal, 17, 20, 59, 60, 120, 128, 158, 170
Coffin, Charles F., 118
Cohn, Adele, 95–100, 246n87
Colbert, Steven, 213
cold-process vulcanization, 9–12, 79
Coleman, D. C., 137
Collins, A. J., 192
Collins, Gail, 64
Cologne, Germany, 128, 148–49, 158
COMECON (Council for Mutual Economic Assistance), 161–62
Communist Party of the United States, 58–59
Compagnia Industriale Società Anonima (CISA) Viscosa, 130, 132
Companhia Nitro Química Brasileira, 114, 176–77
Comptoir des Textiles Artificiels (CTA), 44, 58, 61, 164
concentration camps, xiii, 145–49, 157, 160, 163, 201, 264n178, 268n31
condoms, 9–10
conjunctivitis, 48, 51, 53, 115
Consumers' League of Delaware, 82–83. *See also* National Consumers' League

continuous fiber process, 42
Cote Bérénice, 141
cotton, 23, 27, 42, 46, 59, 63, 89, 113, 120, 126, 136, 138, 140, 148, 169, 180, 216, 220
Courtauld, John, 54
Courtauld, Samuel, 54, 56, 73–74, 152–53, 164–65
Courtauld Institute of Art, 164
Courtaulds. *See* Samuel Courtauld and Company
Coventry, England, 48, 51, 149, 172
Covington, J. Harry, 70
Covington, Kentucky, 204
Covington, Virginia, 87, 94
crew doffing, 123
crimped rayon, 156
Cross, Charles, 24–26, 28, 112, 168
Cuba, 177
cuprammonium, 24, 61, 71, 87, 216
Curie, Marie, ix
CVC Capital Partners, 200

Dacron, 168
Dalton, Hugh, 153
Danville, Illinois, 109, 205, 219
Dariaux, Genevieve Antoine, xiv
Davenport, Iowa, 204
Davies, Ken and Brian, 195
Davies, Robert, 195
Davy, Humphry, 2, 3, 4
Delacroix, Victor, 11
Delaware Rayon Company, 83–85, 108
Delpech, Auguste, 11–13
demonstrative medicine, 38
Devoto, Luigi, 38
Dior, 164
Disraeli, Benjamin, 30
disulfiram, 209–10
Dixon, Ernest, 190
Dr. Clason Viscose Company, 75
Donora, Pennsylvania, 173–74
dope dyeing, 153, 155
doping, 115–16
Dorr, Richard Eugen, 159

Doussalba, 141
Dow Chemical, 171
Downfall (film), 150
Dreaper, W. P., 56
Drinker, Philip, 99, 105–6
Duchenne de Boulogne, Guillaume, 10–11
Dunsmuir, California, 211
duopoly, xiv, 66, 70
du Pont, Eleanor, 66
du Pont, Irénée, 63, 64, 65–66, 83, 121
du Pont, Pierre S., 83
DuPont Cellophane Company, 44, 76
DuPont Company (E. I. du Pont de Nemours and Company), viii, 44, 57–58, 61, 62, 65–66, 69–70, 86–87, 106, 111–13, 121–24, 135, 156, 167–68, 170–71, 172, 204, 212, 255n70
DuPont Fibersilk Company, 44, 46
DuPont Rayon Company, 44, 46, 68
Duvivier, Julien, 178

Earle, George H., 97, 101–4
East Germany, xiii, 161–62, 184, 195
Echo Scarves, 135
eco-friendliness. *See* green marketing
Edens, Bessie, 88–89, 91
Edsell, David, 118
Egypt, 273n108, 276n136
80/20, 210–11
El-Attel, M. M., 276n136
electroplating, 7–8
Elizabeth II, Queen, 169
Elizabethton, Tennessee, 87–91, 193
Elkington, Mason and Company, 7–8
El Paso Natural Gas Company, 193
Enka, 164, 184, 200
ENKA-Vereinigte Glanzstoff, 194
Enos, Victoria, 86
environmental concerns: air pollution, 47, 51, 163, 172–74, 199–200, 219, 273n98; soil pollution, 199–200; water pollution, 51, 172–73, 199, 204

298   Index

Environmental Health Perspectives (journal), 198
Environmental Protection Agency (EPA), 199–200, 202–4, 211–12, 219
Enzone, 212
epidemiology, 178–81
Equifax, 108
Erskine, Lillian, 98–100, 102–3
ethanol, 6
ether, 7
evidence-based medicine, 38
explosive properties, 17, 23, 25, 27, 107–8, 111, 113–14, 119, 177, 179
exposure levels, 99, 105–6, 124, 155, 177, 185, 189–90, 202, 208, 218, 250n136
eye damage, 34, 47–51, 115, 123, 161, 196, 276n136. *See also* amblyopia; conjunctivitis; keratitis

Fabrique de Soie Artificielle de Tubize, 113, 119
factice, 73, 239n83
factories. *See* viscose rayon industry
Factory Act (Great Britain), 32
factory towns. *See* workers' villages
Farley, Eliot, 118, 253n45
Farley, James Aloysius, 63–64
fashion, xiv, 71, 135, 141, 163, 164, 169, 216
Faulkner, Joan, 182
Fay, Peter T., 191
Federal Trade Commission (FTC), 68–70, 75, 83–84, 216
FEFASA (Fabricación Española de Fibras Textiles Artificiales), 259n125
Ferretti, Antonio, 130
Fibranne, 141
Fibro, 152–53
filmstrips, 126
Finland, 139, 280n23
Flexel, 204
Flint, Wales, xii, 34–35, 180, 194–95
flock, 215
Flox, 125, 127

Food Machinery and Chemical Corporation (FMC), 188, 193, 202–3
forced labor, xiii, 141–49, 151, 157, 159–60, 264n178
Formosa Chemicals & Fibre Corporation, 207
Fortune (magazine), 69
Foster, Charles, 21
Four Saints in Three Acts (stage production), 76
Foxconn, 209
France: carbon disulfide studies in, 10–15; sponge production in, 205–6; viscose rayon industry in, 43, 57–58, 74, 140–41, 164; in World War II, 140–41
France Rayonne, 140–41
Franklin Rayon Dyeing, 118
Fredericksburg, Virginia, 171, 203
French Academy of Sciences, 13
"Friedrich Engels" plant, 161
Fries, Pierre, 141
Front Royal, Virginia, 202–3
fumaroles, 3

Gaddy, Marvin, 203
Gahura, F. L., 149
Gajewski, Fritz, 159–60
Gallaher, Christine, 89, 91
Gehrmann, George H., 122–24
Genasco, 28
General Artificial Silk Company, 28
General Hospital (television show), 214
General Mills, 205
Genesis project, 197
Genet, Louise, 13
Germany: carbon disulfide studies in, 15–16; viscose rayon industry in, 41–43, 57–58, 74, 125–29, 142–49, 151, 156–61. *See also* Nazi Germany
Gibsy (film), 145
Gilles de la Tourette, Georges, 15
ginger jake, 122

Glanzstoff, 58, 63, 125, 128, 129, 148–49, 158, 161, 164, 255n74, 280n23. *See also* American Glanzstoff
glos, 44
Goodyear, Charles, 9
Goodyear Tire and Rubber, 209, 284n74
Gordy, Samuel, 104–8
Graham, Doyle, 190
Grahame-White Aviation Company, 116
grain storage fumigants, 210–11
Great Britain: carbon disulfide studies in, 16–17; cellophane in, 61–62; cellulose acetate usage in, 116–17; and postwar viscose rayon industry, 156–58; viscose rayon industry in, xiii, 32–36, 46–56, 57–58, 135, 153, 164–65, 180–82, 199; workplace protection in, xi, 30–36, 46–56, 191–92; in World War II, 135
Greener Ideal, 216
Greenfield, North Wales, 153, 180–81, 191, 195
green marketing, vii, viii, 215–17
Gregory, William, *Outlines of Chemistry, for the Use of Students*, 5–6
Griffin, Mary, 26
Griffin, Roger B., 24
Groms, Adolf, 159
Gross-Rosen concentration camp, 146
Grünenthal, 160, 268n31
Gualino, Riccardo, 129, 134, 257n95, 274n115
Guillain, Georges, 15
guncotton, 23, 112–13

Haff disease, 172
Hallmark, 215
Halston (Roy Halston Frowick), 220
Hamilton, Alice, 78–84, 91–95, 97–103, 105–9, 178, 183, 255n69; *Exploring the Dangerous Trades*, 82, 83, 109; *Industrial Poisons in the United States*, 45–46, 78, 80; *Industrial Toxicology*, 91–92
Hammett, Dashiell, *The Thin Man*, 77

Hancock, Thomas, 8–9
Happy Valley, Tennessee, 88, 90
Hardin, Elizabeth, 91
Harper's Magazine, 80–81
Harvard-MIT School of Health Officers, 117
Harvard School of Public Health, 92, 94–95, 99, 105, 118, 188
Haskell Laboratory of Industrial Toxicology, 123–24
heart disease, xi, 179–92, 195, 208, 209, 276n141
Heath, F. C., 18–19
Heaton, Ida, 91
Hedges, Burke, 177
Hedges, Dayton, 177
Heim (plant director), 158
Heller, Abraham A., 59
Hendon Aeroplane Factory, 116–18
Henwood, W. R., 192
Herbert, Paul Macarius, 160
Herbunot, Louis, 13
Herrick, Elinore Morehouse, 85–86
Hertel, Michael, 36–37
Heunghan Synthetic Fiber Company, 206
Heyman, Edward, "If Love Came Wrapped in Cellophane," 75
Heyme (chief engineer), 158–59
The High Life (film), 177–78, 274n115
Hinsworth, Harold, 182
Hirose, J., 154
Hiroshima, Japan, 154–55
Hiss, Alger, 60
Hitler, Adolf, 128, 150
Hoechst, 61, 124, 161, 194
Hoffman, Arthur, 89
Holland Tunnel, 174
Hoover, Herbert, 88
Hopewell, Virginia, 113–14
hot-process vulcanization, 9, 79
Houseman, John, 76
Howard-Swaffield, H., 181–83, 275n124
Hueper, Wilhelm, 121, 123–24

Humbert, Agnès, 141–43, 158, 261n152
The Hunger Games (book and film series), 220
Hutchins, Grace, *Labor and Silk,* 58–60, 111–13
Huxley, Aldous, *Brave New World,* 74, 220
hydrogen sulfide, 49–50, 105, 127
hysteria, 13–15, 85

I. G. Farben, 61, 124, 126, 148, 159–61, 200, 255n70, 255n73, 255n74, 257n98, 264n178
imitations, 72–74
Imperial Chemical Industries, 156
India, 74, 176, 207–8, 218, 283n60
Indian Rayon Corporation, 176
Indonesia, 176, 207, 218
Industrial Injuries Advisory Council, Industrial Diseases Sub-Committee (Great Britain), 192
Industrial Medicine and Surgery (journal), 165
Industrial Rayon Corporation, 86–87, 185–86, 193
Ing. A. Maurer, 207
Innovia Films, 204, 216–17, 219
insanity. *See* mental disturbance
Insulare, 41
Inter-Industry Committee on Carbon Disulfide of the Man-Made Fiber Producers Association, 188–90
International Association of Machinists and Aerospace Workers, 186
International Monetary Fund, 67
International Oxygen Company, 59
International Publishers, 58–60
International Symposium on Toxicology of Carbon Disulfide (1966), 183–84
International Telephone and Telegraph, 220
Inventions for Victory (exhibition), 135
Irmão, José Cecilio, 177
Israel, 176–77

Istituto Nazionale Fascista per l'Assicurazione contro gli Infortuni sul Lavoro (INFAIL), 133
Italian Society of Occupational Medicine, 133
Italy: occupational medicine in, 38, 41; viscose rayon industry in, 38–41, 43, 57–58, 74, 129–34, 163–64; in World War II, 129–34, 136, 149

J. T. Baker Company, 174
Jääski, Finland, 139
Jakubovic, Alice, 149
Jamaica ginger, 122
Japan: viscose rayon industry in, 43–44, 57–58, 74, 135–36, 150, 154–56, 206–7; workplace protection in, 43; in World War II, 135–36, 150
Jeffers, William M., 138
Jewish forced labor, in World War II, 146, 148
Johns Manville, 199
Johnson, Henry, 68
Jolibab, 141
Jones, Brownie Lee, 87, 243n47
Jones, Hazel, 91
Jones, Thomas P., *New Conversations on Chemistry,* 4
Journal of Industrial Hygiene, 45
Journal of the American Medical Association (*JAMA*), 45, 79–80, 104–5, 107, 248n116, 276n141, 277n153
Journal of the Society of Dyers and Colourists, 23
Joynson, Richard Hampson, 31–33
Joynson-Hicks, Lancelot, 54
Joynson-Hicks, William (later Viscount Brentford), 30, 35, 46–47, 53–54
Judson, A. H., 21

Kaiser (works manager), 158
Kalle and Company, 61, 124, 126
Karelian isthmus, 139, 276n140
Karsch, Johannes, 268n37

Kelley, Florence, 82
Kelly, W. T., 30, 32, 47, 52, 54
keratitis, 49, 115, 161
Kessler, Henry, 246n87
Keun, Irmgard, *Das kunstseidene Mädchen (The Artificial Silk Girl)*, 71–72, 177
Keynes, John Maynard, 137
Khoroshko, Vasily K., 27–28
Kim, Rok Ho, 206
kimonos, 135
Kingsbury, Susan, 94–95
Kinka Spinning, 154
Kiviette, 135
knitting factories, 65
Kohorn, Oscar von, and sons, 120–21, 150, 175–77
Kohut, Heinz, 265n190
Köln-Rottweil AG, 42, 124, 255n73
Kranenburg, W. R. H., 50
Krefeld, Germany, 141, 158
Kruse, Kevin, 64
Kuitu Ltd. (Kuitu Oy), 139
Kummer, W. C., 90–91
Kwon, Kyung-Yong, 206

Labor and Industry series (International Publishers), 58–60
labor reforms, 102
labor unions, 47, 52, 59, 87, 95–96, 104, 114, 186, 194, 244n56
La Cellophane, 61–62
Lampadius, Wilhelm August, 1–2
Lancet (journal), 30–31, 36
Langelez, Albert, 140
Lanital, 130, 155
Lansdowne Borough, Pennsylvania, viscose factory in, 26–28
Lardner, Ring, 76
Laudenheimer, Rudolf, 16
Lauder, A. Estelle, 242n24
Law and Order (television show), 213
lawsuits brought by workers, 18–19, 84–85, 104–5, 191–92, 199
Lazard Speyer-Ellissen, 58

Legally Blonde (film), 213
Legge, Thomas, 33–35, 52, 54–55, 118
Lehman, Herbert H., 90
Lehman Brothers, 58
Lehmann, Karl B., 36, 50, 124, 155, 255n69
lend-lease legislation, 137
Lenzing, 200–201, 215–16, 273n108, 282n56. *See also* Zellwolle Lenzing
Leonard, Louise, 91
Lewistown, Pennsylvania, 94, 97, 102–3
Lewy, Friedrich Heinrich, 98, 106, 160, 166
Liberty Funds, 201
Life and Labor Bulletin (newsletter), 87, 90
Lindol, 122–23
Lindsay, Matilda, 89–90
linters, 113
Little, Arthur Dehon, 24–26, 117
Little, Royal, 117–19, 253n38
Little Jack of All Trades (children's book), 218
Little New Deal, 102, 104
liver damage, 116–18, 210, 276n136
Łódź, Poland, 140, 148
Long, F. Farwell, 110
Long, Horace and Margaret, 199
Longda (JianXi) Differential Fibre Company, 207
Lord Woolton pie, 152
Loriga, Giovanni, 39, 258n111
Lorillard, 122
Loudon, Tennessee, 219
Lovosice, Czech Republic, xiii, 148, 280n23
Lübke, Ernst, *Der deutsche Rohstoffwunder* [The German Raw Material Miracle], 128
"Lucy in the Sky with Diamonds" (song), 214
lung disease, 59, 104, 180, 215
Lustron Company, 117–18, 253n45
Lux Film, 274n115
Lux soap, 72

302 Index

Lyle, William Hugh, 190
Lynch, Gerald Roche, 54–55
Lyocell, 201, 215–16

MacArthur, Douglas, 154
Machinist (newsletter), 186
Mackey, Roger, 202
Madsen, Billie, "My Cellophane Baby," 75
Magnusson, Ragnar, 196
Malamud, Bernard, 128
Maltz, Albert, "Man on the Road," 59
Manchester, Petricha E., 82–84, 92
Manchuria, 135, 150, 155–56
Mancuso, Thomas, 185–87; *Help for the Working Wounded*, 186
The Man in the White Suit (film), 153
Marcet, Alexander, 2, 4
Marcet, Jane, *Conversations on Chemistry*, 4
Marcus Hook, Pennsylvania, 28, 80, 94, 97, 110, 203
Margiotte, Charles J., 104
Marie, Pierre, 14, 15
Marinetti, Filippo Tommaso, xiv, 131–32, 257n104
Marissa, Brittany, 212–13
Martineau, Harriet, 7–8
Masina, Giulietta, 178
Mauthausen concentration camp, 147
Maxton, James, 31–32
Maxwell, Vera, 135
McDaniel, Raymond (Buster), 186
McDonald, Ellice, 121
Medical Research Council (Great Britain), 180, 182–83, 191
La Medicina del Lavoro (journal), 36–38, 39, 134, 165
Meer, Fritz ter, 255n70, 268n27
Mendeleev, Dmitri, 27, 227n88
mental disturbance, ix, 10–12, 17–18, 20–22, 40, 53, 78–79, 95–97, 102–3, 132–33, 144–45, 155
mergers, 114, 118, 124, 171, 193–94, 233n82

metam sodium, 211–12
methane gas, 170
methyl iodide, 212
methyl isocyante, 211
methyl isothiocyanate, 211
Metropolitan Life Insurance Company, 46
Mexico, 136, 175
microelectronics manufacturing, 209
Midas, 212
Midland-Ross Corporation, 193
Midwest Blend (newsletter), 170–71
Miller, Lee, 76
Millon, Auguste-Nicolas-Eugène, 19, 239n85
mineral sulfur, 17, 40
Minerva Medica (journal), 38
Modal, 215
monkeys, experimentation on, 150–51
monopoly, 171
Morgan Stanley, 138, 167–68, 175
Morgenthau, Henry, 137
Mork, Harry S., 117, 118, 253n45
Morris, J. N. "Jerry," 180–85, 188–89, 192
mortality rates, 181. *See also* standardized mortality ratios
Mothwurf, Arthur Franz Felix, 88, 90–91, 157, 267n21
Mueller (chief chemist), 158
Mullens, Frances, 91
munitions, 111–13. *See also* explosive properties
Musée des Arts et Traditions Populaires, 141
museums, in homage to viscose rayon industry, 195–96, 220
Mussolini, Benito, 129–31, 133, 136
Myers, Richard, "If Love Came Wrapped in Cellophane," 75
myopathy, 134, 172
Mytishchi Viskosa, 60

NASA, 214
National Aniline and Chemical Company, 80

National Association of Wheat Growers, 211
National Consumers' League (and state affiliates), 82–83, 86, 94, 242n25, 243n47
National Guard, 90
National Institute of Environmental Health Sciences, 198
National Institutes of Health, 122
National Rayon Week, 63
Nature (journal), 168
NatureFlex, 217
Nawrocki, Arnold, 171
Nazi Germany, 125–29, 134, 141–49, 151
Negro, Fedele, 53–54
Netherlands, 49–50, 164, 194, 233n82
Neuengamme concentration camp, 145–46
neuropathy, 34, 38
New Bedford Rayon, 84
New Deal, 91
New England Journal of Medicine, 92
New Yorker (magazine), 76
New York Post (newspaper), 64
New York Times (newspaper), 66, 90, 102, 113, 168, 216
Nichols, John, 105
Nightline (television show), 209
NIOSH. See U.S. National Institute for Occupational Safety and Health
Nitro, West Virginia, 112–13, 202
nitrocellulose, 23, 61, 111–16
Nitro Química. See Companhia Nitro Química Brasileira
N-methyl morpholine-n-oxide (NMMO), 197, 215–16
Nobel, Alfred, 112
Nobel conglomerate, 112, 194
Nojax, 205
Noro, Leo, 139, 261n139, 276n140
Norris, Howard, 151
North American Rayon Corporation, 214
North Wales, 153, 180–82

nylon, 124, 167, 169
nylon-6, 124, 167, 255n70
nylon-6,6, 124, 167, 255n70
Nylonge Company, 205

occupational disease legislation, 102–5
occupational medicine, 38, 41, 45, 133
Occupational Safety and Health Administration (OSHA), xi, 106, 188–91, 217, 218, 250n136
O-Cel-O, 205
offshoring, vii, xi, 114, 206
O'Hara, Frank, "Second Avenue," 213–14
Old Hickory, Tennessee, 44, 86–87, 112, 170, 204
Olin Corporation, 203–5
Oliver, Michael, 191–92
Oliver, Thomas, 17, 35–36, 40–41
Operation Paperclip, 160
Ophüls, Marcel, 141
Orgatex, 159
Orlon, 167–68
Orlov, Georgii Mikhalovich, 261n139
OSHA. See Occupational Safety and Health Administration
Outlast fibers, 214, 286n90
Ozalid, 256n80
Ozaphan, 126

Padmanabhan, V. T., 283n61
Pakistan Rayon Corporation, 176
Palin, Sarah, 214
parachutes, 118
paralysis epidemic, 122–23
Paris Faculty of Medicine, 13
Parisian artificial silk, 23
Parkes, Alexander, 7–8
Parkinsonism, 52–54, 106, 133, 150, 211, 219
Paske (plant director), 158
Payen, Anselme, 10
Peacock, Edward, 137

304   Index

Pennsylvania: occupational disease legislation in, 102–5; studies of viscose rayon industry in, 94–103
Perfect, Vicky, xii
perfume, 73, 239n85
Perkins, Frances, 104
Perlon, 124
Permanent Commission and International Association on Occupational Health, 183
pesticides, 19–21, 210–12
Peterson, Frederick, 18
Petit, John Read, 28
Petit, Silas, 28
petroleum, 169–70, 172
Phrix, 125–26, 128–29, 141–46, 148–49, 158–59, 161, 195, 255n75, 264n176
Der Phrixer (magazine), 128–29
A Phrix Factory Comes into Being (film), 125–26
phylloxera, 19–20, 212
picric acid, 84
Pieck, Wilhelm, 161
Piergallini (plant director), 132
Pinney, Christopher, 208, 283n61
Platform Specialty Products Corporation, 212
poetry, 131
Poland, 113, 140, 144, 146, 148, 162, 279n8
Politzer, Alfred, 282n45
pollution: air, 47, 51, 163, 172–74, 199–200, 219, 273n98; soil, 199–200; water, 51, 172–73, 199, 204
polyester, 168–69, 271n72
popular culture, 75–77, 212–14
Porter, Cole, 75
Price, Joseph, 189–90
Price, Waterhouse and Company, 68
price cutting, 65
price-fixing, 68, 84
propaganda, 126–28, 135–36, 145
proportionate mortality studies, 181. *See also* standardized mortality ratios

protein-based fibers, 130, 155–56
psychiatry, 13, 53

Quarelli, Gustavo, 52–54, 133–34, 258n111
Quarterly Journal of Economics, 67, 72

Rädel, Siegfried, 161
Raffan, Keith, 195
Rassegna della Previdenza Sociale (journal), 133
Ravensbrück concentration camp, 264n178
rayon. *See* viscose rayon
Rayon: A New Influence in the Textile Industry (Metropolitan Life Insurance Company), 46
"Rayon Goes to War" (booklet), 110–11, 151
Rayonhill, 176
Rayonier Canada, 220
Rayon Manufacturing Company of Surrey, 47, 48, 54–56
Rayon Peruana, 176
The Rayon and Synthetic Fiber Industry of Japan (report), 154–55
Rayon and Synthetic Yarn Association, 65, 67
Redaelli, Piero, 39–40
Red Plenty (novel), 214
reed grass, 130–31, 163
reeling rooms, 84–85, 97, 100, 105, 108
Reinhard, Eli, 70
Reinhard, Isidor, 67–68, 70
Reiter, Hans (Karl Julius), 126–27
Renshaw, Arnold, 36–37, 50
The Retail Credit Company, 108
Rheinische Kunstseide Aktiengesellschaft ("Rheika"), 141, 158
Rheinische Zellwolle AG, 143–45, 262n155
Richardson, Benjamin Ward, 17
Richmond, Virginia, 94, 172, 204
Richter, Richard, 265n190
Riegel Paper, 122

Rimini, Alessandro, 130, 163
Roanoke, Virginia, 44, 91, 156, 173
Robinson, William Cornforth, 32
Rochester, Anna, 60
Rogers, H. R., 48
Rome, Georgia, 114–15, 123
Romney, Mitt, 64
Roosevelt, Franklin Delano, 63–64, 97, 101–2, 137
Rose, H. Wickliffe, 154–55
Rosenblatt, Sigmund, 36–37
"Rose Pâle" (song), 164, 269n47
Ross, James, 16–17
Rostow, Walt, 176
Royal Institution (London), 3
rubber industry, ix, 8–12, 14–19, 21–22, 32–36, 38, 53, 73, 78–81, 209
Russia, viscose manufacturing in, 27. *See also* Soviet Union
Rust, Philip G., 66

Salvage, Samuel Agar, 44, 65, 67–70
Samuel Courtauld and Company, viii, xii, 28, 31, 34, 35, 44, 51, 56, 57, 58, 61–63, 69, 112, 125, 128, 129, 137, 148–49, 156, 158, 164, 172, 174–75, 180–84, 191–92, 194, 197–200, 265n185, 273n105
Sander, August, 144
Sander, Erich, 144–45, 158, 262n157
Sanders, George, 203
Saran Wrap, 171–72, 272n89
Sarvar, Hungary, 113–14, 119
Sateri (Fujan) Fiber Company, 207
sausage casings, 75, 204–5
Sayre, Ronald, 202
scarves, 135
Scheiber, Walther, 147–49, 160
Schenck, Ernst-Günther, 147, 149–50, 159–60
Schilling, Richard, 178–81, 184–85, 191–92
Schmölz, Karl Hugo, 262n157
Schoenfeld, Michael H., 189
Schwarza-Zellgarn AG, 148

Schwenkmeyer, Frieda, 87
Scott, Rachel, *Muscle and Blood,* 202
seaweed, 216
Sette Canne, un Vestito (documentary), 163–64
sexual disturbance, 12, 17, 133–34, 218–19
Shadowtime (opera), 285n84
Shandong Yamei Sci-tech, 207
Sheffield plate, 7
Shemitz, Esther, 59–60
Sherman Act, 171
Sicily, 40–41
Sidlow, Anthony, 173
Siegburg, Germany, 143–45, 158
"Siegfried Rädel" plant, 161
silk, 63
Silk Association of America, 44
Simpson, James, 5
skin blistering, 123
skinless wienies, 190, 205
slave labor. *See* forced labor
Slovakia, 149, 163
Smartfiber AG, 216
Smith, "Cotton Ed," 138
Smith, LeRoy H., 156–57
Smoot-Hawley Tariff Act, 65
SNIA (Società di Navigazione Italo-Americana), 43, 58, 129–34, 136–37, 143, 155, 163, 164, 174, 194
Sniafil, 43, 130
Sniafiocco, 136
Snow, John, 5
Sociedad Nacional de Industrias Aplicaciones Celulosa Española, 136–37
Société Industrielle de la Cellulose, 62
Societé Misr pour la Rayonne, 176
La Société Nationale de la Viscose à Grenoble, 141
sodium tetrathiocarbonate, 212
soil fumigants, 211–12
soil pollution, 199–200
Solfatara volcano, 3
solvents, 6–7, 116

The Sorrow and the Pity (film), 141
Sotgia, Alice, 132
South Africa, 174
Southern Summer School for Women Workers in Industry, 91
South Korea, 206
Soviet Union: viscose rayon industry in, 60; in World War II, 139. *See also* Russia
Spain, 136, 259n125, 273n105
Spanish Civil War, 136, 253n124
Special Yarns Company, 118, 253n45
Speer, Albert, 147, 149
Spinfaser AG, 125
spinnerets, viii, 130, 143, 158, 159
spinning baths, viii, 34, 35, 42, 49, 60, 97, 105, 108, 123, 127, 142–43, 158
sponges, 205–6, 215
Spontex, 205
Sprague, Howard, 179
Spruance, Daniel C., 24–25, 28
spun-dyed staple, 155. *See also* dope dyeing
SS (Schutzstaffel), 146–47
standardized mortality ratios, 188–89. *See also* proportionate mortality studies
staple: blends using, 42, 63, 126, 136; British manufacture of, 152–53; defined, 42; French manufacture of, 141; German manufacture of, 42–43, 124–27, 255n74, 255n75, 257n98; health hazards of, 42; increased production of, 57; Italian manufacture of, 130; Japanese manufacture of, 135–36; manufacturing process for, 42; spun-dyed, 155; Swiss manufacture of, 138; U.S. manufacture of, 113, 200–201; during wartime, 124–27, 130, 135–36, 138–41, 147
Star Trek II (film), 197
Statesman and Nation (magazine), 54
Steel-Maitland, Arthur, 32
Stein, Gertrude, xiv, 76
Stettheimer, Florine, 76

Stewart, Fred G., 104
Stössinger, Felix, 71–72
Strauss, Richard, 121
straw, viii, 126, 145–46, 159, 259n125
Strebel (doctor), 49
strikes, 86–91, 114, 194
Strindberg, Göran, 178
Süddeutsche Zellwolle, 127, 158–59, 194
Sufu, 136
suicide, 17, 34, 41, 97, 143, 186, 201, 204–5, 206–7
Sukarno, 176
sulfite cellulose, 129, 146
sulfur mines, 40–41
Summer Schools for Labor, 94. *See also* Southern Summer School for Women Workers in Industry
Suncook Mills, 118
Surtout, A., 13
Svenska Rayon, 139, 196
SVIT (Slovenské Vizkózové Továrne), 149, 163, 220
Sweden, 138–39, 196
sweet oil of vitriol, 7
Switzerland, 11, 49, 117, 119, 137, 138, 159, 166, 207, 233n79
Sylphrap, 62
Sylvania Industrial Corporation, 62, 70, 171, 202
"synthetic," 72–73
synthetic biofuels, 120
Synthetic Fiber Developments in Germany (report), 156–57
Synthetic Fibers Team, 156–57
synthetic rubber, 124, 160
Synthetic Yarn Federation of America, 87

Taiwan, 176, 207, 282n54
tariff protections. *See* trade protections
Taussig, Frank William, "Rayon and the Tariff," 66–67, 71, 72
Tecumseh, Kansas, 204, 216, 219
Teepak, 190, 205
Teikoku Rayon Company, 58

Teisinger, Jaroslav, 183
Teixeira de Mattos Brothers, 58
Temin, Henry, 105
Tencel, 197, 200–201, 215–16
Tennessee Valley Authority, 173
terylene, 168
tetrachlorethane, 116–18, 117
Textile Foundation, 266n7
Textile Research Institute, 154, 156, 266n7
Textile World (magazine), 42
Textron, 118
Thailand, 207, 218
thalidomide, 160, 268n31
Thanlow, Harald, 222n15
Theresienstadt (Terezín) concentration camp, xiii, 148
thiocarbanilide, 79–81
Thomson, Virgil, 76
3M, 205, 219
Thüringischen Zellwolle, 125, 146–49, 157, 161, 200
Tiller, John, 180–81, 184–85, 192
Time (magazine), 120, 137
Times Literary Supplement (magazine), 71–72
tires, rayon used in, 111, 135, 138, 156, 169, 175
tobacco, 179
Tommy James and the Shondells, 214
Tonawanda, New York, 205, 219
Toray Company, 206
Torviscosa, Italy, 130–31, 163
Toxic Release Inventory, 199, 219
Toyo Rayon Company, 58, 206
Trachtenberg, Alexander, 59
trade protections, 65–67
Trades Union Congress, 47, 52
Transparent Paper Limited, 61–62
Transparit, 62
Transylvania County, North Carolina, 203–4
Trewhella, Robert, 41
tricholorethylene, 92
tricresyl phosphate, 122–23
Triklidou, 141

Trollmann, Johann Wilhelm "Rukeli," 145–46
Truman, Harry, 173
Trumper, Max, 104–8
Tubize Artificial Silk Company, 112–14
Tubize Chatillon Corporation, 114–15, 123
turpentine, 6

UCB group, 204
Ultrasuede (documentary), 220
Union Carbide, 59
Union Club, 65–66
United Kingdom. *See* Great Britain
United States: carbon disulfide studies in, 17–19; cellophane in, 61, 170–71, 203–5; cellulose acetate usage in, 117; occupational medicine in, 45; and postwar viscose rayon industry, 154–57, 159; viscose rayon industry in, 44–46, 64–65, 91–109, 135, 150, 167–72, 188–92, 193, 197–205; workplace protection in, 188–91; in World War II, 135, 137–38, 150
Universum Film AG, 125
Updike, John, *Rabbit Run,* 213
U.S. Department of Justice, 171
U.S. Department of Labor, 92, 97–98, 189
U.S. Division of Labor Standards, 92–93, 97–98
U.S. Field Information Agency, Technical (FIAT), 156, 159
U.S. National Institute for Occupational Safety and Health (NIOSH), 185–86, 188–90, 197–98, 202, 204, 218
U.S. National Retail Dry Goods Association, 44
U.S. Office of Strategic Services, 136–37
U.S. Supreme Court, 70

valley oak, 3
Van Vechten, Carl, 76
varicose veins, 75
VEB Zellstoff- und Zellwollewerke Wittenberge, 161, 195

ventilation, 10, 35, 54, 96, 108, 116, 127
Venturi, Lionello, 134, 257n95
Vertin, P. G., 184, 187–88
Victor, Sally, 135
Vigliani, Enrico, 134, 165–67, 178, 183–84, 258n111
Vikase, 219
Viscacelle, 62
Viscofan, 205, 219
Viscose Ambulatoriums, 75
Viscose Company, 44, 46, 57, 65–68, 102–3, 112–13, 249n122. *See also* American Viscose Company; American Viscose Corporation
The Viscose Continuous and Rayon Staple Fibre Plants of the British, American, and French Occupation Zones of Germany (report), 157–58
Viscose Corporation of Virginia, 44, 91
Viscose Emmenbrücke, 138
viscose rayon: British production of, 24–26, 30–31; in British wartime culture, 135; cellophane vs., 70–71; cellulose as basis of, vii, 126, 130–31; development of, 22–29; economic significance of, viii, 31; in Fascist culture, 129–34, 136; green marketing of, vii, viii, 215–17; in Japanese wartime culture, 135–36; manufacturing costs of, 25; manufacturing process for, viii–ix, 25, 42, 201; marketing of, 62–63; in Nazi culture, 125–29, 134; popular perceptions of, 70–75, 135–36; predecessors to, 23–24; public's ignorance of, vii, ix; "rayon" as trade name, xiii, 44, 61; recent applications of, 214–17; sources of information on, xi–xii; in Soviet wartime culture, 139; synthetic rivals of, 167–72; as threat to natural fibers, 63; in U.S. wartime culture, 135, 137–38; in wartime, viii, 110–51. *See also* viscose rayon industry
viscose rayon industry: advances in, 42; communist perspective on, 58–60; contemporary, 214–17; in East Germany, 161–62; eye damage linked to, 34, 49–51; first-person accounts of, 84–87, 104, 142–45, 147, 196; forced labor in, xiii; in France, 43, 57–58, 74, 140–41, 164; in Germany, 41–43, 57–58, 74, 125–29, 142–49, 151, 156–61; in Great Britain, 32–36, 46–56, 57–58, 135, 164–65; health hazards of, vii–viii, 25–27, 42, 51–52, 60, 78, 84–109, 114–15, 132–33, 140, 142–46, 155, 162, 180–92, 197–209, 218–20; health warnings concerning, 39, 51, 54–55, 91–93, 102–3; inspections of, 34–35, 48; internationalization of, viii–ix, 164, 175–76; in Italy, 38–41, 43, 57–58, 74, 129–34, 163–64; in Japan, 43–44, 57–58, 74, 135–36, 150, 154–56, 206–7; labor actions against, 86–91; museums in homage to, 195–96, 220; popular perceptions of, 127–28, 212–14; postwar status of, 153–77; "rayon" as trade name, 73; rearrangement and dismantling of, 193–97, 200–207; secrecy and obstruction concerning operations of, 18, 82, 98, 101, 124, 151, 181–83, 218; in Soviet Union, 60; Taussig and White's essay on, 66–67; in United States, 44–46, 64–65, 91–109, 135, 150, 167–72, 188–92, 193, 197–205
Viscose Treatment, 75
vision. *See* eye damage
Viskase, 205, 286n90
Visking, 205
viskrings, 217
Vistra, 42, 124, 148, 255n73, 257n98
Vollrath (chief chemist), 158
vulcanization, 8–10, 22, 73, 209
vulcanized cellulose fiber, 72–73
vulcanized vegetable oil, 73

Walker, William H., 117
war crimes, 159–60
Warren, Earl, 171

water pollution, 51, 172–73, 199, 204
Waxed Products Company, 70
Weifang Henglian, 207
Welles, Orson, 88
Werner, Alfred, 119
Whinfield, J. R., 168
White, E. B., 76–77
White, Harry Dexter, 137–38, 259n131; "Rayon and the Tariff," 66–67, 71, 72
Widzewska Manufaktura, 140, 148
Wigton, United Kingdom, 216
"Wilhelm Pieck" factory, 161
Willcox, William, 54–55, 118
Williams, Margie, 211
Williams, Mrs. Harrison, 130
Wir von Glanzstoff=Courtaulds (magazine), 128
Wiślicki, Feliks, 113, 264n176
Wittenberge Elbe, Germany, xiii, 125–26, 145–46, 195
Wodehouse, P. G., 240n95
Wolff, Paul, 127–28
Women's Trade Union League of America (WTUL), 86–87, 89, 91, 94
Wonjin Foundation for Victims of Occupational Diseases, 206
Wonjin Rayon Company, 206–7
wood pulp, 140, 174. *See also* cellulose
Woolton, Frederick James Marquis, Lord, 152–53
Worbs, Helmut, 195
workers: advocacy options of, xi; affected by carbon disulfide, vii; lawsuits brought by, 18–19, 84–85, 104–5, 191–92, 199; strikes by, 86–91, 114, 194. *See also* carbon disulfide—maladies caused by
workers' compensation, 47, 81, 97, 103–5, 138, 199, 202
workers' villages, 28, 125, 149
workplace protections: exposure levels, 99, 105–6, 124, 155, 177, 185, 189–90, 202, 208, 218, 250n136; in France, 12–13; in Great Britain, xi, 30–36, 46–56, 191–92; Hamilton's advocacy for, 80–83, 108–9; in Japan, 43; lack of federal, 105–6, 188–91; modern, 201; in United States, 188–91; during war, 138. *See also* workers' compensation
World War I: cellulose acetate in, 115–17; munitions manufacture in, 119–20; viscose in, 112
World War II, 125–51; crimes related to, 159–60; forced labor during, 141–49, 151, 157, 159–60, 264n178; France and rayon, 140–41; Germany and rayon, 125–29, 142–49, 151; Great Britain and rayon, 135, 137–38; Italy and rayon, 129–34, 136–37; Japan and rayon, 135–36, 150; other countries and rayon, 138–39; research related to, 120; Resistance movement, 141; Soviet Union and rayon, 139; United States and rayon, 135, 137–38, 150
Wright, John Pilling, 84
Wright, Wade, 46, 232n63
WTUL. *See* Women's Trade Union League of America

xanthate, 25, 35, 132, 277n88

yeast, 129, 146–47, 159
Yerkes, Leonard A., 65
Young Men's Christian Association (YMCA), 94
Young Women's Christian Association (YWCA), 87, 91

Zacapu, Michoacán, Mexico, 175
zellwolle, 42–43, 124, 127, 151. *See also* staple
Zellwolle Lenzing, 146–48, 157, 201. *See also* Lenzing
Zenalyser, 210
Zenk, Herbert, 161–62
Zetkin, Clara, 161, 268n35
Zimmer, Verne, 97, 99, 103, 104